Science Assessment
in the Service
of Reform

Gerald Kulm

Shirley M. Malcolm

editors

"Outstanding academic book of 1992"
—*Choice*

Chapters 3, 8, 9, 11, 13-18 are revised versions of chapters that appeared in *This Year in School Science 1990: Assessment in the Service of Instruction* edited by Audrey B. Champagne, Barbara E. Lovitts, and Betty J. Calinger and published by the American Association for the Advancement of Science.

Library of Congress Cataloging-in-Publication Data

Science assessment in the service of reform /
Gerald Kulm, Shirley M. Malcom, editors.
    p. cm.
Includes bibliographical references and
index.
ISBN 0-87168-426-8
1. Science—Study and teaching—
Evaluation.
2. Science and state—Evaluation.
3. Academic achievement—Evaluation.
 I. Kulm, Gerald. II. Malcom, Shirley M.
Q181.S3595 1991
507'.1—dc20   92-37901

CIP

AAAS Publication: 91-33S
International Standard Book Number: 0-87168-426-8
Printed in the United States of America

⊕ Printed on recycled paper

Contents

Preface

One of the most pressing topics within the realm of educational reform is that of assessment. How can we find out what students know, how well teachers teach, whether our schools work, if New York is doing as well in its reform as Maryland, or whether the performance of U.S. students still lags behind that of students in Korea? In fact, "How well are our students doing?" seems to be discussed more than "What should students know or be able to do?"

Some policymakers have used the data that assessments provide to argue for reform. Education in science is "high-stakes" when our economy is increasingly dominated by science and technology. Others dismiss the message that such testing brings because the details are garbled and implications unclear. But the central theme remains: by almost any standard that has been used in assessing science learning in the United States, U.S. students do not measure up, and the implications for the nation are grim. On the other hand, if curriculum and teaching are to improve, new and more effective ways to assess student learning are needed. The central message of this volume on assessment in the service of reform is that there should be some basic givens, whatever system of assessment is used. At a minimum, the system should be free of bias; it should reflect what is being taught and give us information to improve classroom instruction, to diagnose problems, and to identify misconceptions of individual students; it should provide a measurement of the effectiveness of a teacher or a curriculum; and, finally, it should reflect what should be valued and taught about science as both a body of knowledge and a method for discovering new knowledge. Clearly, those givens have the potential to improve the science and mathematics curriculum in every school that understands and heeds the message.

The education reform movement has been driven by assessments. It was the steady decline of SAT scores and the international comparisons between the U.S. students and those of other countries that attracted the attention of legislators and business people. It was the gender and race and ethnic group differences in scores that compelled the educational community to look at differential course taking, uneven treatment of students by teachers, and bias in placement and grouping. And it has been student performance on the state-mandated assessments that have maintained the focus of local school districts on the need for adopting the standards proposed by science and mathematics educators.

We both know from personal experience the difference that assessment can make in opportunities for students. As high school science and mathematics teachers, we were not prepared to address assessment issues, even though it was a requirement to assign grades. As with most teachers, we experienced the prevailing pressures such as preparing students to take state-mandated standardized tests (even though it interfered with teaching inquiry and reasoning), or being told not to assign A's in certain classes, or giving department-wide tests. We also experienced students who were motivated through high expectations, achieving well beyond those who had been identified as being more capable. And those students who selected an easier course than they were capable of in order to get a higher grade.

We have spent much of our professional careers worrying about how to make better distinctions among what students know, what they can do, and what they are **capable** of knowing and doing. Too often, early and inaccurate assessments of children's abilities become self-fulfilling prophesies that limit their opportunity to learn and, later, to contribute as full citizens in this democracy. Assessment can never be the exclusive domain of psychometricians, educators, and cognitive researchers. What we find out, how we find out, and what we come to understand about students all exist in a larger social and political world that affects what we do with this knowledge, once acquired.

This volume, which presents papers by assessment and curriculum specialists, psychologists, researchers, and teachers, is particularly timely. The theme throughout, assessment in the service of reform, relates curriculum to assessment reforms and both to the structural reforms that must be made in the educational system of this nation if we are to remain a world power in the 21st century. We believe this volume will help to inform a discussion that by its nature is – and, in this country, must be – politically based.

Assessment must complement and inform instruction. Readers of this volume will find much to use immediately in making assessment a tool for the meaningful reform of school science. But thoughtful readers will also find meat for many discussions about the future of assessment, and of science and mathematics education, in the United States. What is needed is more than a few minor adjustments in our present system. What is needed is a revolution in outlook that will bring all children the education they need to realize the American dream.

Gerald Kulm
Professor of Mathematics Education
Texas A&M University

Shirley M. Malcom
Head, Directorate for Education and Human Resources Programs
American Association for the Advancement of Science
July 1991

List of Contributors

Jane Armstrong is Director of Policy Studies at the Education Commission of the States in Denver, Colorado.

Ronald K. Atwood is a professor at the University of Kentucky in Lexington, Kentucky.

Joan Boykoff Baron is an education consultant at the Connecticut State Department of Education in Hartford, Connecticut.

Carol Briscoe is an assistant professor at the University of West Florida in Pensacola, Florida.

Joseph C. Campione is Professor of Educational Psychology at the University of California School of Education in Berkeley, California.

Nancy S. Cole is Executive Vice President of the Educational Testing Service in Princeton, New Jersey.

Angelo Collins is Associate Professor of Science Education at Florida State University in Tallahassee, Florida.

Kathleen B. Comfort is an educational program consultant for the California Assessment Program at the Department of Education in Sacramento, California.

Thomas M. Dana is an instructor in the Department of Science Education at Florida State University in Tallahassee, Florida.

Alan Davis is Assistant Professor of Educational Psychology at the University of Colorado School of Education in Denver, Colorado.

Glenn Fay, Jr., is a science teacher at Champlain Valley Union High School in Hinesburg, Vermont.

John Guyton is an assistant professor in the Department of Elementary and Secondary Education at Murray State University in Murray, Kentucky.

Edward H. Haertel is an Associate Professor of Education at the Stanford University School of Education in Stanford, California.

Wynne Harlen is Director of The Scottish Council for Research in Education in Edinburgh, Scotland.

Maryellen Harmon is a research associate at the Center for the Study of Testing, Evaluation, and Educational Policy at Boston College in Boston, Massachusetts.

Karl Hook is a university instructor at Florida State University in Tallahassee, Florida.

Lisa Hudson is an OERI senior research associate at the U.S. Department of Education.

David W. Johnson is Professor of Educational Psychology and Co-Director of the Cooperative Learning Center at the University of Minnesota in Minneapolis, Minnesota.

Michael Johnson is Director of the Science Skills Center in Brooklyn, New York.

Roger T. Johnson is Professor of Curriculum and Instruction and Co-Director of the Cooperative Learning Center at the University of Minnesota in Minneapolis, Minnesota.

Michael Kamen is Assistant Professor of Elementary Education at Auburn University in Auburn, Alabama.

Thomas Kellaghan is Director of the Educational Research Centre at St. Patrick's College in Dublin, Ireland.

Judith A. Kelley is Professor of Chemistry at the University of Massachusetts at Lowell in Lowell, Massachusetts.

Gerald Kulm is Professor of Mathematics Education at Texas A&M University in College Station, Texas.

Kirstin Lebert is a teacher with the District of Columbia Public Schools.

Anthony W. Lorsbach is an assistant professor in the Department of Teacher Education at Bradley University in Peoria, Illinois.

George F. Madaus is Boisi Professor of Education and Public Policy and Director, Center for the Study of Testing, Evaluation, and Educational Policy, at Boston College in Boston, Massachusetts.

Shirley M. Malcom is Head, Directorate for Education and Human Resources Programs, at the American Association for the Advancement of Science in Washington, D.C.

Ray Marshall is a professor in the LBJ School of Public Affairs at the University of Texas in Austin, Texas.

Michael E. Martinez is a research scientist with the Educational Testing Service in Princeton, New Jersey.

Mary Budd Rowe is a professor in the School of Education at Stanford University in Stanford, California.

Donald C. Snyder, Jr., is a physical science teacher at South Philadelphia High School in Philadelphia, Pennsylvania.

David L. Stevens is an earth science teacher at Glenelg High School, Howard County Public School System, in Howard County, Maryland.

Carol Stuessy is an assistant professor of Science Education at Texas A&M University in College Station, Texas.

Mary Thiel is a principal with the Delwood Community School District in Delmar, Iowa.

Marc S. Tucker is President of the National Center on Education and the Economy in Rochester, New York.

Acknowledgements

Undertaking a book project like this requires the work of many people at all phases of the process. The most significant recognition must go to the editors and contributors of the This Year in School Science volume, *Assessment in the Service of Instruction,* upon which this book has been based. It was the work of Audrey Champagne, Barbara Lovitts and Betty Calinger in providing us the original volume that gave us the opportunity to build additional policy inputs into an expanded discussion of assessment.

Our gratitude goes to the authors of the chapters of the expanded volume who put up with our requests for rewrite, references and review. We owe special thanks to those authors who also spoke at the forum and in so doing had to do double duty.

A special note of appreciation goes to the publications staff, especially Kathryn Wolff, who saw this project from beginning to end. Betty Calinger has been the rock that has held the Forum on School Science and This Year in School Science together. She has been liaison and listener, critic and confidant. Special thanks go to Jennifer Hatton for managing Malcom's life so that time and help was found for her to work on this volume. Thanks go to Mike Feuer of OTA for his very important input.

To our friends and family who put up with us when assessment was all we could talk about we extend our deep appreciation, and humble thanks, and our promise to bend their ears with something new.

The policy climate for assessment has moved considerably since this project was begun. We hope that this book provides background and context for the policy choices which lie ahead.

Part I

Policy Issues in Science Assessment

Shirley M. Malcom

The publication of *A Nation at Risk* is viewed by many as signalling the call to arms for educational reform in the United States. The message was clearly sent that our standards had fallen; we needed to regain lost ground by having our schools work harder, our students work harder, and our teachers work harder. In science and mathematics the message was clear: students needed to take more of both subjects. Little was said about what would comprise more, what outcomes we were looking for, or the need to reach all students. The states responded to the first wave of reform largely by increasing the requirements for high school graduation; mostly in the fields of science and mathematics. The second wave of reform was ushered in with the publication of *A Nation Prepared: Teachers for the 21st Century* by Carnegie Forum on Education and the Economy and *A Time for Results* by the National Governors' Association. The arguments here were different: the problem is not the loss of standards we once had, but rather the need to meet challenges in education we had never had to face. Focusing on teachers as key in meeting these new world class standards, the report argued for increased professionalism for teachers and restructuring for schools.

In science and mathematics the major reform projects began around the mid 1980s with a recognition that there needed to be a rethinking of the goals for teaching and learning in these fields. What did students need to know and be able to do? What science was worth knowing and what knowledge, values, and habits of mind regarding science, mathematics, and technology should characterize a high school graduate in the 21st century? What sense of mathematics did all students need and what knowledge and skills in mathematics would allow them to use and apply mathematics in their lives?

As we go about the task of reinventing schooling in America we realize that this is not only a conversation for content experts, but a major aspect of national, state, and local politics. The President of the United States has declared himself to be the "Education President"; in meetings the 50 governors, who have likewise committed themselves to education reform, have joined the President in declaring among the reform goals that we should be first in the world in math and science by the year 2000. Those who started out wanting simply to get the science and math right for what students will need have ended up in the middle of major league politics. Clearly, the policy agenda affects the resources available for reform, the direction of reform, the purposes of reform, and the targets of reform. Part One of this volume addresses this agenda, beginning with discussions of the role of assessment in structural reform, control of testing, and equity issues in assessment. Included among the authors are some of the major architects and voices within the current reform movement. Tucker sets the stage for the discussions of assessment in the book by explaining how and why assessment has assumed such a prominent role in the reform debate. Marshall connects the economic competitiveness arguments – a major engine driving reform discussions – to concerns for equity, making the case that science and technology have become basics in a knowledge-driven economy and that assessment allows us to monitor attainment of higher-order skills by all. Harmon discusses bias in science testing and raises concerns about the multiple roles assessment has been asked to play. Kulm summarizes the discussion of a panel on who should control assessment, noting the practical, political, pedagogical, and ethical concerns that arise as the stakes within assessment increase. Together, these papers detail the major policy issues that must be addressed in the years ahead as this nation strives to meet the year 2000 goals.

Why Assessment Is Now Issue Number One

Marc S. Tucker

There is, I think, some mysterious unseen hand that shapes the consensus in American education. Almost overnight, assessment has become the focal point of a great debate about the purpose, shape, and control of American education, and it seems more likely than not that, within a decade, the United States will have a national examination system. My purpose in this chapter is to offer my personal views about how that extraordinary change happened and what it might mean.

A Nation at Risk

For those of us with a professional interest in American educational policy, modern history begins with *A Nation at Risk,* released in 1983 by the National Commission on Excellence in Education.[1] That report did an extraordinary job of calling the nation to arms, but I consider its analysis to be way off the mark. In essence, the report stated that the problem with American education is one of decline and that the response should be the restoration of lost standards. *A Nation at Risk* recommended a return to the standards that were considered ideal for those who completed high school and went on to college during the 1950s. It recommended restoring the Carnegie units — the time on task, as it were — for the major core curriculum subjects.

The states reacted immediately by focusing on two areas: minimum standards for students and minimum standards for teachers. By 1985 more than 22 states had instituted minimum competency standards for student graduation. Almost as many imposed minimum competency standards for teachers, setting basic literacy standards without which teachers could not continue their employment. This was the first indication of the increased importance of assessment in the new movement. The culmination of the line of thought with which *A Nation at Risk* began was the release of the James Madison High School proposal by William Bennett, then U.S. Secretary of Education.[2] It almost perfectly described the high school that

I attended in the late 1950s. It captured the idea that if only we returned to how things were, the country would be okay.

Two New Reports

In 1986, within a period of only three months, the ground shifted substantially. Two reports — *A Nation Prepared,* released by the Carnegie Forum on Education and the Economy, and *A Time for Results* by the National Governors' Association (NGA) — produced a thorough reconception of both the problem and the appropriate response.[3,4] According to these reports, the problem is not that American schools have fallen from some former achieved standard, but that the world has changed and the United States has failed to respond. The international economy has taken a new direction, these reports said, and for our nation to prosper in the years ahead, we must completely rethink how we develop and use our human resources.

Since the turn of the century, the implicit view guiding American educational policy has been that we could do quite handsomely by educating only one quarter of the population to a high standard requiring independent thinking. Now, however, we are entering an era when virtually everybody needs to be able to think independently, act autonomously, and behave professionally in the workplace. With a whole new set of specifications emerging for American education, we need teachers — not just for a few students, but for everyone — who can prepare, deliver, and teach a thinking curriculum. This situation requires a reconception of the role of teachers and teaching in the United States and the development of a true profession of teaching. The question is how to get these teachers — and how to keep them. I will try to boil down to one paragraph a very complex answer to that question.

Reorganizing American Education

The nation needs to reconceive the organization and management of its educational system. We must restructure the schools in much the same way that leading corporations around the world are restructuring themselves for high performance. We must set very clear goals for students. We must accurately measure student progress against those goals. We must push decisions about how education is delivered to students down to the teachers and the principals, eliminating as much of the intervening bureaucracy as possible, and hold those teachers and principals accountable for the results. Simple to write down, but an extraordinary change for American education — a whole new view about how to organize the educational enterprise.

At the core of the formulation are the pivotal requirements of explicit goals and appropriate measures. The Carnegie and NGA reports were published in the spring and summer of 1986. What happened next was remarkable. All over the country, a new concern for goals arose. Not for the kind of goal statements that we used to produce — vague formulations that we wrote out and then tucked away in the bottom drawer — but real agreements specifying in detail what students ought to know and be able to do. AAAS's Project 2061 report, the Mathematical Sciences Education Board's *Everybody Counts,* and the National Council of Teachers of Mathematics' *Curriculum and Evaluation Standards for School Mathematics* are examples of real standards, evidence not only of a concern for goals, but also of a virtual renaissance in new modes of assessment.[5-7] The National Board for Professional Teaching Standards, the National Commission on Testing, and Project Zero at Harvard — all developed an interest in what came to be called "authentic assessment." The Secretary of Labor's Commission on Achieving Necessary Skills is also looking at work force readiness goals and new measures, and California and Connecticut have invested millions of dollars in the creation of whole new assessment systems and new syllabi to guide them.

The Charlottesville Summit

All of that initial activity culminated in an extraordinary event — the 1989 summit in Charlottesville, Virginia. For only the third time in U.S. history, the President called the governors together to discuss a particular topic — in this case (and for the first time), education. The primary focus of this meeting between President Bush and the governors was the production of a set of educational goals for America. The second focus was how to assess progress toward achieving those goals, and the third was how to restructure the U.S. educational system.

I believe that we cannot achieve the goals without restructuring the system very much along the lines that I outlined above. In short, the issue of assessment arises immediately out of the need to specify goals. Participants at the Charlottesville meeting realized that just assessing results will do little or no good unless new modes of assessment and new items assessed are seen as part of a much larger scheme for restructuring American education. This view was spelled out in the statement emerging from Charlottesville and by ensuing events such as President Bush's January 1990 State of the Union address and a February statement released by the governors.[8] Making it clear that they were serious, the governors and the president appointed a panel, chaired by Colorado Governor Roy Romer, to monitor the nation's progress toward educational goals.

The Romer Panel

The first and most interesting question that the panel faces is whether it will stick with the charge with which it began — that is, monitoring progress toward the goals enunciated by the nation's leaders — or whether it might go beyond that; whether the panel might set up some kind of assessment system that would not simply monitor progress, but would produce it. The other interesting issue with regard to the Romer panel is the conflict that exists about who will carry out this important monitoring function. That conflict is driving both the goal-setting process and the definition of the means by which we assess our progress towards those goals. Members of the U.S. Congress, angry that they were not given a voting role on that panel, offered legislation, Senate Bill 2, to set up another mechanism in which it would have a much stronger role.

Issues of assessment and goal setting, far from being window dressing, have become so important to the future of education in this country that they are worth fighting about in a very serious way. This nation, unlike most others, has no means of determining who is to decide who is going to run the show.[9] Now, for the first time in my memory, this country is groping toward a *real* national policy on education, and the issue of assessment is square in the middle.

Recommendations from *America's Choice*

In June 1990, my organization, the National Center on Education and the Economy, released the report of the Commission on the Skills of the American Workforce called *America's Choice: High Skills or Low Wages![10]* The report's first recommendation was the creation of a national examination system for the United States. Until very recently, few people would have invested even five minutes in developing a proposal for a national examination system, because prospects for its implementation appeared nil. Nevertheless, we proposed a national examination system that would establish a Certificate of Initial Mastery, which most students would be expected to achieve by age 16.

One key idea in the proposal by the commission was that the examination system should be designed to provide incentives to American students to take tough courses and to work hard to succeed in them. The study on which the commission's work was based made it clear that America is almost alone among the industrialized countries in having a system in which, for most students, there is no necessary connection between how they do in school and what they want to achieve later in life. More than 75 percent of America's students need only a diploma to get a job or go to college. Scores on examinations have no bearing on what job they will get, what pay they will receive, what their career prospects are, or even whether they will get into college. That is not true for any other leading industrialized country.[11] What we have managed to do in this country is to rob ourselves of the most important asset our schools can possibly have: the effort of students to succeed in school and to do more than the bare minimum.

Second, and no less important, the proposal called for a system that sets high standards for everyone and in which everyone can succeed. This is a dramatically new idea about assessment. Since the 1920s, American psychometrics — and, more importantly, our thinking about learning and teaching in the United States — has been based on the idea that ability is distributed on a bell curve, that achievement is the result of ability, and, therefore, it, too, will be distributed on a bell curve. Thus, it has been accepted that an important psychometric property of tests is the ability to distinguish sufficiently among the people taking them so that the results will produce a bell curve. In other words, in order for a testing system to be accepted as technically sound, some students must fail. This is a central feature of American testing theory and of American educational practice.

The commission's proposal said, "Let us move to a criterion-referenced system; let us decide what it is that the students need to know and be able to do; let's set that objective very high; and then let's make sure that everyone gets there." A very different idea. But if our economy is to succeed in the future, we must produce *an entire population* that possesses skills that most of us think only professionals or managers need. It is not only morally satisfying to conceive of a system in which everyone can succeed at a high level, it is now essential for the future of the country.[12]

Next, the proposal called for multiple modes of assessment: performance examinations, portfolios, examples of best work. It advocated an examination *system,* not a single examination — a standard to which a number of exams can then be calibrated, enabling us to have one national system that will accommodate more than one national exam. This is not a uniquely American idea; it actually occurs in a number of other countries. It represents an enormous but not insuperable technical challenge. Politically, it is an essential requisite for developing a national examination system to reflect a national set of goals.[13]

Next, the commission recommended moving toward a system that is based far more on the judgment that human beings make about the competence of others based on their work. Given the qualities that we are looking for in students — that is, thinking skills — it is an inescapable fact that we must move toward a system based far more on judgment than the current one, which is based largely on machine scoring.

Finally, the recommendations said that we need to move toward a syllabus-driven examination system. These ought to be exams one can study for. The idea of the whole system is that such exams will drive improvement, which can only come about if one can study for them. This implies a major change in the way we think about assessment. To study for the examinations, students have to know what their goals and objectives are. They have to see most of the questions that were asked each year.

Emerging Educational Policy Criteria

In the following paragraphs, a set of criteria that emerge from the policy environment are described. First, as we move toward new examination systems and new assessment modes, it must be possible to compare the performance of one student to another, one school to another, one school system to another, one state to another, and, without doubt, one nation to another. Second, these examination systems must be syllabus driven; that is, they must be based on explicit statements about goals for students in the domain for which the assessment is being produced. What should students know and be able to do? Third, these systems must be conceived of as driving changes in curriculum and teaching. We are not talking about examination systems designed merely to monitor how the system and the students are doing. The purpose is far more important than that; it is to change the curriculum and the way students are taught. Next, and critically important, the assessment needs to incorporate and be guided by the need to develop thinking skills — not simply routine skills — in students across the board. By the word "thinking," however, I do not mean aridly academic. There is a consensus emerging that what is truly important is the capacity to apply what one knows to the problematic, chaotic settings of real life.

For more than a century, American education has made a fairly sharp distinction between those persons with academic skills who get academic work and those persons who don't have these skills and who therefore are suited only for work that doesn't require much facility with abstractions. My view is that we are moving toward a consensus that that distinction no longer holds. The students who are going directly into the workforce from high school will need a very high level of academic skills, as most of the rest of the industrialized world concluded years ago. Students who are college bound need to experience a school environment in which they are constantly applying what they know to real-world situations, because that's the only way for them to really master the material.

I believe we are moving toward a consensus in this country that assessment systems need to be criterion-referenced for all of the reasons suggested earlier. That is, these systems should reflect a consensus about what students should know and be able to do, and they should also reflect the concomitant and critically important assumption that everybody can successfully learn.

I believe we are also moving toward a consensus that multiple modes of assessment are required and that no single mode will do. In addition to performance tests, the accumulation of work over time is a critically important part of the system that we need to create. Those multiple modes of assessment should include opportunities for trained assessors to make judgments about the competence of the people being assessed. This is not to say that the nation has decided to abandon multiple-choice tests. Most thoughtful people find that neither necessary nor wise. Multiple-choice tests have a role to play, but probably a greatly diminished one.

There is growing agreement that the system used for assessment should be tied to incentives that operate on both teachers and students. That is, we must embed assessment in a policy such that, when students do better, their teachers do better, and when the students do not do better, their teachers suffer real consequences. Those are incentives for education professionals.

Student incentives are also critically important. Clearly, what makes students in other countries invest far more effort in school than most Americans do is that something hinges on it. This is not a complicated idea, but it does mean that the assessment system has to be designed for that purpose. We have to assess skills that colleges think are important for doing well in college and that employers think are important for doing well in the workplace. When students do well on these examinations, it will make a difference later in life because the decisions that colleges and employers make will be based on the results of the examinations.

Next — and at least as important as any item here — these examinations have to be fair. We are already concerned about bias in examinations, but another aspect of fairness is more important. If American schools move toward an examination system at the local, state, or national level in which performance on the examination makes an enormous difference with respect to life chances, and if the U.S. education system moves toward examinations that are based on a syllabus, then it is critically important that *all* students be exposed to a curriculum that will prepare them to do well on the test. That is the central fairness challenge, in my view. If we move toward a system in which examinations make an enormous difference, and students out in the suburbs, particularly the wealthy ones, are those who are prepared for those examinations while students in the inner city are not, then we will have set up a large proportion of students for failure.

Conclusions

I think that any system that attempts to establish a single national examination will not succeed in the United States. We need a *system*; we do not need a *single exam*. We need a standard to which all of us can compare ourselves to see how we compare to others. We need a systemic view of the entire educational system into which assessment fits. One of my primary themes is that this overpowering interest is not in assessment for its own sake, but rather is an interest that flows from a changed conception of the requirements of the American education system. Accompanying that change in outlook is a new sense that the entire system needs to be overhauled.

The design of new assessment procedures must be tied to a new conception of how schools are governed, what the curriculum is going to look like, how the teachers are going to be trained, what the consequences of the assessment are (that is, the uses to which it will be put), and a host of other things. In effect, the restructuring agenda lies just below and is broader than the assessment agenda, and includes teacher professionalization. The latter has very important consequences for the way we think about assessment. In the future, teachers will not be simply the recipients of new assessment systems; they will be creating them and will undoubtedly be doing most of the scoring — because teachers can make judgments and machines cannot. Only teachers can make the system work.

What we do about assessment will have an important bearing on the professionalization of teaching. The conduct of student performance assessment may turn out to be the means by which teachers gain control over their profession. It may be in the end what drives their reflection on their own work, as they assess the work of others. As teachers all over this country assess the work of students in other classrooms, they will examine more closely their own curriculum and the way they teach. That is one of the great hopes of the new approach to assessment.

The events described earlier in this chapter occurred during only four years — 1986 to 1990. The best way for me to convey my amazement at how far we've come is to recount what happened in the summer of 1990 after I gave a speech describing *America's Choice* to a group of superintendents and principals from about 10 school districts in the lower Hudson River Valley. When I finished the speech, a man in the back of the room, unable to contain himself, jumped up and waved his hand. He said,

> Look, I think this idea for a national examination system is absolutely terrific, but it will never fly. It will never fly because of the people in this room, my colleagues. I've been a school board member for 20 years or more. They will never buy it. School boards all over the country couldn't possibly buy it. Think about what you have said, Mr. Tucker. It flies in the face of local control, everything this country has ever stood for, the very basis of school district authority.

And he went on in the same vein. I replied, "Okay, let's put it to the test. How many of you are school board members?" About 100 hands went up. I asked, "Of those of you who are school board members, how many think this idea for a national examination system as I have laid it out is a good idea?" About 70 hands went in the air. I then asked, "About how many of you think it's not a good idea, and would be opposed to it?" Fewer than 15 raised their hands. Everybody in that room looked at each other in complete amazement. They couldn't believe it. A great deal has changed in the past four years.

References and Notes

1. National Commission on Excellence in Education, *A Nation at Risk* (Washington, DC: U.S. Department of Education, 1983).
2. Bennett, W., *James Madison High School Proposal* (Washington, DC: U.S. Department of Education, 1987).
3. Carnegie Forum on Education and the Economy, *A Nation Prepared: Teachers for the 21st Century,* Report of the Task Force on Teaching as a Profession (New York: Carnegie, 1986).
4. National Governors' Association, *A Time for Results: The Governors' Report on Education* (Washington, DC: National Governors' Association, 1986). Four follow-up reports have been issued, with the 1991 report as the fourth and final in the series.
5. American Association for the Advancement of Science, *Science for All Americans: A Project 2061 Report on Literacy Goals in Science, Mathematics, and Technology* (Washington, DC: American Association for the Advancement of Science, 1989); reprinted as, Rutherford, F. J., and Ahlgren, A., *Science for All Americans* (NY: Oxford University Press, 1990).
6. Mathematical Sciences Education Board, *Everybody Counts* (Washington, DC: National Research Council, 1985).
7. National Council of Teachers of Mathematics, *Curriculum and Evaluation Standards for School Mathematics* (Reston, VA: National Council of Teachers of Mathematics, 1989).
8. U.S. Department of Education, *National Goals for Education* (Washington, DC: U.S. Department of Education, July 1990).
9. As a further comment on this point, I believe that the direction of American education ought not to be set exclusively by politicians. Other people — educators, businesspeople, independent leaders of the society — have wisdom to offer and roles to play. It is quite important for us to create some mechanism by which many people in this country from different sectors come to own the goals and own the means by which they are assessed. There are a set of very important political issues that we haven't come to grips with as a country; now, however, for the first time in the history of the United States, we may be groping toward a national education policy. It's not surprising that there is an argument about who ought to control it.
10. Commission on the Skills of the American Workforce, *America's Choice: High Skills or Low Wages!* (Rochester, NY: National Center on Education and the Economy, 1990).
11. We did a survey at the National Center on Education and the Economy to determine whether most employers, including many prestigious ones, required only a diploma. We conducted interviews at some 2,500 firms in the United States and abroad, in every sector of the national

economy, and that is what they told us. When employers use the diploma as their hiring standard, it means absolutely nothing to them with respect to what high school graduates know and are able to do. All it means is that the applicant had the "stick-to-itiveness" to stay to the end and so is more likely to show up for work than those who didn't get their diplomas.

We found that employers very rarely get transcripts. We asked employers whether they wanted transcripts and they said it wouldn't make any difference because they have no idea what "Math III," for example, means. Also, employers profess not to believe in the grades. We surveyed high school students in Rochester, New York, on the same subject. They indicated that if they show up most of the time, turn in something, and don't cause trouble, they'll get a passing grade. The students stated that "most of us know we're failing, and we get passing grades all the time." The employers know that, too. So they don't request the transcripts, they don't know what courses the students took, and they don't know what grades they got. All that's left is the diploma. That is not the case in most of the countries of central and western Europe, in Japan, even in Singapore. In those countries, the courses, the grades, and teacher recommendations are very important in getting an apprenticeship, which is the key to a career, because these indicators really mean something in those countries.

With regard to college admissions, 58 percent of the young people in the United States who graduate from high school go on to college, but to what college do they go? More than half attend colleges that require only a diploma for entrance. What happens to the students after they get in is another matter. What I am concerned about are entrance requirements, especially since community colleges all over the United States are now petitioning their state boards of education to allow them to admit students who do not have a high school diploma.

Imagine how this entire system looks to the students. For more than 75 percent of America's high school students, "if you're not going to college, you're finished." High school makes absolutely no difference and leads only to dead-end jobs. And it doesn't make any difference, they tell us, what courses they took or how hard they worked. It's utterly irrelevant.

In Rochester, we also set up focus groups of college admission officers and the people who do hiring in firms and asked the college representatives what their standards were. They said, "We don't have any standards, we have no requirements, what we have is preferences." "What do you mean?" we countered. They replied, "Look, it's a real world — supply and demand. If we insisted that high school students in this country have the qualifications necessary to do okay — not real well, just okay — we wouldn't have enough applicants and would have to start closing our doors. We're not about to do that." That's what they said, that's what the students know, and that's the system we have in the United States. For more than three-quarters of our young people, the system produces a world in which it is not worth making an effort in school. Do just enough to get by, just enough to get a diploma. We would be delighted to find evidence to the contrary, but so far we have not found any.

Of course, another aspect of a new national assessment is that employers must be convinced to accept it. It has been suggested that it would be much more cost effective — and probably easier — to go directly to industry and essentially ask them to use the assessment data we already have from local schools. Perhaps industry should just ask for more substantiation of student performance, as measured by assessments already in place. If industry would do this and if students saw that performing better did make a difference, perhaps there would be no need to put another assessment system in place.

The answer to that suggestion is, What data are now available that represent not only an honest estimate of the performance of the students but also an *acceptable* standard of performance? Should we use, for example, the school system's data from the Iowa Test of Basic Skills as the examples of student performance to be shared with local employers? Knowing that if the employers then use those data to decide who gets jobs, what the pay rates should be, and what career opportunities will be available, we will, in effect, cement in place the curriculum that is represented by the Iowa Tests of Basic Skills?

The central issue is that the nation needs a new curriculum, a new instructional style, and new objectives for students, and that assessment can play a powerful role in bringing that about. My view is that it can — and will. The implicit or explicit syllabus and the assessment tool that we use makes a profound difference. Students' performance on all the existing measures could be shared with employers, but it would set the nation back substantially by keeping in place a system that is inadequate for today's needs.

12. Some thoughtful people have expressed concern about the impact that comparing students to a standard might have on the tradition of competition in the United States. My response is that, for 15 years or so, we've had one international assessment of mathematics and science capabilities after another. Each one has been more devastating than the last, to put it mildly. What effect has this had on the American people? There are data that show that most Americans think that U.S. schools are pretty bad except for those where they themselves live. How can this be? These two facts — that is, our capacity to ignore the international data and what the poll data show about our general view of the quality of education in the United States versus that in our local school

— come down to the same explanation. If you go to a working-class community and talk to parents whose children are attending the same school they did, the parents will say that it looks like their children are taking the same courses that their parents did, using textbooks that look similar, and going through at about the same rate. So what is the problem? If you go into a fairly well-to-do community in the suburbs where real estate prices are determined by the quality of the schools, parents have two questions: Do the same proportion of kids get into Harvard, Yale, Stanford, and Cornell that used to? and, How does this district compare with other upper-crust districts in the state? If the answers to those questions are "yes" and "just as good," then parents ask, "What do these international comparisons in math and science mean? What exactly is the problem? The chance that my child will go to Yale and become an investment banker in Manhattan is just as good as when I went to school 20 years ago."

Comparing themselves to other districts within their state is similar to what General Motors and Ford and Chrysler did at the beginning of the 1980s. Each competed only with the other big two. We can continue to send the same proportion of students to Harvard, Yale, and Cornell; there may even be the same chance of getting a job in the working-class community as there was 20 years ago; yet, the whole country can fail in the international arena. We need a way of comparing ourselves to other nations in terms that the American people can understand and appreciate. We haven't learned how to do that yet.

Even at the school level, there are numerous ways that the data can be used to make useful comparisons. One issue of particular importance is, How can data from student performance assessment be used to structure teacher incentives? Most people assume that that means a pay-for-performance scheme, with individual teacher bonuses related to the progress of the students in their class. What I have in mind is a system in which staffs of entire schools might receive a bonus at the end of some period, with the bonus dependent on the progress that all of the students in the school make. It might be in the form of pay or it might be increased resources for the school. The point is, if the rewards that went to the school were based on the progress that *all* its students made, there would be an enormous incentive for the staff to collaborate and for the most able teachers to help their colleagues with lesser abilities. These incentives do not now exist in most schools.

One can use these assessment data to structure incentives for competition within the school, or such information could be used to construct an incentive system that would produce more cooperation, more collegial work, and so on. Furthermore, in the assessment system that we have in mind, a large part of the basis for evaluating a student's competence would be project work, much of it done in collaboration with other students. It's possible to produce data about student performance that are designed to encourage collaboration and cooperation among students. Obviously, it's also possible to construct a situation that produces only competition. We do that all the time now. Interestingly, the biggest complaints about that are coming from employers who say, "Once these kids come into the workplace, what counts most is their capacity to cooperate with one another. We don't want to hire people, no matter how good they are, who can't work as members of a team. You educators have a system in which only individual effort counts." Cooperation is now seen as cheating. But one can design an assessment system that encourages cooperation, collaboration, and quality because, in effect, the score obtained is based on the degree to which the work that is produced collaboratively is first class.

13. Calibration is an important step in making possible different forms of exams and avoiding a single national examination. It should be possible to reach a consensus (at least in some areas) on what students ought to know and be able to do. The Mathematical Sciences Education Board and the National Council of Teachers of Mathematics brought us a long way toward a consensus about what students should know in mathematics. We're moving toward a consensus in science and writing. In addition, a number of states are developing exams to take advantage of the emerging consensus in those and other subjects.

As agreements emerge, a number of organizations and states may be expected to develop examinations loosely related to consensus on the goals. That would provide a series of exams with a number of variations, some of which might be quite significant, but which would still be related to a developing consensual view about what students ought to know and be able to do. It should be possible to calibrate these to one another. At present, however, we have not yet developed the calibration mechanism, or the standard, or even the consensus on content. Nevertheless, for a whole host of reasons, the idea seems not only plausible but necessary.

2

Equity in Science Education

Ray Marshall

Three very important questions are related to equity in science education. The first is, Why is science education important for everybody in today's world, and not just for the elite (which is what we've been doing)? The second question is, Why is equity important today, particularly in the United States? Finally, we need to ask, What is the relationship of equity to assessment? I will argue that equity and national performance have merged as joint concerns because we will not achieve one without the other. That means that equity is no longer simply a moral issue, but has become a very important economic and political issue. And while moral issues are fundamental and terribly important, my experience is that ideas cannot be sold mainly on moral grounds to other than moral people. One needs to hook into some kind of economic or political interest. I will argue that this is where we are now in the United States and that equity should sell because it makes economic sense.

Why Science Education for All?

Today, science education is important for all people, not just for a few, for economic and technological reasons that have come to the fore in the past two decades. To understand this, we must look at what has happened to the U.S. economy in this century.

Physical and Human Resources

We know that the United States became the world's leading economy without having all its citizens understand science. This was so because the mass production system that has dominated the American economy since early in this century organized the work so that only managerial, professional, scientific, and technical workers had to have higher-order thinking skills. Most workers could earn a living with basic literacy and numeracy. One reason for that is fairly simple and straightforward: The United States had abundant natural resources. Now, however, natural resources are no longer very important; they have been superseded by science and technology.

Theodore Schultz won the Nobel Prize in economics for demonstrating that the contribution of physical resources to improvements in American productivity is zero for as far back as records have been kept.[1,2] We are continuing to substitute knowledge and information for physical resources — that's what technical progress is all about. For example, Schultz and other agricultural economists have demonstrated that there has been no increase in physical resources in American agriculture since 1925. Today, we use less labor, less land, less physical capital than in 1925, yet we have tripled, even quadrupled, agricultural output. Ideas, skills, and knowledge are substituting for physical resources.

Consider information technology. Today, approximately eight pounds of fiber optics can substitute for several hundred tons of copper. This is knowledge substituting for physical resources, and one cannot make that substitution without the necessary human resources. Schultz's work also demonstrated that human capital (ideas, skills, and knowledge) — not commodities — has *always* been our most important resource. About 80 percent of America's increased productivity since 1929 has been due to technology and changes in human capital, with only 20 percent due to increases in physical (or machine) capital. The return on investment in people is higher than the return on investment in physical resources. Yet too many of our social and economic policies are still designed to deal with physical rather than human resources.

Mass Production and Economies of Scale

Another reason for America's economic success was economies of scale. Mass production made it possible, for example, for Henry Ford to reduce the cost of a touring car from $850 to $360 over six years. He did that via large-scale organization, reducing the cost of production by manufacturing millions of units. The system was organized so that the few people at the top — managers, scientists, engineers — were the thinkers. A laborer did not require much education to make a good living on the line. Most people could go to work in a mass production factory and, because of economies of scale, could greatly improve their standard of living.

The United States used mass production in many different areas, and for a while it was easy to raise the nation's standard of living by doing relatively routine work. The organization of the mass production factory system, called "scientific management," was developed by Frederick Winslow Taylor. He believed that work should be organized so that the least intelligent workers could do it. In fact, if workers had to do any thinking, something was wrong. In his view, workers should be appendages to machines. Thinking would interfere with the routine.

The mass production system served America very well for years, but then two things happened.

First, technology changed the basic organizational structure of mass production, and now, information technology permits many of the advantages of economies of scale without requiring the scale. In other words, we are now reprogramming computers instead of retooling machines. Economies of scope as well as scale result; many different products can be produced using the same technology.

The second thing that happened was the internationalization of the American economy. Henry Ford could produce many cars because he and General Motors and Chrysler ultimately had the American market pretty much to themselves. They learned early on to organize the market, split it up among themselves, and fix prices, thereby achieving economies of scale. But in an American market open to 40 or 50 foreign automobile manufacturers the old economies of scale are gone. One has to be competitive in a different way than Henry Ford and Chrysler and General Motors were earlier.

The Need for a New Education System

The United States can no longer rely on economies of scale and what economists term "interindustry shifts" to provide easy improvements in its standard of living. Production has changed. The way work is organized today, one cannot make a good living without being able to handle science and technology. The new technology is ubiquitous and the economic gap between people with and without technical education is widening (see Table 1). And for the nation's economy to be world class, all its workers must be educated and trained to world class standards.

The American educational system was organized like Ford's factory: its function was to turn out people who did not have to think, educated by teachers who did not have to think. It was assumed that the people in charge of the system would know what everybody needed to do; just like Frederick Taylor, they assumed that there was one best way to do it. The professor of education's job was to find out what that best way was; the job of the men who ran the school system was to impose that method on the women who did the teaching. Prior to the mass production system, more teachers were male. One reason women replaced men as teachers was that women had less power than men. Just like Taylor's factories, people with limited power are needed to make the mass production school system work. In fact, not so long ago, if a woman teacher got married, she got fired. Marriage brought women additional power; they could no longer be as easily controlled, so out they went.

Table 1. Income and Educational Attainment: Median Income for Various Levels of Education (1987 dollars)*

Year	Elementary	High School	1 to 3 Years College	4 Years College or More	Median Income High School Graduate as % of College Graduate
1973	$19,562	$31,677	$37,730	$49,531	64.0%
1979	18,013	30,038	35,902	47,011	63.9
1987	16,094	27,733	36,392	50,115	55.3

Annual Rate of Change in Real Income at Various Levels of Education

1973-79	-1.4%	-0.9%	-0.8%	-0.9%
1979-87	-1.4	-1.0	0.2	0.8
1973-87	-1.4	-0.9	-0.3	0.1

Source: Lawrence Mishel and Jacqueline Simon, *The State of Working America* (Washington, D.C.: Economic Policy Institute, 1988), table 21, p. 11.
*Incomes of families by educational attainment of householders, 25 years of age or more.

Why Is Equity Important?

At the most fundamental level, equity is important because it holds a society together. Any organization depends heavily on internal unity and appeals to higher values to achieve its goals. But people don't give their lives for a few more groceries. Economists miss the boat when they assume that all people want are material rewards. Nonmaterial, or higher, values are what people will die for.

A country that has internal unity will be great. Unfortunately, the United States has been losing its internal unity since World War II. WWII unified us because it was an all-out war, something certainly not possible in the nuclear age. The real challenge for a democratic society is to unify people in the absence of war. What we must search for is what William James called the moral equivalent of war. This is true of every society, and it is especially true in the unique, increasingly multicultural, multiracial society that is America.

Our Multicultural, Multiracial Society

Some time during the next century, non-Hispanic white people will be a minority of the U.S. population. This will happen in California by the year 2000, in Texas by 2010, and in New York by 2015, depending on the assumptions made about immigration and birth rates. It is already happening in the schools, which are a kind of early warning system for society as a whole. Almost all of the growth in the work force between now and the year 2000 will be women and minorities. A majority of minorities. And that is not speculation. Economists have to speculate about many things but not about demographics, because those are fixed by the people who are already here, by their age composition, and by what present trends are likely to produce in the future.[3]

A multicultural, multiracial society has many advantages over a homogenous society. We don't appreciate or celebrate that enough. If we have diversity, we have creativity. The quality of life is much better. One thing this new world is going to require is an abundance of creativity and the ability to deal with diversity that many people — the Japanese, for instance — have trouble understanding. Americans have the real advantage because most of the people in the world are not white, and the number of white people will decline absolutely and relatively as far as we can see into the future — certainly well through the 21st century. Most of the young adults who will enter the work force over the next 50 years are in the Third World, and they are mainly nonwhite.[4] What we learn about our own multiracial society could give us real insights into the rest of the world — if we take advantage of the knowledge.

It is not necessary to emphasize the downside of being a multicultural, multiracial society. Nothing is more deadly than racial, ethnic, and religious conflict. History provides many lessons about the consequences of ignoring equity and not making sure that people are treated justly. And what I mean by equity and justice is doing the most for those who need the most help. In fact, one of the most important equity concepts is that nothing is more unjust than the equal treatment of unequals.

People are going to be either assets or liabilities. We can say — as economists frequently do — that equity costs too much, and that equity conflicts with efficiency. That view is nonsense. Equity and efficiency work together. We must educate people so that they can take care of themselves or crime will escalate — and it costs much more to keep people in prison than it does to educate them. In my state of Texas, $25,000 to $30,000 a year is spent on each prisoner and more prisons are being built each year. The correlation between dropping out of school and being incarcerated is higher than the correlation between cigarette smoking and lung cancer. We can prevent these problems by seeing to it that people are well trained and well educated.

Equity, Performance, and Assessment

What does it mean to say that America is a competitive society, and what does this have to do with assessment? In today's internationalized environment, one can compete in only two ways. The first is by cutting workers' real, after-inflation income, which is what the United States has been doing since 1973. The second is by paying much greater attention to productivity and quality. There are no other options.

In 1990 the Commission on the Skills of the American Workforce (CSAW), which I co-chaired, compared the United States with six other countries (four European nations, Japan, and Singapore).[5] The study found that the United States is unique when compared to these six other countries because most American companies are competing by trying to cut wages rather than by becoming more productive through improving the quality of their products. Less than 10 percent of American companies were trying to compete in ways that make it possible for their workers to maintain and improve their incomes. Most were focusing on reducing wages and costs rather than reorganizing the work to become more competitive.

From a Producer-driven to a Quality-driven System

What kind of organization is required to increase productivity and quality, whether in a school, a government, or General Motors? There are some important principles at work here. First, the organization must be *quality driven*. Why? Because quality really means meeting the student's, client's, or customer's needs. Mass production systems are producer driven. Today, consumers are driving the system, and the principles of a consumer-driven system are very different from those of a producer-driven one. According to Henry Ford, the buyer could have any color car as long as it was black. Today, to sell the product successfully, we need to customize it to meet the customer's requirements.

Second, management has to become much leaner. Whether in a school or a factory, a quality-driven system turns decision-making over to the people who are actually doing the work. In the mass production factory, a large part of the work was directly on the product — putting bolt #35 on the rear left wheel. Workers didn't need much skill and the work was controlled by the line. Today, most workers do indirect work, which means, among other things, using the information that machines make available. Knowledge of science and technology is necessary for understanding and imposing order on information.

Ordered information is necessary for improving the quality of the product, reducing costs, and increasing productivity. For example, computers can make more information about students available to teachers, which can be used to meet each student's unique learning needs. Similarly, computerized information can help teachers understand what teaching methods produce the best results or help schools understand which students are most likely to drop out and why. In the work place, information can be used to help workers understand why defects occur at particular points in the production process or to help companies identify markets with greater precision. Information can help companies understand which products are selling and which are not, as well as how much of a particular product or part is in the supply lines. None of this ordered information would be available or could be used effectively without knowledge of science and technology.

Making Everyone Personally Responsible

It is more efficient to prevent defects than to try to catch them at the end of the process. Making everyone in the school, factory, or government agency responsible for quality, or excellence, prevents defects. This is illustrated by the case of a worker in a California plant that was reorganized to make all workers part of a team and responsible for product quality. When asked if he noticed a difference between the new system and the old, he said,

> Well, everything is different. In the first place, I feel a lot better about it because now they at least recognize that I know something and that I can participate in this process. The second thing is that now I know that it is my job to prevent quality defects and, therefore, I do it — I can stop the line. In the old system, if a three-wheel car or five-wheel car had come by, I'd just ignore it. If I had said, "Look, there goes a three-wheel car," my foreman would have said, "Mind your own damn business. We've got inspectors to catch three-wheel cars. That's their business." But now, it's everybody's business.

The same thing should happen in schools — teamwork to see to it that students learn. It ought to be everybody's business, not just certain teachers, or certain counselors, or certain students. That is a very important difference in the management system.

There are many illustrations of the effectiveness of a team approach to manufacturing. In general this approach improves productivity, quality, and flexibility by reducing the need for supervisors, inspectors, and other support workers; by increasing the number of skills each worker has (thus avoiding the wasted time that was common in the mass production system, when some workers had to wait for others to finish their specialized tasks); and by making all workers in the team responsible for the group's work assignments. The concentration of the group's collective skills greatly facilitates problem solving.[6]

An illustration of a team approach in education was the one used in James Comer's schools in New Haven, Connecticut. Comer and his colleagues took some of the worst schools in New Haven and made them among the best. His model is now being used in over a hundred schools throughout the country. When asked what he and his associates did to produce these results, Comer gives two answers. First, they changed attitudes. They caused everybody — teachers, parents, students — to believe that these poor, predominantly black children could learn. Second, the school management was reorganized on a team basis. The management board consists of the principal, representatives of teachers and parents, and child development (or mental health) specialists. Parents not only serve on the management board, but also work as teachers' aides and perform other tasks in the schools. The main objective of Comer's method is to focus school and family attention on student learning. This contrasts vividly with the traditional mass production model where teachers work in isolation and without the help of child development specialists or parents, and where the main standard for measurement is average daily attendance and time in school, not student achievement. The Comer model likewise has the students learning and working in teams.[7]

Sharing Information

The second major reason for lean management systems in high-performance organizations is that fewer bosses are required if employees can do their own work, impose order, think, and solve problems. The old system — the hierarchy system that still exists in most of our schools and factories — was based on the assumption that workers didn't know much and didn't need to know much. In the old system, foremen withheld information from workers; in fact, that management system came from the Prussian military doctrine of "need to know" (that is, "we [bosses] will not share information with you unless we think you need to know"). The new organization of work depends on everyone using shared information to improve the performance of the whole organization. That also means that management doesn't have to supervise people they don't trust. That's quite different from Taylor's system where management believed that the workers would loaf if they weren't continually watched.

Creating Positive Incentives

A high-performance organization uses a positive incentive system. That's how one gets people to work together. Unfortunately, most U.S. organizations employ either negative or perverse incentives. Negative incentives tell employees that if they don't do right, they'll be fired or reprimanded. Fear is the motivating force. People are not likely to go all out on the basis of fear — rather, they have to *want* to go all out. A perverse incentive system is one in which employers say they want one thing but reward employees for doing something else, which is the case in most of our colleges and universities. We say, for example, that we want the faculty to be good teachers, but they know that good teachers don't go anywhere.

Professors know that they can only advance on the basis of research; they also understand that if they are not good teachers, it won't cause too much damage provided their research is good. But if they are just good teachers, they will not go very far in most universities.

Worse yet, in American factories, most workers believe that if they go all out and increase productivity, they'll *lose* their jobs. If we want workers, teachers, people in our organizations to go all out, we must give them better incentives than to threaten their jobs. That's why IBM and a few other U.S. companies and most companies in foreign countries provide job security, sometimes even offering life-time employment. That's the way to tell a worker that he or she will not get fired for improving the performance of the organization.

All kinds of perverse incentives exist in school systems. Think of the perversity in saying, "we're going to cut the money for programs for lower income kids but leave it wide open for the educable mentally retarded." One doesn't need much insight into human nature to know that the number of "educable mentally retarded" students is suddenly going to increase. Officials will start labeling everyone they can as educable mentally retarded even though one of the worst things one can do to a child is give him or her that label without any basis in fact.

Incentives and Assessment

What does a positive incentive system do? First, it rewards what it says it wants to achieve. That's why assessment is so important. The old system was process-driven rather than outcome-driven. It was assumed, for example, that people who followed all the rules were going to be successful teachers. There were no measures to show whether or not that was true. A positive incentive system employs accurate assessment techniques, then gears rewards to the desired outcome. This doesn't mean simply financial rewards — because people are not motivated just by money — it means recognition. Participation itself is an important part of a positive incentive system. People need to know that they have contributions to make. They need to know that others believe in the value of their work. Then they build on that knowledge to do their best.

A positive incentive system also is concerned about equity and about the internal unity of the organization. Nothing destroys an organization faster than injustice and unfairness. That's the reason that the Japanese do things that Americans consider to be "Mickey Mouse." When the Japanese took over the NUMMI plant in California, they eliminated all the elitist trappings of American managers — the opulent offices, private dining rooms, different pay scales. It's difficult to convince workers that they and management are all in this together and that everyone should go all out when the executives vote themselves a $21 million increase in salary and cut the workers' pay! That won't build internal unity. The Japanese understand this, and that's one of the reasons that they are running rings around the United States in the car manufacturing business.

Another reason why a positive incentive system is so important is that this new kind of work setting gives everybody a great deal of discretion. Since it is difficult to compel excellence, workers must want to go all out to improve quality and productivity on a continuing basis.[8] But this new kind of organization not only requires equity, it requires people with higher-order thinking skills of all kinds. Workers need an experimental frame of mind, which is probably the most important thing that can come from science education. It is not whether one learns biology, chemistry, or physics, it's developing the ability to deal with ambiguity while knowing that, in the real world, there are no answers at the end of the book. I've seen books that state, "We're going to teach you how to solve problems — here's a problem, and you go through these steps to solve it." I say, "Wait a minute — the most important thing I must do in my life is figure out what the problem is." An experimental frame of mind for teachers, for workers, for everyone is what we need if America is going to be world class.

How Are We Doing in This Country?

The CSAW study found that very few U.S. companies were seeking to reorganize work to increase productivity and quality. Most were sticking with the traditional system and were trying to compete by reducing wages, directly or indirectly. It is difficult to change a system that has a history of success. The United States did very well with the old system — and developed a great deal of hubris about it early on. (Hubris is when the learning stops. In 1977, the chairman of the board of one of our major automobile companies told me, "The Japanese can have those little cars — there's no money in that. You know, we're world class. They'll never get the big cars." We all know what happened.)

The same thing has happened in our schools. Our system was designed to turn out people who were literate and polite because that was what employers wanted. The commission study asked employers whether they perceived a skill shortage today. About 20 percent of them said yes. When asked to describe a worker with the skills they needed, three-fourths described people who can read, write, show up on time, and be polite; only 5 percent wanted people who could think and deal with problems. The second thing employers told us was that America doesn't give companies any incentives to work toward achieving a high-income, high-quality country. There is no industrial policy — and no educational policy.

Other countries have plans to help their companies become world class, and that includes plans for quality educational systems. Those plans include educating all their people in science and technology. While all of the other countries we studied differ in economy and culture, they share a commitment to the education and training of their workers and to high-performance work organizations; we do not. They have national consensus on building high-performance economies; we do not. They insist that virtually all of their students reach a high educational standard before leaving secondary school; we do not. They provide high-quality, technical education for noncollege-bound students; we do not. They operate comprehensive labor market information, job search, training, and income support systems; we do not. They support company-based training for frontline workers; we do not.

U.S. schools are not organized to provide that type of training for everyone. We have a very inequitable, two-tiered education system that does a lot for students who are going to college. Those who are not are largely ignored, and, unfortunately, it is determined very early who is going to college and who is not. American educators keep talking about not wanting to track students when they are 15 or 16 years old as the Germans do. But we track them when they are born; we track them when they enter the first grade; we decide that early that some people are not going to make it. In the Year 2000, only about 30 percent of U.S. workers will be college graduates. About a third of our work force growth during the 1990s will be high school dropouts; another 30–35 percent will be high school graduates who do not go to college, and very few of these will receive any formal job training.

Developing a Policy to Change the System

The question then becomes, How should we change the system? The most important initial action is to develop a policy. We need to develop a consensus that America wants to be a world-class, high-income country. If we do that, then we are likely to pay much more attention to quality education for everybody. Once that need is understood, all other actions will fall in line.

The next most important action is to develop strategies for improving our learning systems, with emphasis on equity for all. Improved math and science education is often missed when dealing with a two-tiered system because some students get math and some don't, some get exposed to real science and some get exposed to something else that may or may not be real science. It usually is not.

All children should start school ready to learn — and they don't. The Head Start program can help. Schools should be restructured into high-performance organizations where student achievement is the driving force. Equity, science, and mathematics must be deeply embedded in whatever assessment scheme is adopted because those are the things that are likely to be left out of the system. School governance needs to be changed, teachers must be responsible for teaching — that is what restructuring ought to be about. Special attention needs to be given to minority education. Already, many students in urban districts are members of minorities, but there are fewer minority teachers — about 30 percent minority students, 10 percent minority teachers — and it's going to be 40 percent and 5 percent by the end of the decade if present trends continue. It's worse in science, technology, and math than in other subjects because only about 4.5 percent of the math and science teachers are minorities.

We need alternative learning systems for students who don't do very well in schools or for whom the schools don't do a good job. The Science Skills Center in Brooklyn, New York (described in chapter 15), is a good example of the type of program needed, along with the Job Corps and the Summer Training Education Program (STEP), to help overcome summer loss. Eighty percent of the educational achievement differential between advantaged and disadvantaged students is due to summer loss. We can make a lot better use of summers than we do now by incorporating science, math, and technology into summer programs.

We should pay more attention to on-the-job training. On-the-job training is terribly important because 85 percent of income-producing education — "income-earning learning" — comes on the job, not in school. Who gets the best opportunities on the job? Mainly white males. Minority males get the worst chances on the job, regardless of education. Race doesn't seem to make a difference with regard to women, who get second best to white males. Our study found that only 7 percent of American front-line workers get work-related education and training. Elsewhere in the industrialized world, a majority of the noncollege-bound get some kind of scientific or technical training for work. We need to create a system that makes it possible for the noncollege-bound to keep their options open. They are not going to do that if they don't get science and math in school. Everyone should graduate with the same basic skills, especially in science and mathematics. This is why we should adopt national standards that *all* students are expected to meet before they leave secondary school, as is done in most other industrialized countries.

In addition to a new education performance standard that all students are expected to meet, the CSAW's recommendations include the following:

- States should take responsibility for ensuring that virtually all students meet this new standard. What we need, of course, is a program to prevent and recover the more than 20 percent of our students who drop out of high school — almost 50 percent in many urban areas — who will constitute over a third of our front-line workers. This could be done by creating and funding youth centers as alternative learning environments for those who cannot meet these standards in regular schools. Incentives for schools to improve learning opportunities for these students would be created by allowing the students to take the money allocated for their education with them to the youth centers. In order to provide incentives for young people to continue their formal education, once the youth centers have been established, children should not be allowed to work before the age of 18 unless they meet the basic education standards or are enrolled in a program to meet them.

- We also need to develop a high-quality education and training system for the great majority of our students who do not pursue baccalaureate degrees. We therefore need to develop a comprehensive system of professional certificates and associate degrees for adults and noncollege-bound students. The CSAW recommends the convening of national committees of business, labor, education, and public representatives to define certification standards for two- and four-year programs of technical preparation in a broad range of occupations. This system should be "open ended" in the sense that students would be able to move easily between certificate programs and college. The system could be financed through "GI Bill" type entitlements of four years of post-secondary education and training for all who meet the basic standards for secondary school leavers. The GI Bill for World War II veterans is one of the best investments the United States ever made. It should be reworked and made universal.

- As noted, a very small minority of American employers are moving to high-performance work organizations or investing in training of front-line workers. The CSAW therefore recommends that all employers be required to invest 1 percent of payroll in the training and education of their workers. Those who do not wish to participate would be required to contribute the 1 percent to a general training fund to be used by states to upgrade worker skills. The commission further recommends that public technical assistance be provided to companies, particularly small businesses, to assist them in moving to high-performance work organizations.

The United States is still the world's strongest economy, but we are weakening rapidly. Whether or not we make it depends heavily on whether we give attention to science education throughout life — not just when a student gets to middle school or high school, but when he or she begins preschool. People need to be able to think in ways that make it possible for them to function in this society — not just on the job, but throughout life. Imposing order on information, making choices, and thinking are much more complicated than in the past. Paying attention to equity will avoid the fragmentation and polarization currently rife in the United States.

References and Notes

1. Schultz, T., *Investing in People: The Economics of Population Quality*

(Berkeley: University of California Press, 1981).

2. Carnevale, A., *Human Capital: A High-Yield Corporate Investment* (Washington, DC: American Society for Training and Development, 1983).

3. For a good discussion, see Haub, C., *Understanding Population Projections* (Washington, DC: Population Reference Bureau, 1987).

4. For further discussion of this point, see Marshall, R., "Jobs: The Shifting Structure of Global Employment," in J. W. Sewell and S. K. Tucker, eds., *Growth, Exports and Jobs in a Changing World Economy* (Washington, DC: Overseas Development Council, 1988), 167–190.

5. Commission on the Skills of the American Workforce, *America's Choice: High Skills or Low Wages!* (Rochester, NY : National Center on Education and the Economy, 1990).

6. See for example, Parker, G.M., *Team Players and Teamwork* (San Francisco: Jossey-Bass, 1990); see also Marshall, R., *Unheard Voices: Labor and Economic Policy in a Competitive World* (New York: Basic Books, 1987).

7. Comer, J.P., *School Power* (New York: The Free Press, 1980).

8. For a good discussion of this point, see Zuboff, S., *In the Age of the Smart Machines* (New York, Basic Books, 1988).

3

Fairness in Testing: Are Science Education Assessments Biased?

Maryellen Harmon

This chapter examines the issues of bias and fairness in science testing and their implications for future developments in the assessment of science programs and pupil achievement. It is timely to do this for two reasons. First, pressure for reforms in science and mathematics education at the elementary and secondary school levels has been intensified further by reports from the National Assessment of Educational Progress (NAEP),[1] the International Association for the Evaluation of Educational Achievement,[2] and the American Association for the Advancement of Science;[3] by the implications of *Workforce 2000*;[4] and by President Bush's recently announced "Goals for Education." The resultant consensus for change forces reconsideration of all aspects of science education, including the role of assessment. Second, there has been increasing pressure from people involved in developing new science curricula to have student achievement measured by instruments congruent in philosophy and content with the materials concerned. And while the developers proclaim that current, standardized assessment instruments are biased against their students in the area of content, there are other issues relevant to bias that must also be examined.

The issue of bias in science assessments cannot be explored without looking first at concerns related to bias and fair testing in general. A few years ago, there was considerable activity in connection with the questions raised about assessment instruments whose language and content could militate against the success of certain groups — notably, African-Americans, other ethnic groups, linguistic minorities, and women. At a superficial level, bias in test items was identified with language — dialectic and phonological differences between the spoken language of a minority group and that of the mainstream — and with item examples and content areas less familiar, say, to females, or to members of ethnic minorities, than they were to white males. At this level, the problem was addressed carefully by the testing community; items were reviewed by adult representatives of the groups in question as well as by experts in measurement and were pronounced acceptable or were corrected. The *Code of Fair Testing Practices in Education*[5,6] was prepared by the Joint Committee on Testing Practices of the American Educational Research Association (AERA), the American Psychological Association (APA), and the National Council on Measurement in Education (NCME) (in 1985 the *Standards for Educational and Psychological Testing* had been published under the aegis of the same three organizations). But the problem did not go away.

It is notable that on the most recent NAEP,[7] at all levels above basic arithmetic, girls scored consistently below boys, and African-Americans, while showing significant progress since the preceding examination, still lagged from 20 to 40 points behind whites in problem solving.[8] It is also significant that, on the Scholastic Aptitude Test (SAT) for college admissions in 1988, women averaged 56 points lower than men, and African-Americans and Puerto Ricans some 90 points below whites. Apparently, nothing is changed by mere emphasis on guidelines, rules, and codes unless the codes are related to existing law and reinforced by something with teeth in it, such as litigation.[9, 10]

The matter of bias versus fairness in testing is a complex one and must be viewed from at least three perspectives: the test items themselves, the test-takers, and the uses to which the test results are directed versus those uses for which they were intended by the test-makers. In examining the test items, we must be concerned with issues of the content of the field tested, the language and context in which it is tested, and the match of that content with the curricular and instructional patterns of the classroom as well as with the mental constructs that tie each item to the content knowledge presumably measured. The content of items also must be scrutinized for its congruence with the life experiences of the test-takers. America is a diverse nation and contexts perfectly normal and understandable to college-bound students of Weston, Massachusetts, may be foreign to those from poverty-structured villages of Appalachia, the American Indian reservations of the far West, or the "Black Bottom" of downtown Detroit, not for reasons of lack of schooling, but because of a total cultural mismatch. Construct validity has another and technical meaning to measurement specialists: the match between the item on the test and its significance for, and ability to measure, aptitude for the type of job or educational competence it is supposed to predict. Does it, in fact, measure, and measure effectively, higher-order thinking or some other competence? To what use will the test results be applied in determining the test-taker's future and has the test any demonstrable authority with respect to that use?

The following discussion is divided into three sections. In the first, I shall distinguish the terms "bias" as understood in the psychometric community, "bias versus appropriateness" as popularly understood, and "fairness." I shall then explain some of the sources of bias or lack of fairness in testing, which may throw light on our concerns as we approach a new era in science assessment. For it is a fact that we are at the beginning of a new era. Across the country, science generally is not being assessed by external tests administered in schools except where schools use mandated batteries of which science is a part.[11] Even then, many states require only the reading, language arts, and mathematics portions of the battery. Because concerns about the validity of the tests in question will be discussed, I shall review very briefly the various ways of regarding validity and reliability.

In the second section, I shall consider samples of existing national norm-referenced tests and text-embedded tests[12] in science from the point of view of content and construct validity and any evidence of bias. It is true that statisticians have methods of detecting test bias in their analyses of test results, but this type of analysis is not available in a mere consideration of the text of the items. The current review is an early part of continuing research taking place at the Center for the Study of Testing, Evaluation and Educational Policy (CSTEEP) at Boston College and further analysis will be forthcoming.

In the final section, I shall look at fairness in testing from the viewpoint of single test fragility when it is used as a sole measure of need or achievement, at some of the uses and misuses to which scores are directed, and at what actually is being measured and its relationship to reform efforts in science and mathematics education.[13] I shall then consider a few examples of new efforts to construct alternative modes of assessment and the task that these alternatives will present to all of us to ensure that they do, in effect, measure what we want to measure and are used in ways that are both fair and useful to those who must be informed by them: teachers, students, parents, and policymakers.

Bias and Validity: Defined and Reviewed

Terms

Bias is commonly defined in the psychometric community as the over- or underprediction of performance on a criterion.[14] The key word in this definition is "criterion." Suppose that the criterion is a grade point average or another test. Measurement specialists, using sophisticated statistical analyses of the results of a given test, can point out that correlations between test scores and these criteria have about the same accuracy for minorities as for any other identifiable group taking the test; that is, if the test is biased, it is equally biased for all! Hence, they conclude that the test is not biased. This begs the question because there is no guarantee that the grades used as a criterion were not based on tests or other measures just as tainted (if, in fact, bias is present) as the one in question, and the grades set up a self-fulfilling prophecy. It also begs the question in another way, as we shall see farther on: the issue of the appropriateness of the ways in which the results are used.

Alternatively, bias may be defined as a quality of the test itself that causes it to measure differentially in different populations. I shall consider below examples of language and context that illustrate this type of bias. In the more popular definition of the term, bias means that the test is unfair, leading to faulty predictions even when scores are used appropriately, or excluding students from opportunities in a way that is unfair, given the nature of the test. That is, the student may know more about a particular objective than the test score indicates (or possibly may know much less), but because of the wording, format, speededness (time constraints), or other factors, cannot demonstrate that knowledge as reliably as other test-takers can. A test is biased if its results do not match the actual situation of student competency. Racial or ethnic bias, in this sense, means that the test systematically reads low or high for particular racial or ethnic groups. It predicts inaccurately for these groups as compared with the reliability of its predictions for the norming or mainstream group. Again, there is a key word: "prediction." In both of these senses, we are concerned with the test as a "gatekeeping instrument" and its ability to reliably predict future success in the next grade, in college, in graduate school, or in employment.

We seem to have closed a circle without being much nearer to the issue of bias. That is, someone could argue that if the future into which the test is the gate is "more of the same" as far as lectures, note-taking, and multiple-choice tests are concerned and if it requires the very performance — knowledge of vocabulary, a recall/application-level knowledge of content, competitive attitude, and individualistic work habits — that renders the test unfair to certain ethnic groups, then as a "gatekeeper" it is not biased, even though its outcomes impinge more heavily on one ethnic group than another. Such an argument must lead us into a much larger topic: a definition of what we want the future and its actors to be at the university or in the job market. If, as reformers urge, we want to define future education or performance on the job differently from the way it is being demonstrated at present, perhaps we need to change the gatekeeper.

In using the test as a gatekeeper in this way, we are presumably interested in its ability to classify test-takers into those who will succeed and those who will not succeed in the tasks of the future. It is when the test is used as the sole predictor of success or failure, when an absolute cutoff score is defined, that we open ourselves to a kind of "classification bias." Recently, the National Commission on Testing and Public Policy (NCTPP) noted:

> [W]henever people are classified on the basis of cutoff scores on tests, misclassifications are bound to occur. Some who score below the cutoff score could perform satisfactorily in school or on the job, and some who "pass" cannot perform satisfactorily. The solution to this problem is not to avoid classifying people: such classifications are

essential and inevitable in modern society. Instead, it is to avoid classifying solely on the basis of one imperfect instrument.

When test results are used *independently* of other relevant information about people to classify them in ways that affect their opportunities, misclassification can amount to substantive unfairness.... Moreover, when test results alone are used to allocate opportunities, unfairness often falls disproportionately on certain ethnic, linguistic, and cultural minority groups. By unfairness, the Commission is referring to the fact that disproportionately fewer minorities who could perform successfully had they been given the opportunity will be selected.[15]

Arguably, one of the best criteria for future performance should be present performance on related tasks. The commission goes on to note the discrepancy between the "relatively large *test score* differences between majority and minority candidates" and the "relatively smaller differences between majority and minority candidates on the actual performance that these scores may be used to predict" — in other words, the kinds of tasks that constitute the job or schooling opportunity for which the test is the gatekeeper.

However, there are other issues about the uses of tests that must be addressed. What kinds of judgments are being made, for example, by the state certification systems, based on the outcomes of the National Teacher Examination, and by future employers, based on the SATs or minimum competency tests? What is the relationship between the items on these tests and the tasks in the future job? By naming them "aptitude" or "minimum competency," are we deceived into thinking that that is what they measure so that we do not ask, "Competency to do what?" Fairness in testing is primarily an issue not of the wording of the items, but of the validity of the test for the uses that will be made of the scores. And if the scores are used to predict something they were never intended to predict, then the test is inappropriate, the prediction absurd, and the situation unfair to the test-taker.

Therefore, the discussion of test bias and fair testing cannot be limited to the shape and language of the test questions or to the characteristics discussed below of various societal subgroups, but must extend to the question: Testing for what? What information do we want tests in science to yield now and in the future? Who will use this information and for what purposes? Of course, tests must be robust and reliable, but the primary concern today must be with the multiple aspects of validity.

Sources of Possible Bias in the Test Itself

Mary Catherine O'Connor, in "Aspects of Differential Performance by Minorities on Standardized Tests,"[16] reviews the research literature on the causes of poor performance on standardized tests by certain ethnic groups. Her review focuses on a search for causes of the differential performance. She sets aside the heritability theory[17,18] as well as the sociological theories that see

> "performance on tests as merely one of a number of measures of school success or failure, including grades and teacher evaluations, retention, and the acquisition of degrees" all of which are determined by "macrolevel social factors and the individual's and community's response to them."[19]

Although the sociological position has merit, it ascribes failure on tests to "society's failure to address the basic economic and social structural factors that reproduce and maintain relations of inequality."[20] This position does not address directly O'Connor's immediate concern with the question: Why do members of certain ethnic minorities do poorly on standardized measures where other minority groups are able to succeed? In fact, I might add, the macrolevel approach distances us from the level of the tests themselves and the uses to which their results are directed and seems to require a reform of the total society before we can consider possible correctives for bias. The theme of O'Connor's analysis leads in a different direction:

> [T]he failure of researchers and educators to change test scores is a function of many things; but at least one of them is the possibility that we do not yet understand the correct way to approach the sociocultural differences among groups in the schools with respect to the activity of taking tests, with all that that implies.[21]

Although her excellent paper addresses the characteristics of test-takers and only indirectly the quality of items, it bears upon our current issue. She looks at language, at construct validity as mediated through the language of non-native English speakers (NNESs),[22] and at sociocultural attitudes and practices. For the NNES child, the test may be a test of language and literacy rather than a test of mathematics, science, or social studies.

NNESs (children from homes where the first language is not English) may be so familiar with conversational English as to be considered competent to take tests in English. But their first language has influenced both the ease and the speed with which they read English and the ways in which they interpret certain nuances of syntax or vocabulary.[23] This influence lasts for years beyond the point where they have achieved conversational fluency — for some, a lifetime.[24] Also, familiarity with the language has a lot to do with confidence in one's judgment. For example, if certain words among the distractors on a multiple-choice test are unfamiliar, the NNES, or the child in the process of acquiring English, does not have the same probability of eliminating wrong choices as the native English speaker has.

In addition, reading for understanding is not merely decoding words, but consists of an interaction between the reader and the author such that the reader, competent in a content area, anticipates or infers much more than is printed[25] — and is meant to do so by the author. But the NNESs may make quite different inferences from the ones you or I would make, depending upon the extent of their vocabulary, their cultural background, and what to them are relevant data. Is translation then a remedy? The fact is that it is very difficult to get dialectical and idiomatic accuracy in translations, and unless the translation is validated independently, serious questions of test validity arise. While there is considerable consensus among researchers that language is a factor in test performance, there is no consensus on what to do about it. Several suggestions for remedies are addressed by O'Connor,[26] which go beyond the scope of the discussion here.[27]

Another area, possibly even more important, is that of cultural differences between minority and mainstream groups. Among cultural differences that seem to be significant in test-taking, O'Connor includes the attitude toward time,[28] the perceived nature of the outcome in the test-taker's mind, the test-taker's belief in his or her ability, past experience with whites in authority, attitudes toward competition versus collaboration,[29] culturally specific interpretations of a task, the amount of personal anecdotal data that should be brought to bear,[30] early experiences with verbal directions, and cultural practices with regard to children speaking in the presence of adults.

D. McShane gave the following example in testimony at a hearing on The Effects of Testing on American Indians:[31]

> A young Ojibwe (Chippewa) student was tested by several professionals and classified as having certain learning and behavioral problems. In part, this occurred because he stared vacantly into space, completed tasks very slowly, and gave "nonreality-based" responses. As it turned out, the boy had a special relationship with his traditional Ojibwe grandfather, who encouraged his dreaming whether it occurred by day or by night and often discussed the nature of dreams with him.
>
> In Ojibwe thought and language, ganawaabandaming, which means seeing without feeling (objectivity), carries less value than moozhitaming, which means feeling what you do not see (subjectivity). The Pascua Yaqui, Northern Ute, and Red Lake Chippewa Nations have enacted education codes that include the ability to daydream as a criterion for identifying gifted and talented students.[32]
>
> Thus, far from being a negative symptom of "nonreality-based behavior," day dreaming is supported by the culture as a means to, and sign of, creativity.

One further example of the influence of early childhood experiences on school success will have to suffice here. O'Connor contrasts two parental teaching styles during a child's first few years.[33] One, characteristic of middle-class educated whites in this country, is verbal. The child is given directions step by step or "talked through" a task such as making cookies or riding a bike. The parent may demonstrate the behavior at the same time, explaining each step and querying the child to be sure he or she understands both the step and the reason for it. In contrast, Native Americans, Hispanics, and some others teach by modeling, but with very little verbalization. The child watches, tries the task, is helped to correct mistakes, and tries again.[34] It is easy to see which style matches school learning and school test-taking modes and, consequently, by familiarity, prepares the child for success in school tasks.

Thus, tests can be unfair in themselves: not measuring what they purport to measure or testing it in ways less accessible to certain populations than to others. In some cases, the test becomes not so much a test of mathematics or science as a test of reading ability. Some of the ways in which items can be biased are by the use of language that is not part of the everyday language of certain minorities, item contexts where people of another culture bring a special framework,[35] and items that call upon background information not equally available in everyone's background.[36] The items are constructed out of a white, middle-class context and are most suitable for that context.[37]

Tests can violate the Code of Fair Testing Practices in Education even more in the way in which they are used. The first and most appropriate use for test scores is to supply information to the teacher on which to base decisions about curricula and instruction. Have the students learned what I thought I was teaching? If not, what evidence is there about where their misconceptions lie? How do these test results compare with my own assessment of students' understanding and ability,[38] with other test scores and grades, with daily work, class discussion, and other indicators? If there is a problem, is it the test, the student, or my teaching that needs to change?

A second appropriate use for the scores of well-constructed, valid tests is to permit the monitoring of programs or pupil progress by policymakers. If the information yielded by the test is indeed representative of the situation, then such policy decisions as providing more money for materials, more time for instruction, and more in-service training for teachers can be informed thereby. But to use test scores for appropriate policy decisions requires that those using them know what the scores mean. It is not possible to "raise the mean score in every state above the national average." Nor can tests, nationally normed for groups of students, be used to make decisions about individual students in defiance of the test publisher's caveats, as was done in one rural school system.[39] Over and over, test publishers, measurement specialists, and other educators emphasize that decisions should not be made on one indicator alone, even if that indicator can be demonstrated to be free of bias.[40]

Tests intended to monitor programs could rightly sample a somewhat different domain from the specific curriculum of a given school: the universe of content that a consensus of science and science educators believe ought to be in the curriculum. If tests are used for monitoring purposes to validate the adequacy of a local curriculum, it is imperative that (i) the universe of concepts and skills they assess be truly adequate according to the best national consensus available and (ii) they assess in ways that are unbiased, unambiguous, and directed to probing the process of thinking.

Teacher decisionmaking is clearly not the primary use for which the scores from standardized tests are intended. They are used for purposes of sorting and stratifying pupils,[41,42] for gatekeeping, for evaluating programs, and even for monitoring teachers' judgments and teaching performance:

> The education reform movement in state legislatures has produced even more pressure upon school administrators and teachers to demonstrate success in terms of test performance...it is becoming all too common for school districts and school personnel to use a numerical score on some standardized instrument to make a "yes or no" decision about an individual student, or to place the student in an instructional level or grouping, without adequate understanding or recognition of: (i) what the test is actually measuring; (ii) the strength of the relationship between the test content and the local instructional program; or (iii) the appropriateness of using the test instrument to make that kind of individual decision at all.[43]

As has been noted already, such use often fails to take into consideration the caveats and warnings of the test publisher until a lawsuit forces such consideration.[44]

Validity Concerns

The issue of appropriate use is a question of test validity and part of a much broader issue. The need for content validity of standardized tests for science reform curricula is one of the driving forces behind the alternative assessment movement.[45] "The issue of validity is the most important consideration in test evaluation. The concept refers to the appropriateness, meaningfulness, and usefulness of the specific inferences made from test scores."[46]

How valid is a particular test for the judgments that may be based on it? To clarify the technical aspects of validity: Tests may be considered from the viewpoint of content validity, criterion validity, and construct validity. There is also the question of alignment of the curriculum and instruction with the test content, sometimes called instructional/curricular validity, which Mehrens[47] considers test preparedness,[48] but which the courts considered of first-level importance in the Florida case Debra P. v. Turlington.[49] The court ruled:

> We hold, however, that the State may not deprive its high school seniors of the economic and educational benefits of a high school diploma until it has demonstrated that the SSAT II [Secondary School Achievement Test] is a fair test of that which is taught in its classrooms.[50]

That is, for a test to be fair, both curriculum and instruction, and not simply the goals and objectives defined by the state for that subject (content validity), must match the content coverage of the test. If the content of a test is drawn from a specifically defined domain and samples that domain adequately, then the test is said to have content validity. A score on such a test could be said to indicate a certain level of mastery of the concepts and processes appropriate to that universe. For example, if the experimental, "hands-on," problem-solving aspects of elementary school science are considered to be as important as the conceptual network and are integrated with it, then assessment must sample experimental, "hands-on" competency as well as verbal conceptual content if it is to have content validity. It is to ensure content validity that test publishers make great efforts to review the curricular guidelines of all 50 states and to include in tests only those items that presumably are being taught everywhere. The fact that such an approach, compassing a broad range of topics in limited space and time, produces only very limited and watered-down items is to be regretted.

There must be another way to achieve content validity, perhaps by abandoning the "national curriculum" that is thus created by textbooks and allowing again individualized local curricula and tests in the good old American tradition of local control. To do so also abandons the possibility of comparing one state's achievements with another's (a goal of the Chief State School Officers). Until a national consensus is reached on what constitutes the domains of school science in the various disciplines and whether or not achievement should be measured externally (perhaps only at exit points like 8th and 12th grades?), true local control may be the only way. Such a consensus seems to require at least 10 years of exploration and experimentation with both curricular and measurement innovations. Premature foreclosing of experimentation and debate could result in "top-down" content and pedagogical decisions, to be discarded ultimately as were many of the new mathematics and new science efforts of the 1960s. Content validity, which must be established as the tests are constructed, is judged only after the fact by someone competent in both the subject matter area and test construction.[51] It is a primary determinant of whether the test is fair.

Criterion-related validity evidence demonstrates that the items being tested do measure, in fact, those criteria judged to be necessary for a future task: college work; the practice of law or medicine; competence to handle safely and maintain big guns; readiness to pursue higher studies in science; ability to solve problems or to design, conduct, and interpret experiments; or ability to think as mathematicians do. If a test has criterion validity, it is predictive of future performance in a specific domain. Obviously, to evaluate this kind of validity requires a clear description of the tasks and underlying conceptual structures of the future performance in question.

The above seem fairly obvious, although predictive validity is harder to attain than content validity. It was for this reason that, years ago, the United States military, as well as those for whom artists audition and many other organizations since, moved to performance assessment, wherein the candidates actually demonstrate the competencies that constitute their future tasks.

To establish criterion validity, the limitations of testing have to be recognized. At best, a test is a small sample of the possible range of items in a much larger domain, a sample taken on a particular day and hour, in a particular set of physical and emotional circumstances. In addition, the information it accesses is mediated by the limitations of the printed word and even the multiple-choice format. Recently, I spoke with two of the developers of the National Assessment Project for the new National Curricula in Science and in Mathematics in England. They told me that they had piloted a particular set of items on a test in primary school mathematics in three different forms: by word descriptions of the problem, by word descriptions accompanied by pictures, and by word descriptions with manipulatives which the children could use in answering the question. There was about a 20-point differential between the scores for each of these approaches, with those using manipulatives scoring some 46 points higher than those who received the problem in word form alone. That is not a surprising result, but a graphic one, indicating that the items using words alone might have been assessing reading ability, or ability to create the scene imaginatively, or even the developmental level of the child, rather than the child's actual grasp of concepts and procedures — a grasp that the child could demonstrate easily using manipulatives, even while recording the answer in written form.

The issue in question is that of construct validity: What inferences can we make from the test item to what the child actually knows and can do? In every test question, there is the assumption that a correct response on this item indicates "right" thinking and an incorrect choice indicates the opposite. But just how close is the connection between this mediator (the words of the item) and the actual concept in a student's mind? Haney conducted a study of test ambiguity with first-grade students and reports many samples of "unusual and perceptive interpretation."[52] For example, when students were confronted on a standardized test with pictures of a geranium in a pot, a cabbage, and a cactus in a pot and asked, "Which plant needs the least amount of water?" 2 out of 11 children chose the cabbage:

Q: Which one needs the least amount of water?

A: That does.

Q: Why is that?

A: 'Cause it's a cabbage. Doesn't need as much water. Only when you clean it.

This child apparently reasoned that the cabbage needed less water than the two potted plants because it was not depicted as a potted cabbage. From this point of view, the logic of the child's reasoning is hard to deny. A head of cabbage once harvested can be kept well and long without water — water is needed only when you clean it.[53]

In another example, when shown a carrot, a cabbage, and a tomato and asked, "Which one is the root of a green plant?" only 3 out of 11 children selected the carrot. Others either rejected the carrot because "it's the root of an orange one" or disregarded the word "root" and selected the cabbage as the only thing green in sight. What was being measured here may have been concrete sequential reasoning or the "child's-eye view," but it was not the ability to identify carrots, tomatoes, and cabbages.

To put the question differently: Does a student's inability to solve a "word problem" involving addition indicate that he or she has no concept of addition, cannot read, cannot visualize the problem, does not know the procedure, or what? What exactly is being measured? How valid is our construct that ties this particular item to that skill? It is arguable that the only valid assessment of people's ability to do something is to watch them do it and/or have them explain why they did what they did.

Mehrens discusses the validity required in tests used for different purposes.[54] For instructional decisions, we require content and construct validity, with content being broadened to include not just "what ought to be in the domain of eighth-grade science," for example, but what is actually in the curriculum.[55] For guidance decisions, such as choosing electives and colleges, we require some measure with predictive or criterion-related validity. For placement in a remedial or gifted program, the test should show us that this is the *least* restrictive environment that matches the student's present concept, skills, and needs levels and, therefore, is the one in which he or she could be expected to grow.

Standardized Tests in Science: Is There Bias?

In science education, if we are defining unfairness in tests by the way they are being used, this becomes a moot question because few states mandate externally developed science tests and in none that we have been able to find is science a part of the minimum competency tests required for graduation. However, some states require two or three units of science for graduation and an external, end-of-course examination in each that counts as a part of the final grade. So, while the results of elementary school science tests seem to be "low stakes" (that is, do not prevent promotion or graduation, or have other serious consequences) everywhere, at the high school level in some states the stakes are very high. Many states that do not mandate any external science testing leave such decisions to the districts, many of which have more stringent criteria for promotion and/or graduation. But the fact remains that, at the K–8 levels, science is largely untested. And what is the outcome? It is a lowering of motivation to teach science. When time is short to "prepare" students for tests in reading, language arts, and mathematics, science untested can also be science untaught.

Having acknowledged the state of affairs, it is still worthwhile to look at the issue of bias in the traditional sense. Are there test items that, by their content or structure, militate against success for members of a certain racial, sexual, or ethnic group? This is an area that test publishers take very seriously and, as indicated above, they make strenuous efforts to keep test items racially and sexually unbiased. However, there are still some items whose context a given population would have less opportunity to be familiar with than would others. For instance, farm content very familiar to rural students, and adequately familiar to well-read, middle-class suburbanites, might be so far outside the experience of urban, inner-city youth as to place them at a disadvantage.[56] The issues of language access mentioned earlier — ease of reading, likelihood of appropriate inferencing, as well as home attitudes toward cooperation and social interaction — are as likely to be operative in science testing as elsewhere and are not entirely limited to NNESs.

The biggest issue involves the content validity of the tests and the pupils' opportunity to access the content in meaningful ways through the curricular and instructional patterns of individual schools and districts. This issue has an added power in elementary school science. Science is generally taught in elementary schools by homeroom teachers, most of whom do not have majors or minors in the sciences. If teachers are relatively insecure in their knowledge of science, to what extent can science testing have instructional validity? There is a strong likelihood that teachers will use precious time for other, tested, curricular areas or will unknowingly introduce misconceptions.

I reviewed several examples of standardized tests in an effort to detect bias in the items themselves but, in fact, did not find it at the language level of native English speakers except, possibly, with respect to inner-city students. For a few items on each test, I believe that inner-city students would not have background information equal to that of other test-takers[57] (see Table 1). What I did find was an astonishing range of domains being assessed on a single test at each level: from astronomy and biology to light, sound, heat, motion, earth science, machines, and more. The questions called for responses chiefly at the levels of recall (facts, information) and learned application. A need for analysis was found mainly in questions based on reading and interpreting charts or graphs or decoding carefully the text of a question expressed negatively: "Which of the following is NOT a cause of...?" and, with the exception of the material of one publisher I reviewed, even these were few.

A large part of one of the tests actually measured reading ability in the science context. This test eliminated much of the recall by providing scenarios containing facts from which inferences and implications are to be drawn. It is a contemporary implementation of Ralph Tyler's recommendations of some 50 years ago.[58] Thus, the test-makers really are assessing: (i) a student's ability to read and interpret text and to see the relationship between the choices and the text — certainly, a general thinking skill suitable to all subject areas; (ii) some processes, such as identifying variables or controls, weighing evidence (to eliminate unjustified statements), or interpreting data; and (iii) the ability of the student to analyze an experiment to identify the question under investigation, the tools required for data collection, and/or the congruence between the experimental design and the conclusions drawn.

Although highly sophisticated in the context of the rest of the test, these are certainly valid items for the kind of science now broadly recommended. It must be noted that this test was revised in 1985, and its newness gives it an advantage over others available for analysis. In fairness, it also must be noted that most test publishers are in the process of revising their instruments and are exploring such options as performance assessments.[59]

A few more comments on the 1985 instrument are in order. As in the other tests sampled, it also surveys a broad but more limited domain of content (9 fields as compared with an average of 13 for the others). For a teacher to teach so broad a range of subjects in any one year must almost ensure that it will be at the expense of depth in any topic and will preclude significant time for "hands-on" discovery and probing for higher-order thinking. Only the surface can be touched in a two-week (or less) exposure. For such coverage and pedagogy, the large number of recall and verbal application items found across the tests is congruent. But what is the impact on science instruction if the type of test reinforces a superficial, time-constrained, nondiscovery type of teaching or reduces science education to learning to read scientific subject matter?

Table 1

Analysis of the Content of Standardized Science Tests

Topics	Recall Facts No./%	Concepts No./%	Application No./%	Analysis No./%	Problem Solving No./%	Synthesis No./%	Evaluation No./%
Test A, Intermediate Level							
13	31/52	20/33	12/20	0/0	0/0	0/0	0/0 (+2 ambiguous, 1 erroneous)

Unless specifically taught, questions such as those calling for "best farmland" in different locales, life cycle of frogs and snakes, and desert plants that are sources of water for animals would be differentially difficult for populations in various geographic areas.

Topics	Recall Facts No./%	Concepts No./%	Application No./%	Analysis No./%	Problem Solving No./%	Synthesis No./%	Evaluation No./%
Test B, 8th Grade Level							
9	13/32	6/15	13/32	11/28	0/0	0/0	0/0 (1 ambiguous)

All questions seemed to be on "learned material" and most provided the data if a process judgment were to be based thereon. The only potential for bias might be in a question related to growing conditions for molds. The answer: "It depends on the type of mold" was not among those given, but it would influence the choice for certain populations.

Topics	Recall Facts No./%	Concepts No./%	Application No./%	Analysis No./%	Problem Solving No./%	Synthesis No./%	Evaluation No./%
Test C, 5th/6th Grade Level							
9	11/37	1/3	13/43	9/30	0/0	0/0	0/0

Bias: One question asking which parts of plants are grown underground had potential for bias against inner-city students.

Note. Examples and preliminary findings from research in progress at the Center for the Study of Testing, Evaluation and Educational Policy, Boston College, Chestnut Hill, MA.

At present, staff members of CSTEEP also are exploring a number of science tests provided by publishers as an additional resource accompanying texts (text-embedded tests).[60] It is no surprise to learn that the content on the tests matches beautifully that in the text, often in the same words. In many texts, there are end-of-chapter materials with "extension" questions that take students beyond the "givens" of the text and invite deeper thinking. These are optional and, usually, are not found in the accompanying tests.[61] There is a market reason for the exclusion. At present, test-makers are trying to satisfy the large majority of test-consumers, and these still follow "traditional" rather than inquiry approaches to teaching science. Hence, current text-embedded tests are unlikely to drive curricular changes. Careful reading and study of the text do provide an adequate base for passing the publisher's tests. Again, there is no evidence of particular bias in language or content against members of a minority group who have been able to decode and analyze the text, but the comments made above still apply. On almost every test, examples were found of item contexts that could be unfamiliar to inner-city youth, as indicated above, although it is conceivable, in some cases, that some familiarity could be achieved by careful reading of the text. No items were found favoring inner-city students inequitably. In addition, while individual items do not reveal content bias, the impact of the overall, multiple-choice format, speededness, and the contexts of certain questions may bear unequally upon members of certain groups more than others.

From all of the above, I would conclude that existing science tests are not seriously biased in the usual sense of the word, but, by their speededness, format, extensive content coverage, and limitations on process and critical thinking coverage, they are biased against thinking, concept networking, problem solving, integrating and synthesizing knowledge across domains, evaluating, and, above all, against spending instructional time in experimentation. They honor facts and information, but their present format, coverage, speededness, and machine-scannable answer sheets render them incapable of testing adequately the innovative curricula resulting from reform efforts in science education. The multiple-choice format encourages one correct answer only, a situation not found in science explorations where so much of what is found also is found to be dependent on circumstances.[62] A great deal of analytical thinking is required if all the distractors not only are plausible in certain circumstances, but also represent various current misconceptions. However, present methods of scoring do not allow for good thinking that leads to "wrong" answers. And, certainly, the multiple-choice test rarely honors good thinking that leads to no answer, a situation with which scientists are most familiar!

The testing community is on the threshold of creating new tests that include performance, portfolio, and writing-across-the-curriculum components. What are the problems that will have to be addressed? How can we be sure that the new tests will be better than what we have? We believe that the most significant problems will emerge in the areas of adequate content coverage, potential for bias, issues of subjectivity and unreliability in scoring, cost effectiveness in view of the labor-intensive nature of administration and scoring, and interpretation of the results; that is, content and construct validity, reliability, and economic effects. It is beyond the scope of this paper to address economic effects.

Care must be taken not only in the choice of language and context for test items, but also in the use of the results. In order to avoid the bias that comes with the misuse of a single assessment mode, we recommend that assessment incorporate multiple modes and never rely on a single-day, single-instrument measurement. Both fixed-forced-choice and open-ended responses as well as uneven matching columns, creation and/or interpretation of graphs and charts, essays and performance assessments including portfolios of work, projects, simulations, and "circuses"[63] may be utilized. For formative assessment, it is advisable that the lines between assessment and instruction be blurred as much as possible; for example, by instructionally embedded assessments and extended problems. Such a blurring facilitates the continuous use of assessment results in planning and instruction by the classroom teacher, without "taking time out for tests." But, in order for this to happen, teachers will need assistance in "reading the situation" while students are working, in interpreting aspects of their performance, and in recognizing underlying concepts and misconceptions. Teachers also may need help in classroom management and in dealing with the risk level involved in not intervening while a student is working.

During the past three years, CSTEEP staff members have been experimenting with alternative assessments. A summary of their findings may be found in *Assessment in the Service of Instruction,* pages 52-59.[64]

Issues of Fairness and Bias with Alternative Assessments

At first glance, it would seem as though the primary issues in using alternative assessments would cluster around the potential for "subjectivity," that is, unreliability in the hands of different scorers. In fact, while this is a concern, it is solved rather easily if care is given to developing scoring rubrics with clear criteria for acceptable responses and indicators of various levels of performance. Several researchers are engaged in this task. For performance assessments, the same conditions apply: published criteria for acceptable performance and indicators by which various levels of performance can be recognized are needed. Still more important is the training of teacher-observers to use the criteria as they observe performance because this is a new method of assessment for most teachers.

However, alternative assessments are not an automatic cure for all of the issues that have been raised above. Such assessments also must be examined for bias in language and content and fairness in use. They are very sensitive to nuances of language and context, even within a culturally homogeneous group. While they remove the barriers created by speeded-ness on today's standardized tests — barriers that are problematic for certain cultures and learning styles — they require far more time to administer and to score, and it may never be cost-effective to administer them on a large scale using external examiners.

The various areas of content recommended by science reformers, such as problem solving, discovery or open-ended experimentation, problem-posing, designing and carrying out experiments, and higher-order thinking, are particularly well matched with performance assessments and would seem to be assessed better in this mode. But if such contents have not been taught or have been taught only in a didactic and nonexperiential way, these tests, by their very nature, will fail the test of curricular and instructional validity. Thinking skills can be measured by alternative assessments with greater security when the person doing the thinking explains his or her thoughts in essay or "think-aloud" form than when evaluators have to rely on one prestructured choice out of four to provide a window into that thinking process. But essay responses raise the spectre of language again. Apart from the obvious concerns for NNESs,[65] there is the serious inability of many students in America to write competently in English at the third-grade level or above. As a result, scoring becomes very time-consuming or itself an exercise in problem solving.

It is possible that the decision to use performance assessments may motivate policymakers, administrators, and teachers to change the ways in which science is scheduled and taught in schools. It also is possible that the regular use of essays on tests may serve as a motivator to teach writing earlier and in more effective ways than is currently done. It even is possible that the criteria developed to score assessments may become valuable tools for teachers in the daily assessment that feeds instruction and may, in fact, help them to clarify the results of instruction. Thus, many of the negatives that could cause alternative assessments to be biased or unfair seem to have other aspects that promote needed improvements in education. Rather than waiting for a national consensus on age-appropriate outcomes in science education, could assessment incorporating some of those outcomes, as they appear, help to elicit such a consensus? Could an upward spiral be created by the dialectic between assessment reform and curricular reform to achieve the excellence in teaching and learning that is so much desired?

References and Notes

1. Data from this study have been published in a number of other reports, among them *The Underachieving Curriculum* by C. McKnight, F. J.

Crosswhite, J. A. Dossey, E. Kifer, J. O. Swafford, K. J. Travers, and T J. Cooney (Champaign, IL: Stipes Publishing, 1987) and "Sex Differences in Science Opportunities and Achievement in the Second IEA Science Study" by W. Schmidt in Larry E. Suiter (Chair), *Understanding Results From the Second International Science Study* (symposium conducted at the annual meeting of the American Educational Research Association, San Francisco, March, 1989).

2. International Association for the Evaluation of Educational Achievement, *Science Achievement in Seventeen Countries: A Preliminary Report* (Oxford: Pergamon Press, 1988).

3. American Association for the Advancement of Science, *Science for All Americans: A Project 2061 Report on Literacy Goals in Science, Mathematics, and Technology* (Washington, DC: American Association for the Advancement of Science, 1989); reprinted as, Rutherford, F. J., and Ahlgren, A., *Science for All Americans* (NY: Oxford University Press, 1990).

4. Johnston, W. B., and Packer, A. H., *Workforce 2000* (Indianapolis: Hudson Institute, 1987).

5. American Educational Research Association, American Psychological Association, and National Council on Measurement in Education, *Code of Fair Testing Practices in Education*, Report of the Joint Committee on Testing Practices (Washington, DC: American Psychological Association, 1985), 9.

6. The work of the Joint Committee to Develop Standards for Educational and Psychological Testing also is sponsored now by the American Association for Counseling and Development, the Association for Measurement and Evaluation in Counseling and Development, and the American Speech-Language-Hearing Association.

7. Mullis, I. V. S., and Jenkins, L. B., *The Science Report Card* (Princeton, NJ: Educational Testing Service, 1988).

8. Dossey, J. A., Mullis, I. V. S., Lindquist, M. M., and Chambers, D. L., *The Mathematics Report Card* (Princeton, NJ: Educational Testing Service, 1988).

9. Test litigation reaches the courts only when a question of civil rights or due process is involved. In general, the courts will not deal with fairness in test content or construction because this is a problem for the education community and education is not seen as a fundamental right protected by law. See N. J. Chachkin, "Testing in Elementary and Secondary Schools: Can Misuse Be Avoided?" in B. R. Gifford, ed., *Test Policy and the Politics of Opportunity Allocation: The Workplace and the Law* (Boston: Kluwer Academic Publishers, 1989), 167–176.

10. Haney, W., "Making Sense of School Testing," in B. J. Gifford, ed., *Test Policy and Test Performance: Education, Language, and Culture* (Boston: Kluwer Academic Publishers, 1989), 51–62.

11. Examples of such batteries include the Iowa Test of Basic Skills, Stanford Achievement Test, California Achievement Test, Science Research Associates Tests, Educational Testing Service/Addison Wesley Test, and others.

12. By text-embedded tests, I am referring to the end-of-chapter reviews, tests, and supplementary test materials supplied by publishers for use with their texts.

13. See also the recent study of existing tests in elementary mathematics by the National Center for Research in Mathematical Sciences Education (NCRMSE), Romberg, T. A., Wilson, L., and Khaketla, M., *An Examination of Six Standardized Tests for Grade Eight* (Madison, WI: NCRMSE, 1990).

14. O'Connor, M. C., "Aspects of Differential Performance by Minorities on Standardized Tests," in B. R. Gifford, ed., *Test Policy and Test Performance: Education, Language, and Culture* (Boston: Kluwer Academic Publishers, 1989), 129–181.

15. National Commission on Testing and Public Policy, *From Gatekeeper to Gateway: Transforming Testing in America* (Chestnut Hill, MA: Boston College, 1990), 10–11.

16. O'Connor, "Differential Performance." See reference 14.

17. Jensen, A. R., "How Much Can We Boost I.Q. and Scholastic Achievement?" *Harvard Educational Review*, 39 (1969), 1–123.

18. See also the review of refutations of Jensen by R. J. Samuda in *Psychological Testing of American Minorities: Issues and Consequences* (New York: Dodd, 1975) and others.

19. Ogbu and others in O'Connor, "Differential Performance," 130. See reference 14.

20. Ibid.

21. Ibid., 131.

22. In this group, she includes Black English Vernacular because there are a number of significant syntactic and phonological differences between dialects spoken at home and "school English." She does not advocate the use of so-called Black English in schools and seems to concur with its opponents that to do so would be to perpetuate institutional racism. See the original for further elaboration.

23. An example cited by O'Connor from a reading test: "If a broncobuster wants to win a rodeo contest, he must observe the contest rules. One of these rules is that the rider must keep one hand in the air. A rider who does not do this is disqualified.

    Question: In a rodeo contest, a broncobuster must keep one hand _____ (under, still, free, hold)" (O'Connor, "Differential Performance." See reference 14.).

24. Recent studies in England, on the influence of age of school entry as a measure of maturity and a predictor of school success, indicate that

the lag in success created by entry at age 5, as compared with entry at age 6, is still evident in the 13-year examinations (staff members of the National Foundation for Education Research, personal communication, March 6, 1990). If this is so for native English speakers, one could anticipate a parallel effect for NNESs who are immature and inexperienced in the use of English.

25. Contrast the speed with which one can read and comprehend a treatise on nuclear physics with that evident in reading a well-written novel.

26. O'Connor, "Differential Performance." See reference 14.

27. Suggestions that have been tried include eliminating speededness and using as a test administrator a person of the same language and culture as the test-taker. The latter approach yielded ambiguous results depending on the age, sex, and culture of the test-taker. The effectiveness of translated tests has been studied both in Africa and in the United States and the problems with validity have been reaffirmed.

28. Native Americans, African Americans, Pacific Islanders, some middle easterners, and others have very different attitudes toward speed and "not all people will work on the tests with equal interest in getting them done in the shortest time possible" (Peterson, 1925, in O'Connor, "Differential Performance," 151. See reference 14.).

29. Several studies have documented the difficulties of children from other cultures with the isolated, silent, independent, competitive style of working on a test. Such a way of working is in sharp contrast to practices in those cultures where all tasks are shared collaboratively and any hesitancy on the part of a learner is detected immediately by a brother, sister, or friend who offers help.

30. See also in this context W. Haney, "Talking With Children About Tests: An Exploratory Study of Test Item Ambiguity," in R. O. Freedle and R. P. Duran, eds., *Cognitive and Linguistic Analyses of Test Performance* (Norwood, NJ: Ablex, 1987), on first-grade students' interpretations of test questions.

31. NCTPP, *From Gatekeeper to Gateway: Transforming Testing in America,* 13. See reference 15.

32. The testimony was cosponsored by the National Commission on Testing and Public Policy and the Native American Scholarship Fund, Inc.

33. O'Connor, "Differential Performance." See reference 14.

34. It is noteworthy that many Chinese parents, like educated middle-class white Americans, also use verbal, direction-giving strategies in child-rearing.

35. O'Connor cites an example from de Kohan about an Argentinian test-taker with the Stanford-Binet of 1937: "The judge said to the prisoner, 'You are to be hanged, and I hope it will be a warning to you.' (Verbal Absurdities, XI). Many children answered, 'Oh! This is silly!'

But when we inquired why it was silly, they said proudly, 'Of course it is foolish. No judge could give anybody the death punishment in Argentina, this would be against the law' ." O'Connor, "Differential Performance," 154. See reference 14.

36. On one achievement test, students are asked, "In places where there are mountains and valleys, where is the best farmland usually found? (a) Near the tops of the mountains, (b) Along the streams running down the mountains, (c) On the mountainside closest to the ocean, (d) Toward the bottoms of the valleys." Without direct instruction (rendering the item largely recall), an American inner-city youngster would be unlikely to select the correct answer.

37. To make this point, in the 1960s there were a number of "tests" in circulation that were written in the dialect and common parlance of North Carolina African Americans. It was nearly impossible for northern whites to "pass" these "tests."

38. Kellaghan, Madaus, and Airasian showed in the Irish Study that teachers confident of their own judgment about students' achievement used a "low-stakes" test to confirm that judgment and that, surprisingly, there were only small differences between the test results and the teacher's judgment. However, there is some evidence that even "low-stakes" tests can produce a "Pygmalion effect." See G. F. Madaus, "The Irish Study Revisited," in B. R. Gifford, ed., *Test Policy and Test Performance: Education, Language, and Culture* (Boston: Kluwer Academic Publishers, 1989), 63–89, and T. Kellaghan, G. F. Madaus, and P. W. Airasian, with the assistance of P. J. Fontes and J. J. Pedulla, *The Effects of Standardized Testing* (report submitted to the Carnegie Corporation of New York and the Spencer Foundation, 1980).

39. Chachkin, "Testing in Schools." See reference 9.

40. See also in this context the recommendations for evaluation in the National Council of Teachers of Mathematics, *Curriculum and Evaluation Standards for School Mathematics*, report of the Commission on Standards for School Mathematics (Reston, VA: National Council of Teachers of Mathematics, 1989), and in the NCTPP, *From Gatekeeper to Gateway: Transforming Testing in America*. See reference 15.

41. Weiss, I., *Report of the 1977 National Survey of Science, Mathematics, and Social Studies Education,* NSF Publication No. SE 78–72 (Washington, DC: U.S. Government Printing Office, 1977).

42. Weiss, I., *Report of the 1985-86 National Survey of Science and Mathematics Education,* NSF Publication No. 1, SPE-8317070 (Washington, DC: U.S. Government Printing Office, 1987).

43. Chachkin, "Testing in Schools," 167. See reference 9.

44. See Chachkin, "Testing in Schools," for numerous examples. See reference 9.

45. Alternative assessment is not to be construed as performance assessment alone, but includes all modes for demonstrating knowledge and skills other than the machine-scorable, standardized tests; that is, free-response, written modes, portfolios, exhibits, debates, simulations, and single-task and extended performance assessments.

46. American Educational Research Association, American Psychological Association, and National Council on Measurement in Education, *Standards for Educational and Psychological Testing*, report of the Joint Committee to Develop Standards for Educational and Psychological Testing (Washington, DC: American Psychological Association, 1985).

47. Mehrens, W. A., "Using Test Scores For Decisionmaking," in B. R. Gifford, ed., *Test Policy and Test Performance: Education, Language, and Culture* (Boston: Kluwer Academic Publishers, 1989), 93–114.

48. This is an important issue. If a school or district has authorized experimental curricula in science or mathematics, it becomes an issue of fairness if these students are tested with instruments validated for traditional curricula covering a broader and different spectrum of content. In effect, this may inhibit the use of innovative curricula according to the developer's intent or result in a teacher's attempt to combine two philosophically different approaches.

49. Debra P. v. Turlington, 408 Florida 5th Circuit Court (1981).

50. Madaus, G. F., *The Courts, Validity, and Minimum Competency Testing* (Boston: Kluwer-Nijhoff Publishing, 1983), 72.

51. Mehrens, "Using Test Scores." See reference 47.

52. Haney, W., "Talking With Children." See reference 30.

53. Ibid., 339.

54. Mehrens, "Using Test Scores." See reference 47.

55. That it is in the curriculum is no guarantee that it will be taught, as Mehrens points out, but, certainly, a test is not valid for use in a district if it does not match the mandated curriculum. Whether that mandated curriculum should be decreed from on high or modified by the teacher to suit the unique needs of a particular class is a different question. See Mehrens, "Using Test Scores." See reference 47.

56. "The evidence (in the Debra P. trial) focused upon a number of items on the test that required familiarity with situations and information not a part of the background of many black or low-income students. Expert testimony identified a number of [questions] black students were unable to answer correctly and that involved unfamiliar material or items where the 'correct' answer to the test developers was not the same answer black students would consider." See Madaus, *Courts, Validity, and Testing*, 9. See reference 50.

57. See the Debra P. case for the finding of the court on background information. See Madaus, *Courts, Validity, and Testing*. See reference 50.

58. Madaus, G. F., and Stufflebeam, D., *Educational Evaluation: Classic Works of Ralph W. Tyler* (Boston: Kluwer Academic Publishers, 1989).

59. However, at the publishers' presentations during the National Council of Supervisors of Mathematics' meeting in Salt Lake City in April 1990, participants reported that all publisher-presenters denied that there would be significant changes in assessments. The alternative communication comes to us through requests for assistance from individual publishers in developing performance assessments as part of their current revision.

60. The Center for the Study of Testing, Evaluation and Educational Policy (CSTEEP) at Boston College is conducting a study of "text-embedded" tests as well as "high-stakes," externally developed tests in science and mathematics. More complete data will be forthcoming.

61. See, for example, at the secondary level, very interesting Prentice Hall materials for biology and chemistry.

62. A question about the growing conditions for mold on one of the science tests seems to expect the best-choice response: that mold would grow in the container *not* kept in the refrigerator, but the question would be answered differently by a student who had been experimenting with the kinds of molds that grow best at 35–40° Fahrenheit.

63. By "circus" is meant a set of performance assessments arranged at stations around a classroom or laboratory. Students move around the room from one station to another as they complete each task.

64. Champagne, A. B., Lovitts, B. E., Calinger, B. J., *Assessment in the Service of Instruction* (Washington, D.C.: American Association for the Advancement of Science, 1990).

65. One project the author is working on allows students to respond in their native language and uses native language readers for scoring. However, the items were construed and developed by native English speakers.

4

The Control of Assessment

Gerald Kulm

The complex issue of who should control assessment impacts on nearly everyone, from the individual student and teacher to the nation as a whole. Recently, the arena for discussion about the fine line between local and national control of education in this country has shifted from curriculum issues to assessment. Although states and local school districts continue to exert nominal control, the science and mathematics communities, governors, and the President have embraced the notion of national curriculum goals and standards. If the need for national curriculum standards is generally accepted, then the idea of a national achievement test naturally follows. But if national curriculum standards and high-stakes national assessments were put in place, a radical shift of control would occur, throwing the educational system out of equilibrium. It is impossible to analyze completely the implications of a national test, much less predict the possible outcomes. Before accepting the idea, however, the issues of control should be examined carefully in the light of such a test.

In this chapter, the issue of who should control assessment is addressed from a variety of perspectives, ranging from philosophical to practical and from purposes to expected outcomes. The discussion, based in part on a symposium held in November 1990,[1] is intended to open the way to careful consideration of the issues before this nation moves even farther along the path of mandated national tests in science and mathematics.

The Role of the Student

In any discussion of assessment, it is critical to remember that it is the individual student who is being assessed. In a very real sense, the individual student is the key person in control issues. Whatever the intentions or actions of teachers, schools, states, professional organizations, or national leaders, their success depends solely on the response of individual students to the assessment. That point is made painfully clear when the stakes involved in a test are different for the student and for the policymaker or school administrator. The administrator's job may be at risk if students do not perform well on a normed standardized test. But students may have no reason to perform well if the test is not a part of their grade or graduation requirements. For the student, only minor stakes, such as placement in a particular ability track, may be involved. In recent years, administrators and policymakers have tried to raise the stakes for students with, for example, "no pass, no play" policies, or high school graduation tied to passing state-mandated achievement tests. The proposed national achievement test would raise the stakes even more for students by encouraging employers to consider the test results in hiring high school graduates.

Students realize very early that their assessments and grades are subjective and mainly in the control of the teacher. Although some enlightened teachers may allow student discussion and input, control at the classroom level is fairly well defined. The teacher may not have complete control, however. Required tests, school district or departmental policies, pressures to assign grades in a particular distribution, and other factors remove some control from the teacher. Moreover, each individual, group, organization, or agency that has even minimal involvement in a student's education has some control over assessment, thus removing some control from both teachers and students. The more removed from the classroom that assessment decisions are made, the greater the concern becomes for control issues. The following sections discuss these issues in some detail and from several perspectives. In each of these perspectives, the fundamental question ought to be whether the control of assessment, or the lack of it, serves to enhance students' learning of science.

Control and the Purposes of Assessment

Lee Jones, coordinator for science assessment for the National Assessment of Educational Progress (NAEP), has pointed out that knowing the purpose of assessment is central to answering the question of who should control it.[2] For example, a classroom test to decide whether or not students are getting the point of a science concept being studied is likely to be much different from a test intended to monitor the status of the nation's students' understanding of the same concept. And even when the purpose of an assessment is specified, deciding exactly what should be assessed is tough. Then there is the question of what instruments are best for showing whether students know something or have acquired some ability. Finally, we have to address how the results of the assessment will be used.

To answer questions about purpose and control of assessment, we must consider who should be involved in answering the other questions along the way. The groups that should be involved in the three questions underlying the purpose of the assessment — specific content, type of test instrument, and intended use of test results — are the groups that should be in control of assessment on an operational level.

Senta Raizen and her colleagues have summarized some twenty-two reasons for assessment.[3] One important reason is to convey expectations to students about what they should know. At that level, control of assessment can only be in the hands of the teachers. Another example is advanced placement (AP) exams, which have a different purpose: that of certification. What should be assessed and who should control AP exams? Again, the questions — of content and control, of structure and purpose — are easy to answer, since they all go back to the college teacher who sets prerequisites for college courses.

On the other hand, for an examination such as the National Assessment of Educational Progress, the purposes, although spelled out by law, are open to interpretation, and certainly identifying who should be involved in control is very hard to do. The enabling legislation states that NAEP's role is to monitor progress, or provide objective information about student progress, in different content areas and in a reliable fashion. However, even though the intended purpose is to monitor the status of students across the nation, simply administering the test conveys expectations about what should be taught in science to students, to parents, to school teachers, and to administrators.

Identifying the purpose of the National Assessment returns us to the questions, What should be assessed? How should it be assessed? What groups should be involved? My answer is that no single group should control it because so many people need to have input into the process. Decisions about what should be assessed should involve the best scientific minds in the country and leaders in science education, business, and industry. Answers to How? should come from these same groups, but with strong involvement of teachers, who know from their classroom experience how to find out what students actually understand. For monitoring educational progress over time, particularly at the national level, the involvement of education and measurement specialists is essential.

Given this context and the numbers and kinds of groups that have reason to be involved, a consensus process is the only possible answer. When the NAEP science test is implemented in 1994, many people will have been involved and each should feel that it has been a consensus process. To make the NAEP science exam more valid, more reflective of what kids know and can do in science, three major goals have been set for the 1994 assessment. One goal is to emphasize depth of understanding of conceptual themes rather than just breadth of science knowledge. A second goal is to improve the current multiple-choice test to emphasize conceptual understanding. The third goal is increased emphasis on thinking skills, probably by incorporating several different alternate assessment strategies.

The NAEP process that Jones describes for deciding on the content to be assessed, determining how to test that content, and deciding how to report results is one of the best examples of a consensus process, working in response to a reasonably clear purpose. What implications does this process have for enhancing students' learning of science? Currently, the NAEP consensus process itself includes wide input from a diverse group of educators and other professionals. The diversity tends to remove responsibility from any one person or group, leaving much of the final control over how the content is assessed to item writers and a few NAEP and Education Department staff who have final approval of the test. This process illustrates the possible pitfalls of a consensus process.

Even if the consensus process was perfect, the purpose of monitoring national progress is far removed from the reality of individual students' day-to-day science classroom work. Student learning is affected only indirectly by the results of the NAEP. For example, a local school board member or administrator might be motivated by NAEP reports to press for changes in the local school science program. The most visible effect that this type of national consensus assessment can have is to serve as a model of resource for state or local assessment efforts. These local or state tests have higher stakes for teachers and students, but they have had the effect in recent years of narrowing the goals of student learning. Tests that had the intended purpose of sampling and monitoring achievement have become the standards for defining essential outcomes. Reaching a consensus on the science that everyone agrees upon as important has the possible effect, therefore, of limiting and narrowing science content. Instead of enhancing and extending students' science learning, this approach to solving the control issue may have the opposite effect.

Ethical Issues

According to Bruce Goldberg, codirector of the Center for Restructuring of the American Federation of Teachers, control of assessment is a moral and ethical matter.[4] The question of control puts the issues of assessment and education into the appropriate context. The answer to the question is not a particular person, nor a matter of who has the power or who is entitled by law or by statute. Rather, determining the purpose of assessment is a moral question that can only be answered by examining the moral reasons behind it. Of course, any inquiry is value laden, this one no less than any other. One might be tempted to say that, if there was agreement about the purposes of assessment, there would be agreement about who should control it because the moral reasons for these purposes would have been given and agreed upon. As it is, there are many different purposes and points of view, each with its own underlying moral reasons.

For testing in general, the most notable purpose is accountability: reporting to the public the results of its investment in the education of its students. The second major purpose of assessment is the improvement of teaching and learning. Teachers must have adequate knowledge to perform their diagnostic function and to make adjustments in their pedagogical strategies. Students themselves can be reflective about their own work and learn from taking tests.

A number of arguments can be made on moral grounds about who should be in control of assessment for these two purposes. Here is one example: A democratic society is founded on the principles of justice and equal opportunity. Since assessments are high stakes and are used as a means of establishing individual opportunities, assessment should be controlled by independent public scrutiny; that is, by panels of citizens representing the interests of those whose future opportunities will be affected. This argument uses justice and fairness as criteria for determining who should control assessment.

A second argument is the following: The economic well-being of this society depends on a work force that is competitive in a global economy. To be competitive, we need an educational assessment system designed to reward those who succeed. Because of the vital interests of the nation, the control of such a system must rest with a national governing board. This is an argument from utilitarianism; it says that the consequences of our actions ought to be the criteria for deciding who should control assessment and its purposes.

Yet another argument is that the goal of schooling is the development of individual autonomy. This goal requires teachers to control the assessment process at the classroom level because only classroom teachers can fashion the continuous ties among curriculum, instruction, and assessment needed for development of individual autonomy. In the hands of teachers, assessment could simultaneously illuminate student accomplishments, point out strengths and weaknesses in the instructional program, and sharpen curriculum goals. This argument focuses on the intrinsic worth of human beings and their potential for self-realization.

Note that these arguments are not necessarily consistent with each other. The moral principles on which each is based could conflict, and so could the practical results. Suppose, for example, that present trends continue and artificial intelligence is developed to such an extent that robots or computers make the nation globally competitive without requiring the education of large segments of our population. The argument from utilitarianism would be in conflict with the argument for self-realization.

Many rational arguments are possible, and each could be posed so that the conclusion follows from the premises. If the arguments are based on incommensurable premises, however, they will lead to paradoxical conclusions. One must conclude that, in those societies where the question, Who should control assessment? is a meaningful one, it cannot be answered by specifying any single person or group. And, conversely, only in those societies where the question never arises is there an unambiguous answer. A society that can answer that question is a society with a shared sense of purpose. Because in our society there is not that shared sense of purpose, all of these differing moral criteria are really incommensurable — no overriding moral criterion holds them together.

How do we deal with a situation in which we have incommensurable moral premises in a liberal democracy dedicated to the notion of progress? In a society in which there is no progress, there is always a sense of purpose. People know what they are supposed to do; it is historically or spiritually embedded in the nature of their society. Our society believes in progress, in opportunity. Problems are solved by securing consensus on some moral principle that underlies the political process. In the absence of a shared sense of moral purpose, we must rely on political compromise. What overriding moral principle should undergird a political compromise on the control of assessment? The answer might be in what John Dewey has to say, for example, in *The School and Society*. He writes, "What the best and wisest parent wants for his own child, that must the community want for all its unlovely. Acted upon, it destroys our democracy." That is, assessment should be based on the expectation that all children, all students, will achieve a high level of competence or mastery — but that in itself would imply that standards were uniform for children whether they lived in Florida or Oregon or anywhere else in the country. So there would have to be national standards. If there are national standards, they can incorporate the utilitarian argument about the need to compete in a global economy. The standards must also inform and help teaching and learning. And the standards would have to be incorporated in an assessment process that was cumulative over time. Projects, performance-based assessments, portfolios — all the things that now go under the guise of alternative or authentic assessment would be used and calibrated to these national standards.

Thus, according to Goldberg, while there cannot be an immediate answer to the question of who should control assessment, there can be development of moral principles that would undergird the arguments made for any particular mode of control. If this happened, there would be national standards created, there would be local initiatives calibrated to these national standards, and all of this would inform teaching and learning.

It is important to keep ethical considerations in mind during discussions of control issues. Goldberg's points about expectations for all students certainly have direct implications for individual student learning. The question is whether these ethical points remain academic discussions or if they become a part of national, state, and local policy decisions for high-stakes tests. Persons who support standardized national tests may argue from ethical grounds that such an assessment would guarantee uniform science education for all students. A similar argument was given a decade ago for state minimal competency tests, resulting in lowered standards and a skills-oriented curriculum.

The idea that the acceptance of moral principles implies the creation of a national standard is interesting. The recent national standards effort seems to be driven by a policy that has economic concerns foremost, and is being carried out by professional groups who have childrens' learning at the center of their goals. While this has been a fortunate combination of interests, it is not clear that the moral considerations themselves would carry the work forward, if American achievement scores were at the top, even if the needs of some segments of society were not met. It remains to be seen what effect these standards, whether undergirded by moral principles or not, will have on individual student learning in science. It seems critical that the groups that have developed the standards, primarily on ethical and moral principles, continue to be involved in the issues of assessment control. All of the moral ground gained by insisting on curriculum reform for all students can be either solidified or lost in the assessment process.

Fairness Issues

Issues of control and fairness, according to Monty Neill, associate director of the National Center for Fair and Open Testing, are tied closely to the effects of testing.[5] He lists specific conditions, effects, and observations about testing that must be considered:

- High-stakes assessments can and will control curriculum and teaching, dominating other modes of assessment.
- A decade or more of increasing intensive and extensive high-stakes, multiple-choice testing has brought little improvement in education for most students and clearly no improvement in higher-order thinking skills.
- Standardized testing reinforces the current educational system that combines assembly-line approaches with behaviorist psychology.
- Testing is not the solution to educational problems that require curricular change and new approaches to teacher education.
- To be instructionally useful for a student, assessment must be individualized, detailed, and based on what is in the curriculum.
- Good assessment requires a good curriculum. A good portfolio is not a collection of photocopies.
- The richness and complexity of individualized assessment cannot be aggregated.
- Teachers, by and large, do not know how to assess learning based on inquiry-centered curricula.
- Performance-based external tests, such as those in California and Connecticut, can be used to encourage improvements in curriculum and instruction. However, there are risks of narrowing, limiting, and trivializing good practice because of the serious and profound

contradiction between in-class needs and what can be actually be aggregated.

Neill believes that there are both positive and negative aspects to assessment reform. He makes the following points about fairness issues. In Vermont, for example, teachers designed much of the approach and teachers are going to be educated across the state in how to use it. In Arizona, open-ended questions are being introduced, such as the math question that asks students to design a playground. The students have to explain what they are doing and show their reasons for doing it. This is not a hidden, surprise test, and teachers are expected to teach to it. There are obvious risks involved in that. But the teachers will have explanations of how to solve complex problems from each student in the class — potentially very valuable information for the instruction process.

As an example of a possible danger, consider the 20-minute essay that was proposed but not actually implemented in the new SAT examination. Had the essay been required, students would walk into a high-stakes test not knowing what they would be asked to write about; would have no time to prepare, do research, or to outline; and would have no time to revise. That sends a very bad message about what writing is. It may be better than absolutely no writing, but we have to be very careful about these kinds of dangers.

Assuming that these observations are accurate, Neill believes that some conclusions follow. First, since the most important areas of assessment are in the classroom and the assessment must serve the learning process, most assessment must be controlled in the classroom, primarily by teachers, but to some extent by students. That does not mean that parents, the community, and the school administration should be excluded. They should clearly be involved in determining what is important and how it affects the community.

Next, the primary task of districts, states, and the federal government should be to help teachers do good assessment work as part of their good teaching. These agencies should model, as much as possible, good assessment practice such as portfolios, open-ended questions, and performance-based tasks. Teachers should be involved in devising these and certainly in learning to grade and rate them. Most important and perhaps most difficult, government and other external groups should not subvert in-class instruction and assessment by making the stakes too high on any given test. Sampling should be done and there should be a general effort to minimize comparisons of individual students and schools. Clearly, there should be no national test, because the only mechanism available for full national tests in five subjects and three grades is a multiple-choice test, and under the assumptions above that will be a disaster. The demands coming from the centralized apparatus are overwhelming the teachers, leading them to simply look at outcome and achievement. Centralized control is not likely to be helpful and is very likely to be harmful. On the other hand, establishing uniform procedures for collecting indicators, Neill says, could be helpful in some areas, such as having similar dropout information across the country.

Finally, the economy should not control assessment. That is not to say the economy is unimportant; it obviously is. But the goal should be healthy, happy, creative, sociable communities of people in which the economy is just one part of the lives of the people and not the overwhelmingly determinant factor. And if that is what people want, then the purpose of schooling must transcend the economy.

Neill concedes that society as a whole requires more than we have often expected our schools to deliver. In order for schools to meet the needs of society, empowerment must come from the bottom up; from the students, the teachers, and the parents. That is where we need to start, and the other things that we do should be useful and helpful and not sabotage and destroy the good things that need to happen at the bottom.

The issues of fairness are particularly relevant concerns for individual students. The success of the American system of local control of education has always depended upon direct involvement by parents and others in the community who have the interests of students uppermost in mind. Recent intervention by external groups, government agencies, professional organizations, and others into the science curriculum and assessment arena has come about partly because parents and local groups have lost interest or have become too busy with other issues. Traditional family support systems for children have changed, leaving more responsibility to outside groups and agencies. Neill's concerns for fairness have arisen primarily because many children, especially those who are in families in which both parents work or in which there is only one parent, no longer have someone — traditionally a parent — who will act as an advocate to monitor and support their learning.

Although outside agencies may intend to have the children's best interests in mind — for example, by requiring them to take a national science test, they cannot be directly responsive to the needs and development of each individual child. Once this control over the individual child's education is given over, it is difficult to regain. Even if the current efforts by many schools to involve parents in their children's education are successful, the impact of high-stakes tests can make that involvement only superficial. In fact, it might be argued that parents' decline of interest and involvement in schools coincided with the implementation of the numerous mandated tests. Parents may have been lulled into thinking that the tests would guarantee that things would improve. Or they may have felt that there was little they could do to influence a testing system put in place by the school district or state "authorities," even if they thought it was unfair to their child. As the nation moves toward even more intense and expanded efforts in assessment, parents and other advocates for children should be involved and unintimidated by external policies.

Teaching Issues

As a classroom teacher and member of the National Education Association (NEA) Executive Committee, Reg Weaver knows that for effective teaching, the days of 30 kids and 30 textbooks should be seriously reconsidered.[6] This means that the days of teaching methods requiring regurgitation of isolated facts should be on their way out. A thematic approach to teaching science is needed, in which students learn scientific reasoning, how to integrate new and changing information, and how to apply abstract information with creativity and invention.

One of the key issues in developing a thematic approach is to define and set national goals for what should be taught, and to discover and implement new ways to enhance student learning. To achieve this, assessment techniques must be tied to the thematic approach. Weaver outlines some examples of concerns that must be resolved. First is the question of breadth versus depth. In other words, what is the essential information that needs to be taught, and how is it to be assessed? (Project 2061 addresses these concerns in its *Science for All Americans*[7].) A second issue is defining the most effective methods of teaching science. How do we teach inquiry and a spirit of discovery and at the same time teach the importance of rational analysis? The third major concern involves reaching agreement about what science content is to be taught at each grade level. Fourth, we must determine how to deal with the issues of knowing versus doing. And finally, there is the need to create interdisciplinary courses, instruction, and assessment.

In a reformed curriculum it will be difficult to separate instruction from curriculum from assessment. We need to learn how we determine a student's depth of understanding. There are many ways to determine a student's mastery of the subject area, but schools are driven by standardized test scores and by report cards that compare school A to school B. We lose track of what students actually are learning because the outcome is reported only in terms of test scores.

Some states are using assessment approaches other than multiple choice. Some are using more direct observation and are placing more emphasis on application skills as opposed to content skills. But in the classroom it is very difficult for some students to move from content skills to application. On standardized tests there are few applications. If content skills are not taught and application is being taught at the high schools, the students are going to fail at the high school level even though they may pass the standardized tests.

Weaver believes we must continue to look for various kinds of approaches for determining students' abilities. We need to think about broader applications of assessment. Teachers will continue to need help in sorting out all of the results of research in the areas of assessment and learning. But if science education is to meet our national goals, we have a lot of work to do.

Teachers must function in schools restricted by rules and standards that have been set by somebody else. If teachers are to be held accountable for what the students learn and if they are to be required to make decisions about what science is to be taught, then they should decide the best way to assess the results. Mixed signals about what to teach and how to assess are just not going to work. That means that teachers must have means to assess other than just paper, pencil, and filling in the bubble.

Weaver's points about the classroom teacher bring home the point that teachers are the key to individual student achievement and to the success of any assessment program. Even with the advances in instructional methods and technology, the old idea is still valid that once the classroom door is closed, the teacher is the one responsible for what kids learn. At present, only scattered efforts are being made to build authentic assessment into teacher education and inservice programs. Until teachers learn how to assess learning that results from reformed curricula such as thematic approaches to science, it is unlikely that there will be much classroom implementation. And unless the material is assessed, students will not believe that it is important. It is at this point that issues of enhancement of student learning become the key. Students will be the victims if they are tested in science classes on only the material that teachers know how to assess, yet are evaluated and ranked on external high-stakes tests with different, higher-order performance items. In order to enhance student learning of science, it is essential that the same effort that is given to developing and implementing external assessments be given to helping teachers learn to use appropriate assessment in their classrooms.

The Need for New Assessment Approaches

Two possible responses to the question of who should control assessment are "no one" and "everyone," according to Tom Pettibone, associate dean for research at the University of Tennessee.[8] Both responses, in reality, are equivalent — if everyone controls, no one controls. And this is the answer that the question ought to have. First, however, it is important to list those who might have some control over assessment. These are (not in any particular order) students, individual teachers, school principals, superintendents, school boards, parents, business community, general public, test makers and sellers, teacher associations (both general and subject-matter specific), universities, research laboratories, other professional associations, politicians, courts, funding agencies, and others. In addition to the

Whos, there are some "Whats" to consider as well. For instance, the curriculum is an important What. To what extent does the curriculum control the assessment process? For that matter, *should* the curriculum control assessment? Curriculum-based assessment requires a close relationship between teaching and testing. There are those who maintain that the situation is just the opposite, that the assessment controls the curriculum. It has been suggested that one way to move toward a national curriculum is to institute a mandated assessment process. Another What is economic development. Should concerns for economic development (that is, jobs) drive the assessment process?

Of the many possible Whos and Whats, there is not one that should be entrusted with sole control of the assessment process — not because any are necessarily untrustworthy or trivial, but because assessment is so important to so many groups that such control must never be given to any single Who or What. Indeed, "control" may well be what is wrong with assessment now. No one denies that assessment has many faces. On the one hand, we want to know how students perform in relation to some norm or criteria. Teachers, parents, and students all need to know this. On the other hand, the same information — probably in a different format — is needed on numerous levels for making decisions to improve instruction. And some of that information is needed for policymakers at levels almost too numerous to list. The public, especially in recent years, wants to know what the educational return is on their heavy tax investment in the schools. The one aspect of control that should be emphasized is control of the assessment process itself. Obviously, for assessment to have validity, there must be consistent standards of administration and data collection, regardless of the particular measurement strategy. This is probably as true for "authentic" types of assessment as it is for so-called "standardized" assessment. Other forms of assessment, including portfolios, can and should be "standardized" as much as the old forms.

Because of the need to serve multiple purposes, audiences, and masters, assessment must be controlled by the many and not by the few. An example is the team approach being used by the University of Tennessee to evaluate innovative schools.[9] These schools come in all sizes and shapes, from across the United States, serving diverse populations of students. Some are elementary, others middle, and still others are high schools. The programs vary from a recess math program at the elementary level to a performing arts program at a high school serving predominately minority students. One West Coast elementary school has a student body speaking seven languages and some 20 dialects. A middle school in Brooklyn has established minischools as a type of magnet school program. A preschool in North Carolina is pairing affluent children with poor ones.

The assessment needs and requirements of these 15 schools are immense and extremely diverse. The assessment process will utilize a variety of sources and techniques and will involve a number of groups to help determine the best and most meaningful assessments to satisfy the diversity of interested participants in these projects. The RJR-Nabisco Foundation wants to know the return on its large investment and also wants to document the characteristics of the successes so replication can be attempted. Interested others include all of the Whos and Whats mentioned before.

The University of Tennessee is working to establish an assessment center to develop new assessments to serve the measurement and evaluation needs of America's schools and school districts. It is the center's contention that, in spite of faults, there are already measurement and assessment tools for individual students and, through devices such as NAEP, for national and statewide assessments. Certainly, these can and will be improved over time. There are not, however, adequate sources of performance feedback and related services needed by schools and school districts to measure progress toward their goals. That's a big order and one which cannot be filled if any one Who is allowed to control the assessment process.

Pettibone suggests an idea different from the consensus process, that many, rather than one or a few groups or individuals, should control the assessment process. The need for multiple types of assessments, beyond tests of achievement or student performance, is an excellent justification for this view of control. A full discussion of the kinds of data needed for these purposes has been provided by Murnane and Raizen (see reference 10), who also focus on individual students as the key to assessment. It is easy and direct to ask students what they think of a recess math program, or to simply count the number of students who participate. The difficult task is to determine the kinds of questions that are appropriate for assessing complex programs, while being creative and open enough to look at simple, observable indicators. There is a danger in assessments of complex programs to test everything that might be a contributing factor in the students' learning. This approach usually means that huge blocks of student time are spent taking tests. The amount of school time given to testing is already excessive.

As the reform process gets under way in an extensive implementation phase, program evaluation will become a more central focus of assessment efforts. It is important that this work be unobtrusive when possible, contribute to learning, and economical in the use of student time. If this aspect of assessment is not carefully monitored and controlled, there is the risk that any added benefit of reformed curriculum or methods will be neutralized by the time spent testing the participating students. On the other hand, if assessment is built into instruction, and simple, observable indicators are used for overall program assessment, the reforms will have a chance to take root and grow.

References and Notes

1. This chapter is based in part on a panel discussion, "Who Should Control the Assessment Process," which took place at Forum 90: Assessment in the Service of Instruction, sponsored by the Directorate for Education and Human Resources Programs of the American Association for the Advancement of Science with the support of The Carnegie Corporation of New York, 10 November 1990, Arlington, Virginia.
2. Jones, L., remarks delivered at Forum 90. See reference 1.
3. Raizen, S. and Kaser, J., "Assessing Science Learning in Elementary School: Why? What? and How?" *Phi Delta Kappan*, (1989) **70**, 9. See reference 1.
4. Goldberg., B. remarks delivered at Forum 90. See reference 1.
5. Neill, M., remarks delivered at Forum 90. See reference 1.
6. Weaver, R., remarks delivered at Forum 90. See reference 1.
7. American Association for the Advancement of Science, *Science for All Americans: A Project 2061 Report on Literacy Goals in Science, Mathematics, and Technology* (Washington, DC: American Association for the Advancement of Science, 1989); reprinted as, Rutherford, F. J., and Ahlgren, A., *Science for All Americans* (New York: Oxford University Press, 1990).
8. Pettibone, T., remarks delivered at Forum 90. See reference 1.
9. Murnane, R. J., and Raizen, S.A., eds., *Improving Indicators of the Quality of Science and Mathematics Education in Grades K-12* (Washington, DC: National Academy Press, 1988).

Part 2

Science Assessment and Curriculum Reform

Shirley M. Malcom

It is very important that papers addressing curriculum, instructional, and policy concerns related to assessment in science be presented in the same volume. Content experts and teachers may have the greatest interest in the educational roles of assessment; the use of assessment as a policy and political tool has come to dominate the public discussion of testing. The question is not whether the public has a role in requiring accountability for use of its tax dollars but rather what should be the relative balance between the public needs and the educational needs that assessment must satisfy. In a knowledge-based economy a high premium must be placed on mastery of skills, demonstration of competencies, and comfort with the content provided in science, mathematics, and technology education. So a separate goal for performance by American students is stated, and the search has begun for appropriate indicators of student achievement.

A story that is circulating from the hearings conducted by the goals panel says that a testifier told a story about a farmer who was weighing livestock to take to market. When the farmer came across an animal that was too small, it was put to one side. After completing the weighing of the other animals, the farmer picked up the small animal and weighed it again; not so surprisingly, the animal weighed the same as before. The moral of the story is straightforward: weighing does not change the size; the animal must be fed to effect a change. The search for the right assessment must be first and foremost a search for mechanisms that help us feed the minds of students. Their food must be filling and nutritious, giving them an appetite for lifelong learning that includes studying science because they like it. Students should be able to take such knowledge, make it their own, and be able to use it in other aspects of their lives.

Part Two of this book addresses the issue of science assessment and curriculum reform. The chapter by Kulm and Stuessy sets out the major national proposals for curriculum reform and the challenges these projects will face in the reform of assessment. Rowe raises the stakes for reform (to support our democratic ideals) and assessment (taking a student's "mental temperature") as she argues for finding and holding fast to the big ideas within science. Cole explores the issue of reconciling the measurement and instructional roles of assessment in science. The chapter by Hudson provides the reader with information on the existing assessments, while Davis and Armstrong review the use of various science tests currently used in state testing. Comfort reviews some of the state efforts to develop and test performance assessments as alternatives for standardized tests. Chapters by Harlen and by Madaus and Kellaghan give us windows into testing in Europe — the first focusing on the use of performance testing in England and Wales and the latter providing an historical perspective on examination systems in the countries of the European Community. The Madaus and Kellaghan paper was done as part of a larger study on assessment by the Office of Technology Assessment, the full report of which will be available in fall 1991.

5

Assessment in Science and Mathematics Education Reform

Gerald Kulm and Carol Stuessy

National goals for science and mathematics education have attained nearly unprecedented attention. Not since Sputnik has the nation's attention been focused so intently and comprehensively on elementary and secondary education. The President has declared that the United States shall be number one in the world in mathematics and science by the year 2000. A national panel of governors and nearly every government agency have responded in some fashion to this goal. In this chapter, we will focus on the issues of assessment in science education reform in this broad context. We will also deal with the role of mathematics in the science reform effort, since science and mathematics are so interrelated, the goals for improving their learning so similar, and the barriers they face in implementation so comparable. Finally, we will consider the challenges of a changing demographic mix in the school population and the continuing problem of science opportunities for women.

Of course, the big questions are, "What science content are we talking about?" and "How do we measure whether American students have learned this content?" In a country in which decisions about educational goals and achievement are local matters, these questions raised at the national level have generated considerable discussion and debate. Some answers to the science content question have been proposed by professional organizations, which have published standards, guidelines, and blueprints for what students should know by the time they finish high school.[1-3] Even though there is not a unanimous agreement on these proposals, they have provided direction and examples for state departments of education and local school districts to adapt or build on in developing their own curriculum reform. Some states and districts have already begun reforms drawing directly upon these reports. Large projects have been funded to pilot-test curriculum planning and implementation efforts that reflect the recommendations of these reports. Evaluation of these substantial changes brings assessment clearly into the picture. Curriculum changes of this magnitude demand assessment approaches that are appropriate and powerful and sensitive enough to reflect the new kinds of learning that are expected of students.

Policy and Politics of Assessment Reform

How to measure national attainment of science and mathematics superiority is an especially sensitive question. There is reason to believe that success or failure of reform in science education rests heavily on the outcomes and directions that assessment takes in the next few years. As a result, the assessment issues in the reform movement have attracted substantial political attention and involvement, both nationally and locally. Some of the panels and groups that have been formed to study directions for assessment are dominated by persons with interests primarily in the policy and political area, in contrast to the earlier curriculum and teaching groups that were mainly under the leadership of educators and scientists.

For a number of reasons, the issues surrounding assessment generate intense opinions, discussions, and reactions. First, many people object to the seemingly logical notion of a national examination. While alternative approaches might exist for estimating the level of science achievement of the nation, it will require the best thinking of the educational community to develop them. Second, even if some national plan for assessment were agreed upon, what would be the actual content of the test? The traditional types of items and mass-testing methods do not seem to match the new goals for science and mathematics education. Assessing problem solving, reasoning, and making and testing conjectures involves measuring individual and small group performances on in-depth problems, an approach that seems poorly suited to mass testing. Finally, some educators fear that a national high-stakes test may subvert progress toward changing the curriculum to emphasize higher-order thinking processes. Even though it is clear that classroom assessment approaches should be process-oriented and closely tied to instruction, the influence of mandated external tests is overwhelming. Teachers and school districts will build their students' learning experiences around the test that has the greatest impact on them.

Proponents of assessment reform have focused mainly on procedures and strategies for changing how specific subject matter is assessed, and science assessment reforms are under way in a number of states. In some cases, these reforms have built upon similar work in writing, mathematics, or other subjects. Once holistic scoring of essays is accepted, scoring open-ended questions in science or mathematics also seems viable. If students use measuring instruments to do lab performance tasks in science, it seems reasonable to let them use calculators on math tests. The direction of assessment seems most likely to change in disciplines where it is natural for assessment to follow the expected performance. For example, writing stories and essays is the expected result of learning to write, so assessment of actual writing samples is a natural thing to do. Doing laboratory observations, making hypotheses, and doing experiments are the activities most people expect to result from teaching and learning science. So it seems natural, especially to those outside of schools, that assessment should focus on the performance of those kinds of science processes.

The atmosphere within the educational community, especially at the national and state level, is shifting toward acceptance of such alternative approaches to assessment. There is still a long way to go, however, in working with district measurement specialists, superintendents, principals, and teachers at the school level. A great deal of work needs to be done to change the way school districts evaluate how well they are doing. Many districts perform quite well on the current standardized tests and have many students who achieve high scores on SAT exams. If these districts send a large percentage of their students to good colleges, they see no reason to change either their curriculum or the way they assess students. It is very difficult to convince them that there may be improvements that lead to a higher level of science learning for all their students. And if reform is to take place, more than schools, teachers, and administrators must be involved. The current testing and evaluation system has created expectations and assumptions in a broad array of people and organizations, from parents to colleges and the business community. As assessment approaches are changed, all of these constituents must be informed. For example, we must deal with the parent who expects a letter grade on a report card and receives, instead, a profile of student proficiencies and a list of areas in need of further work or study, and the personnel office or college that expects a transcript and receives a written summary of a portfolio of student work.

Assessment and New Learning Goals for Science

Changes in curriculum goals require concurrent changes in approaches used by teachers in improving learning, by schools in reporting to parents and school boards, and by agencies in monitoring progress toward broad goals. For these reasons, restructuring assessment is one of the most significant challenges facing science and mathematics education reform. National groups such as the American Association for the Advancement of Science, the National Science Teachers Association, and the National Research Council have convened committees and conferences to address the issue. Since recent experience in American education seems to verify that the material tested is the material taught, science and mathematics educators have called for new approaches to assessment and provided some directions for changes.[4-6] Promising work has begun to appear at the national, state, and local levels.[7-9] Improvements are being made in some state tests;[10] however, more development is needed to help teachers to identify specific approaches that work in the science classroom and to learn how to use them. Teachers have been legitimately concerned that if they "fight the system" and teach inquiry and problem solving, their students will suffer on mandated tests that focus mainly on terminology and computation. There is a wide consensus that unless assessment is changed, reform in science and mathematics education will not be comprehensively implemented at the classroom level.

What implications do the new goals for science education have for assessment? The major reform documents have stated some general themes and common ideas that challenge current instructional approaches and assessment practices. The most fundamental goal is that science and mathematics learning should focus on broad conceptual understanding, problem solving, and habits of thinking. This view of learning is the major force driving new thinking about assessment. Current tests — one-shot, multiple-choice, and focused on mastery of simply stated behaviors — are not appropriate for assessing this kind of learning. Problem solving and conceptual understanding are not skills that can be learned, practiced, mastered, and tested in the short term. They develop gradually and contextually while connections are made with related concepts and experiences. When learning is individualistic and dependent upon experience and motivation, assessment must be continuous, sensitive to individual differences, and open-ended enough to be capable of reflecting deep and broad understanding. Although some separate testing may be necessary, most assessment should be nonintrusive and integrated with learning activities.

Most educators agree that science learning should progress from extensive early hands-on explorations, to informal experiments and qualitative understanding, and finally to more formal and quantified work in the later grades in high school. This view of learning requires assessment capable of monitoring and reflecting the growth of ideas and concepts. Much traditional testing has focused on final outcomes of learning. Even pretests or formative tests are constructed in the same way as summative tests; only the time in the learning sequence at which they are administered is different. New assessments are needed to provide information about what students learn in the early informal settings. They should show to what extent and in what ways students have integrated experimental knowledge with known or developing concepts. They should give insight into how students formalize information gathered through explorations and hands-on work into images and symbol systems. New assessments are needed to measure development of students' quantitative and qualitative understanding and show how they are related. Finally, approaches are needed that monitor and assess students' readiness to progress from exploration to informal experiments and then to more formal work. In this type of learning, end results are less important than the processes and experiences that lead to the desired outcomes.

Another goal of a reformed curriculum is to reduce the sharp lines between the sciences, and to reflect science as an integrated domain of knowledge. Science is unified through broad concepts that link the content of the separate disciplines. When the focus is on broad themes that link important content and ideas, tests of isolated concepts and procedures are not sufficient. Even broader or more "authentic" assessments of knowledge in specific subjects such as physics or biology may not be curriculum appropriate. Although there is no agreement on the extent to which the curriculum itself should be integrated, at least some portion of science assessment must let students show how well they can apply an integrated knowledge of science. Long-term projects, investigations, or reports seem to permit assessing this type of knowledge or ability. Since these kinds of activities often involve cooperative work by two or more students, approaches are needed that assess the knowledge gained and demonstrated through these activities, both individually and collectively. Other innovative approaches are needed to assess how well science knowledge is integrated and available for problem solving. The goal is for students both to gain in-depth content knowledge in specific subjects and to develop the ability to apply that knowledge across fields. If they do not acquire sufficient content understanding in specific subjects, "less is more" can become "less is less."

A final goal common to many curriculum reform proposals is that the relationships among the sciences, mathematics, and technology be emphasized and developed. Currently, even in elementary grades, science is most often taught separately from mathematics. The separation increases through the middle grades and high school. The issues are not settled as to whether or how to develop an interdisciplinary curriculum; however, if teachers are expected to emphasize connections and students are expected to understand and apply concepts and ideas across science, mathematics, and technology, new assessment approaches will be needed. The next section explores in more detail the issues of integrated science, mathematics, and technology and the implications for assessment.

Assessment and Integrated Science and Mathematics/Science

Current curriculum reform efforts by the American Association for the Advancement of Science (AAAS), the National Science Teachers Association (NSTA), and the National Council of Teachers of Mathematics (NCTM) focus on making mathematics and science relevant. Explicit connections among the disciplines of science, the domain of mathematics, and the real world, often made through the use of educational technology, are suggested as essential in the development of mathematical and scientific literacy. AAAS Project 2061's *Science for All Americans*[11] proposes scientific literacy for all through an integrated understanding of science, mathematics, and technology as interdependent human enterprises. The Scope, Sequence, and Coordination of Secondary School Science (SS&C) developed by NSTA promises "a content core of scientific knowledge that every educated person should have at the end of his/her secondary education, knowledge that allows a person to make informed decisions about scientific matters rather than be swayed by pseudoscience, and to understand the scientific side of modern life where it impacts on the individual."[12] Both science reform projects involve development of curricula that reduce the amount of content that is covered ("teaching less that is also more"), that weaken or eliminate rigid subject-matter boundaries, and that pay more attention to the connections among science, mathematics, and technology.

Although no one definition of "integration" satisfies all educators, the concept is becoming a recognized option for science curriculum reform in the United States. The trend is toward integration, even though there is a long history in the United States of compartmentalizing knowledge into "subjects" and teaching bits of "content" in mathematics, science, and calculators/computers in discrete "units" of instruction. In the rapidly paced, changing world of the 21st century, however, the problems are not subject-specific or compartmentalized: world hunger, poverty, disease, war, and environmental pollution are "integrated" by their very nature. Although technology has exacerbated many of these problems, technology can also be used to solve them. A knowledge of the complex negative/positive interactions that occur among science, technology, and society is necessary for all citizens to make decisions in a democratic society. Citizens must also have the mental capabilities to conceptualize problems holistically, to evaluate the relative importance of information gathered from a number of sources, and to see the entire picture composed of many overlapping pieces.

For those who have not had integrated problem-solving opportunities in school, the transfer from traditional classroom learning to real-life problem solving may never occur. The question of "When will I ever need to know this in my real life?" takes on an ominous tone as we look at educating 21st century Americans for a technologically driven world of work as well as supplying the mathematicians, scientists, and engineers to maintain the nation's competitive advantage in the world marketplace. The question also is relevant when literacy is understood as education that enables human beings to be life-long learners, fully capable of experiencing and comprehending a world that is rapidly changing and often driven by factors associated with the scientific and technological enterprise.[13] Integration may allow us to produce classroom scenarios that engage, motivate, and encourage learners to be creative and elaborative in their learning about the natural and technological world in which they live.

Although the integrated viewpoint appears attractive for classroom teaching, both science and mathematics educators agree that integration should not replace all separate teaching of mathematics and science. Mathematics, science, and technology educators who met at Wingspread[14] in 1991, for example, agreed that students need to understand that the domains of mathematics and science, which are indeed different, also complement one another. Much of what we call "scientific knowledge" could not exist without the language of mathematics to simplify and explain the complex relationships that exist in nature. Deep conceptual understandings of mathematics involve seeing connections between the abstract mathematical conceptualizations and the concrete realities of the real world. Although mathematics and science complement one another, mathematical concepts (for example, "number") represent discrete understandings, as do scientific concepts (for example, "evolution"). Nevertheless, science can often provide a contextualized foundation for learning mathematical concepts, and mathematics provides the language that quanitifies and simplifies explanations of natural phenomena.

A number of innovative curriculum projects explicitly integrate mathematics and science and/or technology (e.g., AIMS, CTG, TIMS, TERC).[15] Technology is used in many of these innovative projects for contextualizing instruction and providing visual scenarios from which real-life problems are posed (or identified by students) and solved. An advantage that has been hypothesized in the use of technology is that it provides a context for "authentic activity" that facilitates student problem solving and increases intrinsic motivation.[16] Technology allows presentation of problems that are posed within a real-world, personally relevant context. The technologically presented context enhances motivation, active mental processing, and elaborative processing, as compared with more traditional contexts that rely heavily on the reading capabilities of children who also must be intrinsically motivated to learn.[17–19] Although the advantages to the learner of integrating technology with mathematics and science have been suggested, most positive outcomes are only intuitively understood. A strong justification from the educational research community for using innovative, integrative curricula is lacking.[20]

Student assessment, which provides specific information about learning, may reveal unique advantages of the integrated approach to teaching mathematics and science. Performance assessment, which by its very nature "blurs the edges among curriculum, assessment, and instruction,"[21] makes an ideal beginning point for the development of authentic assessment for contextualized, innovative, integrated curricula. In performance assessment, students learn by engaging in the assessment-curriculum-instruction enterprise, which consists of a number of tasks or activities within units labeled as "problems." The tasks provide opportunities for students to synthesize their knowledge, make connections, deepen their understanding of concepts, and create for themselves learning situations that foster the development of deeper levels of understanding. Tasks also provide opportunities for teachers to evaluate evidence from a number of sources — written, oral, visual, and computer-based. Evaluations that are multidimensional exceed the one-question, one-answer notion often used for assessing a student's acquisition of scientific facts. Issues relating to the design, selection, and evaluation of learning tasks or "activities" have been relatively neglected in educational research and scholarship.[22] It may be that analysis of learning activities occurring within the integrated context, such as the analysis structure suggested by Brophy and Alleman[23] or others[24,25] may reveal needed information about the unique aspects that an integrated mathematics/science curriculum offers to learners.

Assessment Concerns for Traditionally Underrepresented Groups

Students who are minorities, speak English as a second language, are poor, or have physical disabilities are currently less likely than others to receive the instructional and social support necessary to excel in science and mathematics. These students are often placed "at risk" for entering science- or mathematics-related careers and occupations. Of particular interest to many science and mathematics educators is the traditionally underrepresented groups of females and minorities, who historically do not pursue advanced course work and careers in science-related fields. A recent report describing the representation of women and five racial and ethnic groups among B.A., M.A., and Ph.D. degree-holders in quantitatively based disciplines indicates that at any given degree level, a group's share of quantitative degrees reflects two factors: persistence in the pipeline and field choice.[26] This report suggests that the scientific/mathematical pool from which quantitative Ph.D. graduates ultimately derive first appears in elementary school and emerges strongly before the 9th grade. Before 9th grade, membership in the pool is defined more by quantitative interests than by skills. The talent pool appears to reach its maximum size before high school, but some migration into the pool continues to occur during grades 9 through 12. Thus, to increase participation in this group, strategies should be implemented to increase the size of the initial scientific/mathematical pool of minorities and women before and during high school.

At-risk students' science and mathematics achievement scores continue to be one of the most visible and troubling factors in the nation's educational efforts. Current assessment approaches indicate that science and mathematics deficiencies are present for minority students compared to whites from the first testing point. The performance gap increases over time, resulting in a cumulative deficit that excludes many minorities from post-secondary study in science, mathematics, or engineering.[27-30] The reasons for the gap begin before school entry. More African-American and

Hispanic children than white children come from economically disadvantaged backgrounds. They appear to be behind their peers and are therefore placed in remedial classes. Students who speak English as a second language are at an even greater disadvantage in a classroom where teachers speak and teach only in English. The content-specific vocabulary associated with traditional science and mathematics teaching makes the subject of science seem more like a foreign language than a subject that fosters curiosity, problem solving, and assurance that the world is indeed a place that is understandable and predictable. Still other factors appear to be working to discourage females in mathematics and science, including teacher and classroom variables, and motivational and social factors.

While there have been slight improvements, the continued gap between whites and African-Americans, Hispanics, and other minorities is cause for concern about the way testing (and, hence, teaching) is done.[34, 35] Fundamental changes are needed in light of research indicating that minority and other at-risk students learn science and mathematics best in environments that involve hands-on work, meaningful applications, and cooperative group interaction.[36-38] If assessment were also aligned with these learning styles, at-risk students might reveal knowledge and performance in science and mathematics greater than that estimated by more traditional standardized tests. As yet, however, little work has been done to investigate how specific alternative approaches to science assessment work with various student populations. Research and development are needed to answer questions such as:

- How well do current tests measure the scientific knowledge, skills, and understanding of at-risk students?
- Will assessment items that focus on meaningful tasks and projects reveal unseen scientific and mathematical abilities in at-risk students?
- Will assessment tasks that are given one-on-one, in small groups, and/or which allow students to respond in their first language reveal unseen mathematical abilities?
- Will alternative approaches to science assessment be fair and accurate in their treatment of at-risk students?

The most promising assessment approaches appear to be those that use hands-on work and focus on student performance. Negative expectations may be overcome by giving all students the opportunity to include their best and most creative work in portfolios, thus providing a wide variety of positive, successful experiences in science and mathematics.

Teachers and Assessment Reform

One of the most important issues in assessment has not received much attention. That is, how can science teachers be helped to learn how to use alternative assessment approaches? If they make changes in assessment, how will their teaching be affected? And how many added demands will assessment reform place on an already less than adequate system of inservice teacher education?

A start in our discussion about teachers leads us to consider current investigation in cognitive psychology that concerns the ability of people to reason, understand, solve problems, and learn on the basis of these cognitive activities.[39] Current researchers view learning as an interactive process among prior content knowledge, new knowledge, and students' cognitive abilities to think and solve problems. Ideally teachers should know about the prior knowledge and learning strategies of their students in order to provide diagnostic, flexible, and interactive instruction that matches students' entry level knowledge. In particular, teachers must sequence instruction to take into account the changes in learner needs at various stages and provide appropriate support or "scaffolding"[40] in the form of verbal, visual, or physical cues or other assistance that enables students to do as much as possible on their own. Assessment, which must accurately ascertain students' knowledge and skills at different points in a learning task, provides a mechanism for the teacher to make decisions about the type and amount of scaffolding to be provided. Teachers also manipulate the instructional methods associated with the task, and in the problem-solving learning situation, they may employ a cooperative learning structure, such as cooperative problem solving,[41] cooperative grouping,[42, 43] paired problem solving,[44] or variations of reciprocal teaching.[45, 46] Other more appropriate instructional methods might include the use of advance organizers,[47] concept mapping,[48] or explicit instruction in metacognitive strategies.[49] Problem-solving instructional sequences of this kind have flexibility built into their structure. The assessment is authentic to the instruction and the learners; it is not something "done" to students by outsiders unfamiliar with the unique learning qualities and capabilities of the classroom learners. Ultimately, the success of the learning experience relies on the authentic interactions of students and teachers in the learning process, and the interactions allow for instruction to be altered to produce ultimate learning outcomes in problem solving.

Outcomes associated with innovative curricula are often multidimensional and complex. For instance, a scheme may be required for assessing the degree of "conceptual understanding in science," graphing interpretation skills and laboratory procedural skills of students who have worked with an interactive computer simulation.[50] Innovative tracking and scoring methods are being developed for measuring the learning outcomes of students who are engaged in innovative, integrated learning experiences. Many assessment teams across the nation are working with teachers, curriculum specialists, and others to develop holistic scoring protocols, portfolios, and checklists for evaluating student performance on interviews, presentations, and written work. For those developing authentic assessments, an instructional activity is viewed as a multidimensional interactive learning experience that results in a number of instructional outcomes.[51]

For the teacher, a number of information sources are necessary for the complex decision making needed to maximize instructional outcomes. Simple "grades" are not useful for deciding what to teach, how to teach it, to which students and for how long. Altering instruction systematically as a result of outcomes associated with a particular learning activity or sequence is not practical without a planning, monitoring, and assessment tool. The tool should provide access to and methods for teachers to manipulate data about student entry knowledge and skill levels, learning activities and their associated levels of cognitive and skill demand for learners, and the instructional context in which the learning activity is offered. Interpretations regarding choice, sequencing, and modification of instruction must be easily made and probably should be represented visually. Methods must be designed for teachers to document and track student progress through particular learning activities, as well as to ask evaluative questions regarding the efficacy and/or efficiency of their choices of activities and their position within instructional sequences. These methods would assist teachers working with heterogeneous classes of students who exhibit differing levels of entry knowledge and strategies.

Computers can be invaluable for manipulating data sets that pertain to cognitive and skill demands of learning activities, entry-level student knowledge and strategies, instructional contexts, sequencing, and eventual learning outcomes of students. Computer technology should allow the teacher to ask questions about individual and class performance over time, as well as about the effects of adjusting instructional activities within a particular dimension (such as a cooperative learning structure) on other aspects of student performance (such as acquisition of content-specific concepts or strategies). Lesh has explored computer-based assessment possibilities including reports that are graphic-based, interactive, intelligent, and multidimensional.[52] This suggests methods for producing visual reports via "learning progress maps" that resemble three-dimensional topographical maps — idealized models of potential student knowledge that show the conceptual terrain that all students are encouraged to explore in a targeted curriculum, as well as multidimensional profiles of development of individual students in various conceptual regions. The overall goal of the learning progress map is to capitalize on the dynamic, interactive, and intelligent nature of computer-based graphics to generate rapidly updatable reports to inform the teacher on a wide range of decision-making issues. The reports reflect the nature of the interactive model of student/teacher/instruction discussed previously — a far cry from the percentage or letter grades that traditionally represent the terminal effects of school and provide no indication as to the richness and multidimensionality of the classroom experience.

Although the learning progress map reporting scheme provides data for teachers to use to alter instruction and evaluate programs, Lesh emphasizes that "instruction-relevant learning progress maps must be designed so that we do not try to inform all decisions and all decision-makers using only a single type of report, test, or data source. That is, the information that is collected, filtered out, interpreted, aggregated, and displayed varies according to the decision and decision-maker that is being addressed."[53] The dangers of using only one or two data sources in making decisions about students and the curriculum cannot be overestimated. If by the Year 2000 the school is indeed viewed as "the site of reform,"[54] the advice provided by Lesh is indeed as appropriate at the level of the school, district, and nation as it is at the level of the teacher. Assessments should empower teachers and schools to make the changes in instruction and curriculum that will bring all children to the goals of literacy for the citizens of this nation. Assessment attached to external rewards and punishment is in direct conflict with a model that empowers teachers to diagnose, modify, and implement learning activities to maximize learning. In that sense, one can view the quality of the learning environment as directly related to the assessment that its decision-makers put in place.

The assessment reflects the educational and philosophical goals of the educational community responsible for making decisions about the learning of its students. Assessments that guide responsible decision making in the classroom directly affect the learning of children. For that reason, the development of assessment strategies that enhance and support decision making at the level of the classroom teacher should have highest priority in educational reform. Ideally, teachers in the future will seldom tell students to put away their books and notes to take a test. Assessment will be an ongoing and integrated part of teaching. Students will have the opportunity to provide input to their assessment records, making sure that their best work, produced without pressure and anxiety, is included in the evaluation process.

Families and Communities

Children learn science and mathematics best when there is coordinated support and encouragement from the school, the home, and the community. Traditionally, this support has been available to students from stable and educated families, attending well-financed schools, and living in communities with a variety of easily accessible resources. Traditional teaching, assessment, and reporting approaches have a reasonable chance of succeeding in this environment. Now, however, societal and family structures are making it necessary for schools and communities to provide additional structure and support to students. Assessment plays a major role in this effort, in reporting student progress to families and in helping communities to determine how well they and their schools are doing. The most important role for assessment, however, is to communicate to families and communities the new goals for science and mathematics education.

One of the factors that often differentiates the achievement and opportunities of students who are less successful in school from those who are more successful is the amount and quality of experience and involvement in science- and mathematics-related activities outside of the school. This difference is due to a number of factors, including lack of home and community resources, unfamiliarity with how "the system" works, lack of confidence or expertise in dealing with science and mathematics content, and the impact of the low science and mathematics expectations of many parents for their children, especially female or minority. Many of these factors have their roots in a fear of science and mathematics. This fear can be traced to traditional testing and assessment approaches in which there is one correct answer or procedure that must be memorized and recalled on a timed test.

The new approaches to assessment can have a positive and dramatic effect on perceptions about the nature of science and mathematics. Families can actually see examples of performances and projects rather than seeing numbers and letters on report cards or assignment sheets. Students can do science assessment activities at home, in the neighborhood, at museums or exhibits. Parents or other family members can be involved as partners in the assessment process, providing information about the child's current and developing interests and motivations. These approaches can provide powerful and effective messages about science, its goals, and the ability of students to do science and mathematics.

Summary

Student learning and attitudes in science must be viewed, and therefore assessed, from multiple perspectives. It is not sufficient simply to reject or reform the standardized tests now used by teachers, schools, and other groups. Students learn science in other school courses, especially in mathematics courses. As the use of technology grows, students will learn science in a multitude of contexts in which technology appears. Science is also learned outside of school; sometimes more than in school. Assessment approaches that only include school-like or textbook-like or laboratory-like settings for science may not provide the opportunity for many youngsters to demonstrate their knowledge and ability in science. And, conversely, there is ample evidence that high achievement on school-based standard science assessments does not guarantee that students will not continue to have misconceptions about fundamental science concepts.

Above all, assessment in science should promote and encourage learning and should focus on what children can do rather than on what they can't do or on what they have only memorized. Science educators must be involved actively in the reform of assessment and ensure that policy-driven assessments generate the fewest possible barriers to the types of classroom assessment that are so needed for an effective science curriculum.

References and Notes

1. American Association for the Advancement of Science, *Science for All Americans: A Project 2061 Report on Literacy Goals in Science, Mathematics, and Technology* (Washington, DC: American Association for the Advancement of Science, 1989); reprinted as, Rutherford, F. J., and Ahlgren, A., *Science for All Americans* (NY: Oxford University Press, 1990).
2. National Council of Teachers of Mathematics, *Curriculum and Evaluation Standards for School Mathematics* (Reston, VA: NCTM, 1989).
3. Aldridge, B. G., "Essential Changes in Secondary Science: Scope, Sequence, and Coordination," *The NSTA Report* (Washington, DC: National Science Teachers Association, 1989); Aldridge, B. G., "Scope, Sequence, and Coordination: A Rationale," paper distributed by the National Science Teachers Association (Washington, DC: National Science Teachers Association, 1990).
4. Burstall, C., "Alternative Forms of Assessment: A United Kingdom Perspective," *Educational Measurement: Issues and Practice,* 5(1), 17–22.
5. Kulm, G., ed., *Assessing Higher Order Thinking in Mathematics* (Washington, DC: American Association for the Advancement of Science, 1990).
6. Shavelson, R. J., Carey, N. B., and Webb, N. M., "Indicators of Science Achievement: Options for a Powerful Policy Instrument," *Phi Delta Kappan* (May 1990), 692–697.
7. Council of Chief State School Officers, *Assessing Mathematics in 1990 by the National Assessment of Educational Progress* (Washington, DC: Council of Chief State School Officers, 1988).
8. National Assessment of Educational Progress, *Learning by Doing: A Manual for Teaching and Assessing Higher-order Thinking in Science and Mathematics* (Princeton, NJ: Educational Testing Service, 1987).
9. Baron, J. B., "Performance Assessment: Blurring the Edges among Assessment, Curriculum, and Instruction," in A. B. Champagne, B. E. Lovitts, and B. J. Calinger, eds., *Assessment in the Service of Instruction* (Washington, DC: American Association for the Advancement of Science, 1990).
10. See Davis and Armstrong, Chapter 9 and Comfort, Chapter 10, this volume.

11. American Association for the Advancement of Science, *Science for All Americans: A Project 2061 Report on Literacy Goals in Science, Mathematics, and Technology* (Washington, DC: American Association for the Advancement of Science, 1989); reprinted as, Rutherford, F. J., and Ahlgren, A., *Science for All Americans* (NY: Oxford University Press, 1990).

12. Scope, Sequence, and Coordination Project, "Content core to be available in May/June 1991," *Currents Newsletter,* March/April, 1991, p. 3. National Science Teachers Association, Washington, D.C.

13. Rutherford, F.A., *Science Education Reform: Implications for Teacher Education Reform and Project 2061* (College Station, TX: Center for Mathematics and Science Education, 1991).

14. Wingspread Conference, a network for integrated science and mathematics teaching and learning, sponsored by School Science and Mathematics Association, National Science Foundation, and The Johnson Foundation, April 26-28, 1991, Racine, Wisconsin.

15. AIMS: Activities that Integrate Math and Science, Aims Education Foundation, Fresno, CA; CTG: Cognition and Technology Group at Vanderbilt University, TN; TERC: Technical Education Research Center, Cambridge, MA; TIMS: Teaching Integrated Math/Science, University of Illinois at Chicago.

16. Brown, J. S., Collins, A., and Duguid, P., "Situated Cognition and the Culture of Learning," *Educational Researcher,* **18** (1), 32-42.

17. Ibid.

18. Collins, A., Brown, J. S., and Newman, S. E., "Cognitive Apprenticeship: Teaching the Crafts of Reading, Writing, and Mathematics," in L. Resnick, ed., *Knowing, Learning, and Instruction: Essays in Honor of Robert Glaser* (Hillside, NJ: Erlbaum, 1989), 453–494.

19. Weinstein, C. E., "Elaboration Skills as a Learning Strategy," in H. F. O'Neil, Jr., ed., *Learning Strategies* (New York: Academic Press, 1978), 31–55.

20. Berlin, D., "Establishing a Research Agenda for Integrated Science and Mathematics Teaching and Learning," paper presented at the Wingspread Conference of the School Science and Mathematics Association, National Science Foundation, and The Johnson Foundation, April 26–28, 1991, Racine, Wisconsin.

21. Baron, "Performance Assessment," 247.

22. Brophy, J., and Alleman, J., "Activities as Instructional Tools: A Framework for Analysis and Evaluation," *Educational Researcher,* **20**(4) (1991), 9–23.

23. Ibid.

24. Collins et al., "Cognitive Apprenticeship." See reference 18.

25. Pressley, M., Goodchild, F., Fleet, J., Zajchowski, R., and Evans, E. D., "The Challenges of Classroom Strategy Instruction," *The Elemen-*

tary School Journal, **89**(3) (1989), 301–342.
26. Berryman, S. E., "Who Will Do Science? Minority and Female Attainment of Science and Mathematics Degrees: Trends and Causes," a special report to The Rockefeller Foundation, mimeo, 1985.
27. Beane, D., *Mathematics and Science: Critical Filters for the Future of Minority Students.* (Washington, D.C.: The American University, Mid Atlantic Center for Race Equity, 1985).
28. Besag, F., and Wahl, M., "Gender, Race, Ethnicity and Math Avoidance,"paper presented at the annual meeting of the American Educational Research Association, Washington, DC, 1987.
29. Bradley, C., "Issues in Mathematics Education for Native Americans and Directions for Research,"*Mathematics Teacher,* **74**(7) (1984), 96–106.
30. Dossey, J. A., Mullis, I. V., Lindquist, M. M., and Chambers, D. L., *The Mathematics Report Card* (Princeton, NJ: Educational Testing Service, 1988).
31. Oakes, J.,"Race, Class, and School Responses to 'Ability': Interactive Influences on Math and Science Outcomes," paper presented at the annual meeting of the American Educational Research Association, Washington, DC, 1987.
32. Matthews, W., Carpenter T. P., Lindquist, M. M., and Silver, E.A., "The Third National Assessment: Minorities and Mathematics," *Mathematics Teacher,* **74**(7), 165–171.
33. Valverde, L. A., "Underachievement and Underrepresentation of Hispanics in Mathematics and Mathematics-related Careers," *Mathematics Teacher,* **74**(7) (1984), 123–133.
34. Davis, J., "Blacks and Mathematical Literacy: Balancing the Equation: A View from NAEP," paper prepared for conference, Middle School Mathematics: A Meeting of Minds, (Washington, DC: American Association for the Advancement of Science, 1987).
35. Kulm, *Assessing Higher Order Thinking in Mathematics.* See reference 5.
36. Besag and Wahl, "Gender, Race, Ethnicity, and Math Avoidance." See reference 28.
37. Malcom, S. M., George, Y. S., and Matyas, M. L., *Summary of Research Studies on Women and Minorities in Science, Mathematics, and Technology* (Washington, DC: American Association for the Advancement of Science, Office of Opportunities in Science, 1985).
38. Spanos, G., Rhodes, N., Dale, T. C., and Crandall, J., "Linguistic Features of Mathematical Problem Solving: Insights and Applications," in R. Cocking and J. P. Mestre, eds., *Linguistic and Cultural Influences in Learning Mathematics* (Hillsdale, NJ: Erlbaum, 1989).
39. Glaser, R., "Education and Thinking: The Role of Knowledge," *American Psychologist,* **39,** 33-104.

40. Collins, Brown, and Newman, "Cognitive Apprenticeship." See reference 18.
41. Brown, A. L., and Palincsar, A. S., "Guided, Cooperative Learning and Individual Knowledge Acquisition," in L. Resnick, ed., *Knowing, Learning, and Instruction: Essays in Honor of Robert Glaser* (Hillside, NJ: Erlbaum, 1989), 393–452.
42. Slavin, R., *Cooperative Learning* (New York: Longman, 1983).
43. Slavin, R., and Madden, N., "What Works for Students at Risk: A Research Synthesis," *Educational Leadership* **46**(5), 4–13.
44. Lockhead, J., "Teaching Analytic Reasoning Skills Through Paired Problem Solving," in J. Segal, S. Chipman, and R. Glaser, eds., *Thinking and Learning Skills Volume 1: Relating Instruction to Research* (Hillsdale, NJ: Erlbaum, 1985), 109-131.
45. Brown and Palincsar, "Guided, Cooperative Learning." See reference 41.
46. Collins et al., "Cognitive Apprenticeship." See reference 18.
47. Anderson, C. W., "Strategic Teaching in Science," in B. F. Jones, A. S. Palincsar, D. S. Ogle, and E. G. Carr, eds., *Strategic Teaching and Learning: Cognitive Instruction in the Content Areas* (Elmhurst, IL: North Central Regional Educational Laboratory 1987).
48. Novak, J., and Gowin, D., *Learning How to Learn* (New York: Cambridge University Press, 1984).
49. Pressley, M., Goodchild F., Flest, J., Zajchowski, R., and Evans, E. D., "The Challenges of Classroom Strategy Instruction," *The Elementary School Journal,* **89** (3), 301-342.
50. Stuessy, C., and Rowland, P., "Electronic Data Acquisition, Computerized Graphing, and Microcomputer-based Labs: Can Electronic Devices Enhance Graphing Skills and Concept Acquisition?" *Journal of Computers in Mathematics and Science Teaching,* **8**(3) (1989), 18–21.
51. Brophy and Alleman, "Activities as Instructional Tools." See reference 22.
52. Lesh, R., "Continuous Progress Assessment and Accountability: The Key to Meaningful School Reform" (College Station, TX: New Directions in Education Conference, June 10-11, 1991).
53. Ibid., 32-33.
54. Alexander, L., *America 2000: An Education Strategy* (Washington, DC: U.S. Department of Education, 1991).

6

Assessment Implications of the New Science Curricula

Mary Budd Rowe

Assessment in the service of instruction is a particularly important concept because it can represent a way of taking a student's "mental temperature." One kind of assessment can happen in every science class if teachers practice "wait time" instead of "talk time." By talking less and listening more, teachers are more likely to hear what is going on in children's heads.[1] Of course, when one hears what is inside their heads, it might be a little disturbing and difficult to deal with. But as we begin to move toward an integrated curriculum, we are going to need to be able to respond to students in a conversational mode and in a far more complex and involved way than perhaps we've been accustomed to in the past.

I might give an example of a group of children I worked with in Harlem. Four students were assigned to care for the guppies in an aquarium. One Monday we came to school and most of the guppies were dead. A big fight began over who owned the guppies that were left and how they were able to identify whose guppy had died. I was convinced that their observations were very discriminating. It was interesting that in Englewood, on the other side of the Hudson River, the same thing had happened to their aquarium. The schools had been in the habit of turning off the power over the weekend, there had been a cold snap, and the aquaria were too small to provide enough of a temperature buffer for the fish to survive the cold spell. What was interesting was that in one group the students' attention was on who owned what was left — and in the other the attention was on how it happened and what could be done to save the ones that were left. So, very often, hearing the discourse is important for a teacher to gain insights into what to do next.

On another occasion, there were four second grade boys gathered around their aquarium and giggling. So I wandered over to take look at what was happening. Their attention was focused on the aquarium; that, at least, was good. One little guy saw me looking and said, "What's that?" He pointed to a fish that was swimming along and had something coming out of its back end. They all looked at me to see what I would say. I said,

"Now, you know what that is. You tell me." There was a long pause, they got a little serious, and finally one of them said, "Shit." I said, "That's what it looks like to me." They got very serious at that point; then one little boy said, "Do they piss, too?" Right then I could not think how fish did that. So I said, "Well, that would be harder to see, wouldn't it?" They all stared at me; then another little boy said, "You mean to say they swim around in all that shit and piss?" Actually, I had never thought about it that way before. But what was really remarkable was what happened next. New York City was having a garbage strike, and one little boy suddenly said, "It's kind of like us, isn't it, with the garbage strike?"

Now, that's a kind of thinking that some developmental theorists think doesn't happen at this age. It seems to me that the point here is to realize that there are many effective stimulus materials if you are open to hearing what children are saying. It's an invitation to take part in conversations that are very rich and provocative for more than just science — children are building stories about life itself.

I look at science as a special kind of story-building activity, whether it's done by a professional scientist or by a young student just beginning to learn it. Science provides multiple entries for conversations with students. You can hear on multiple levels, and students are usually communicating multiple features of themselves as they talk. Let me give you another example:

I took a couple of chickens in a common cage into another second grade classroom. We had been working with operational definitions and the importance of good observations. I left the chickens in the classroom for a week because time to contemplate is very important for understanding. Children at this age — and at many ages — may not engage with a stimulus right away. When I came back a week later, the children were all gathered around the cage. I sat down with them to get on eye level to join their conversation — I am really talking about assessment here. Then one little Puerto Rican boy said, "This one is the mother and this one is the father. Do you want to know how I know?" I said, "Yes, I do." He said, "This one is the father." He frowned and continued, "This one, when it pecks, it tries to hurt. That's the father." And then he put his hand on the other one and said, "This one, when it pecks, its peck is more like a kiss. That's the mother."

I still get a bit of a chill when I think of this again. Now, as a teacher I have really a difficult decision to make: am I going to teach him about pecking order? Scientists talk about pecking orders, but they don't say much about how a chicken knows enough to peck on this one and not peck on that one. But this boy told another kind of story, and it fit. He had an operational definition and a useful simile.

These stories address some very important issues about assessment in the context of instruction. I am very committed to a much larger view of what science and math instruction is all about. I am committed to the importance of getting standard data of all kinds, which do tell us things about patterns, but I don't think we observe and hear enough. Learning to have conversations instead of inquisitions is a very powerful way of starting to get data in context. And for those of you who are methodologists and have some doubt about this as a way of doing research, I would say that this is a kind of action research in which all of us who have a concern for youngsters can participate with interest and enthusiasm. It is the kind of research that you see in medicine and the clinics these days.

Of course, I'm talking about a kind of story telling in which you are informed, then you develop a story to tie the observations into a plot to test, so you can figure out what the real plot is. With youngsters, I am interested in what world view they are building. Their world view has a lot to do with the behaviors they will try and the amount of effort and persistence they will show. I want to quote something that Sumner wrote in *Perspectives in Biology and Medicine* concerning the importance of really attending and hearing on multiple levels. He said,"Beyond the mnemonic effect, anecdotes have an important epistemological function as well as the general social value of story telling in the professionalization of the young and in the communal life of all physicians. Anecdotes play an intellectually respectable part not only in introducing established information to new learners but also in advancing the new knowledge of experienced learners. They have a use in the scientific investigation of disease. They suggest where primary research attention needs to be turned, functioning as a preliminary critique of current therapies and staking out new knowledge, new subgroups in the clinical spectrum of disease."[2]

I think that there are parallels for science teachers. Science is a very powerful vehicle to open up what is going on in children's heads; also to help us in developing curricula which we think will begin to meld together in some ways that will be powerful for youngsters. But how are we going to find out whether or not we are successful teachers? Not by a lot of separate little tests. We really want to find out what sort of coherent plot — story line — students are making for themselves about the world and their place in it. The issue, in a way, is learning to converse — to make some sort of image and to talk about it.

We're doing a lot of work on curriculum. The intended curriculum gets negotiated through our conversational interludes with students, then something walks out of their heads — the learned curriculum. But we really don't know much about what it is that they are learning. And for many the curriculum isn't very successful; otherwise, we probably wouldn't have as many dropouts as we do. We need to find ways to converse with these students about what we think are the problems that are confronting them.

Part of learning to hear youngsters on multiple levels — like the child in the chicken story — is knowing what other levels of processing are going on and deciding on that basis when we should talk or in what ways we're going to talk together. I've compiled a set of questions — as have researchers, especially clinicians, working with children all over the world — that, I believe, address some of the central issues in kids' heads as they start to build their stories about the world and their part in it (Figure 1).

One thing that you overlay on what you are doing in curriculum or instruction is to ask how what you are teaching is contributing to developing answers to those questions. It is particularly important to start this analysis prior to middle school. The United States has the earliest age of onset of pubescence of any nation in the world. Teen-age pregnancy — at 13 or even younger — and very high-risk infants are of great concern. The things that are on young people's minds at that age are, I think, not at all what we are teaching them in school. How can we bring those agendas together?

I happened to ride home on the bus one day and overheard two little girls who were sitting in front of me. One of them said to the other, "Are you on the pill yet?" The other said, "No, my mother thinks I can wait six months to a year." There is an energy and discourse and thinking about phenomena that we have to get children to share with us. We have to be able to merge our science with that discourse in a way that will help them make fewer dangerous decisions than so many of them are making now. We must do this in a way that builds hope in children about a future in which they can and will want to be an important part.

The last question is, Who cares? A template that you can lay on planning and what you are listening for in the assessment is, What are the signs that the students cared about what went on in class? The signs to look for are that they liked it, that they see some consequence, and that they would like to engage in doing something about it.

How might we organize a way to get at what's happening with students based on what we're doing? What story are they building for themselves? Another template that you might consider as you develop instructional units is the knowledge that what you learn and the way you learn it is closely tied to the likelihood that you'll ever use the knowledge.

An example of that is an experience I had during a trip to China some years ago. I always like to try "wait time" experiments wherever I get a chance. On short notice, I was given a chance to do some science with a seventh grade group in Shung Hi that had been studying English. The only equipment I had with me was a stack of Fresnel lenses, so I handed these out to 60 nice, well-disciplined students. Many of these kids had never seen magnifier, had never looked through one, so a little skill training was in order. As a check of what they were seeing, I asked them to draw their forefinger. Then I asked them what their observations were. Pretty soon they began to say, "Oh, little red and yellow and blue dots."

I've done this activity all over the world, but I'd never before had anyone tell me that they saw little red, yellow, and blue dots. So I went over and

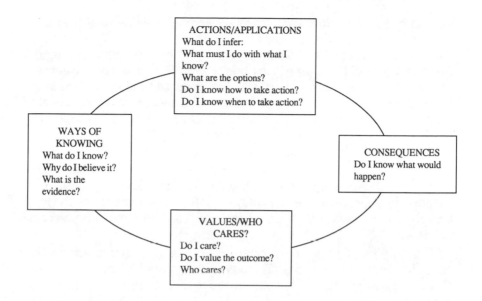

Figure 1. Questions that address some central concerns of individual students.

1. What kind of country is this?
2. What values control activities?
3. Where do I fit in?
4. Do they expect me to succeed or fail?
5. How much effort do I need to make?
6. Is success worth the effort?
7. Can I get help?
8. Do I have the energy and endurance?
9. What happens if I do not make the effort?
10. What am I up against? What is the competition?
11. What difference can I make?
12. Do I care? Does anybody care?

Source: Rowe, M.B., "Science Education: A Framework for Decision-makers," *Daedalus* **112** (2) (1983), 123.

said, "Let me see." Sure enough, there were red, yellow and blue dots. Now, what was the reason? Are Chinese born with red, yellow, and blue dots on their hands? No. Was the lens dirty? No. Was it the printing ink from their books? No, most of their books were black and white. But many of the schools have a factory attached to them and the kids work in the factories. The factory for that particular school happened to produce all of the chalk for the Shung Hi school district. The important part of the story is what happened next. One of the students said, "That's why they keep wanting us to wash our hands!"

You couldn't see the chalk dust with the naked eye, it was that ground in. Now my theory is that if you learn something, and there is observational support for a practice you are told to perform, you are more likely to comply. There is a relationship between what you know and how you came to know it.

The story building that kids are always doing in relation to themselves and school goes on every day. There is science story building, but there is also story building on the bigger questions. For example, a teacher was practicing wait time and a student said to her, "Well, Mrs. Buster, how come you aren't repeating things anymore?" There was a pause — she didn't know what she was going to say — and another kid said, "I know, she knows that we know by the tone of her voice which answer she likes and which she doesn't and we can stop thinking." So you're not the only one doing assessment. The kids are assessing, too. Similarly, when you really start to listen — begin to have conversations with the kids — children will go home and say to their parents, as one fifth grader did, "This is the first time in all my years in school that anybody really cared about what I actually thought about something."

I did a site visit for a National Science Foundation project in which the teachers were working very hard on new curricula. I knew they were really caring teachers, but a lot of science teachers and math teachers are introverted folks. So how they feel about something doesn't get out and visible. The perception of the kids was that the teachers only cared about science, that that was all they thought about. So the students' ability to hear and profit and their willingness to engage in doing the science was far less than it might have been. When I gave this feedback to the teachers, it came as a shock to them because they were as committed a group of people as I ever met. I suggested that we would have to find ways to indicate that we're alert to other parts of the students than just the part doing science. We identified five or six things that even a beginning teacher might do, like standing at the door about twice a week and picking out a couple of kids who seem unusually happy or unusually down. It's a way of engaging — it's like you're sort of a thermometer, but you can't read a system if you cannot get in and get calibrated in that system.

Another important development in science education is the opportunity for science teachers to have more productive conversations with their colleagues in mathematics. They have had a longer history of doing research and thinking about teaching and learning than we've had in science. The new book by Lynn Steen, *On the Shoulder of Giants,* is beautifully done, but you would think it was written by scientists if you looked at the chapter titles. There is a real opportunity to make progress if we can develop assessments that preserve the directions that have been set. Do we know how to properly preserve in an assessment task the important features of the connections between, for example, structure and function? I doubt it!

I was flying from Oregon to San Francisco and two youngsters were sitting next to me. One was ten, the other was eight. The stewardess came along and handed them a book with a punch-out-and-assemble model of the jet we were on. The weather was very foggy, and we could not see much out of the windows. The older girl assembled the plane and said, "Oh, these are the wings, this is a tail," and so on. I thought, wouldn't that just delight the heart of some measurement person? The girl sat quietly for a bit and then said, "Does this plane we're on, ah, does it have wings?" She couldn't see the engines and wings out in the fog. She knew things had wings but she didn't understand the relationship between structure and function. All she had was just a formal level of understanding. So hands-on work is not sufficient; we need a way to get at the deep patterns and to find the holes in the patterns. The drive to cover material quickly makes it difficult for students to pick up patterns and for us to have the kinds of conversations with kids we need to undercover the patterns.

Now, as we consider assessment, we must know what our goals are. For me the most important goal is this thing called democracy. I think it's the most difficult type of social organization to maintain. It is the most demanding for teachers whom I look on as the "social genes" for democracy. We have to think about the Chinese philosopher Chuang Tsu over 2000 years ago who said that the pheasant has to take ten steps for a mouthful of food and a hundred steps for a bellyful of water. If it were kept in a cage, it would not have to do that; it would be treated like a king. But it doesn't want to be kept in a cage. Why? Because it would not be happy. Why would it not be happy? Because it would not be free.

It seems to me that what we're trying to do is teach our people to live outside a cage, to be fundamentally happy, or at least fundamentally free with a modicum of happiness.

The way we're doing testing and assessment right now isn't going to do it. Sadly, testing in its most common form is driving content presentations in texts. We are mutilating science instruction with textbooks loaded with facts but empty of science's most powerful feature — patterns and connections – all in the service of accountability. We cannot really fault the publisher if that is what sells. I believe in democracy and I don't want it to fail. Programs that reflect some of the major values and processes of science will serve that goal well. We must move ahead with new approaches to assessment, but our methods and the messages that they contain must not be trivial or we undermine our goal!

References

1. For a discussion of "wait time" and a review of related research, see Rowe, M. B., "WaitTime: Slowing Down May Be a Way of Speeding Up," *American Educator,* 11(1) (1987), 33–43, 47; see also Rowe, M. B., "Science Education: A Framework for Decisionmakers," *Daedalus,* **112** (2) (1983), 123–142; reprinted in L. L. Motz and G. M. Madrazo, eds., *Third Sourcebook for Science Supervisors* (Washington, DC: National Science Teachers Association, 1988).
2. Sumner, *Perspectives in Biology and Medicine*

7

The Impact of Science Assessment on Classroom Practice

Nancy S. Cole

People who teach students science, people who make science tests, and, increasingly, the U.S. public share many common concerns about science education. They share concerns that students need to learn science more than ever before in our nation's history, concerns that our tests of science need to focus on the most important aspects of the science we want students to learn; and they share concerns that tests today need to help educational improvement by having a positive effect on classroom practices in science.

Many of these concerns are not new. Teachers always want students to learn more and better. Test makers have always sought to have their tests reflect the most important aspects of the subjects tested and to be useful educationally. And the U.S. public has looked before to education in science to keep the nation strong. Thus, we consider a set of problems and concerns involving assessment that are not new. One might reasonably ask why they have not yet been solved and why today this set of concerns is receiving such renewed and central attention in calls for educational reform.

In this paper I will discuss some of the conflicts between the purposes of assessment for large-scale measurement and assessment for improving instruction—reasons these old problems are with us unsolved. In addition, I will examine some of the reasons these old problems seem to have an unusual chance of being solved in this decade.

The Traditional Large-Scale Measurement Goal

Large-scale multiple choice testing in this country has developed over the last half-century in response to some important needs. Current widely used tests are efficient and often effective ways to measure many important types of skill and knowledge. In spite of much discussion of how tests should change, current tests are almost surprisingly good measures of what students are learning in school.

Such tests are designed to measure and to produce scores that are meaningful measures for individuals and/or groups, scores that are valid and reliable measures. Since these tests are given to large numbers of persons, they are designed as well to produce that measurement in ways that minimize cost in time and money; that is, to be efficient. These driving design characteristics — effective and efficient measurement for large-scale use — have guided the nature and characteristics of the tests. These are the purposes for which they were built and for which they have been used, presumably, until recently.

The tests that best met the goals of effective and efficient measurement have been largely multiple choice (machine scored); centrally mandated, developed, administered, and scored; formal and independent of classroom uniquenesses with delayed feedback of scores; and designed primarily for users external to the classroom. The dominant concerns in their design have been validity (how to measure what we intend to measure), reliability (how to measure consistently), objectivity (how to produce scores that are not affected by the scorer), and efficiency (how to test large numbers of students quickly and inexpensively). They were not designed for a single curriculum or a particular classroom situation and were assumed to be neutral indicators of student progress, largely isolated from classroom-level concerns. Efforts were made to provide helpful information to teachers where possible, but such efforts were subsidiary to the driving design characteristics.

The New Instructional Goal for Testing

In the 1970s, some new situations arose. There was concern nationally about basic skills. People moved to enumerate the separate skills they wanted students to learn and to test them one by one (at least at lower grade levels and for less complex skills). This thinking lent itself nicely to school achievement testing, and there was something of a national exuberance over so-called "criterion-referenced" testing. Statewide school testing was being implemented by states attempting to ensure that students were being taught these basics, and test scores were published in newspapers for school districts and schools in an attempt to hold the schools accountable for what students learned. These supposedly neutral indicators of student performance were becoming involved in a high-stakes enterprise.

It was only a small step to recognize that such high-stake tests could themselves be a lever, an action, to change the educational system. Until this period, the test content always was presumed to follow the educational content. By the early 1980s, it had become socially acceptable to design a test for the purpose of producing education change. Test content was being used to lead educational content. Increasingly, policymakers were thinking about tests as tools to produce educational change rather than measurement tools.

As states increasingly tried to use tests in this fashion in the 1980s, it became clear that the tests were not having the desired effect. Test scores were not going up dramatically, if at all.[1] Although teachers reported spending more time on the topics covered by the tests,[2] there was increasing concern about whether the covered topics were all of the important ones on which time should be spent and about whether time was being spent, not on learning the general topics of the test content, but on drilling particular item formats or even particular items of the tests.

Increasingly, measurement researchers began to think of tests as a part of the educational system in which they are used — not separate, isolated, independent, neutral measures. Tests are a part of the system and they affect and are affected by the system. Concepts of validity came to include this much broader perspective of testing.[3] Systemic validity was identified as the validity of a test in helping to produce desired instructional goals.[4]

These experiences bring us to the early 1990s. Now we find many educators seeking ways to use tests more effectively within the educational system as a supporters of and agents for positive and constructive change. Further, there is increasing recognition that, if we are thinking of tests as tools with which to try to create positive change, then we need to design them explicitly for that purpose.

Considering tests as educational or instructional tools rather than measurement tools has enormous implications for their design. In 1988 I contrasted the characteristics that would drive the design of tests differently if the driving goal was improved instruction rather than effective and efficient measurement.[5] Table 1, adapted from that paper, provides that contrast. In the left column of Table 1 are characteristics that drive the test design when large-scale measurement is the goal. These characteristics begin with validity, reliability, objectivity, and efficiency. In the right column are the characteristics that would be dominant concerns for instructional purposes. From this perspective, the judgment of test quality would depend on the impact of the tests on instruction and learning. Design features that promote good instruction and student learning would be sought.

As shown in the left column, tests for measurement purposes are usually mandated by the school system or the state. On the right, it is the teacher who mandates the assessment to meet a particular classroom need. Tests designed for measurement have generally been designed to be widely applicable and suitable for central data processing, and that often results in a multiple-choice format. When tests are designed for instructional purposes, the format of the tasks is determined by their instructional value.

When guided by educational impact, other features would almost surely be different. Tests designed for instructional purposes would likely be adapted by the local teacher to the local context. Issues of time and cost efficiency would be far less dominant if students were learning from the assessment process. Local scoring by teachers and/or students to give quick feedback would likely be desired features for instruction and learning, and

Table 1. Characteristics of Tests Designed for the Different Purposes of Measurement and Instruction

Assessment designed for measurement	Assessment designed for instruction
Valid	Quality judged by effect on instruction
Reliable	
Objective	Design determined by instructional purpose
Efficient	
Centrally mandated	Teacher mandated
Widely applicable	Adapted to local context
Multiple choice	Format determined by instructional value
Machine-scored centrally	Locally scored
Delayed feedback	Immediate feedback
Used independently	Used with other information
Stable scores	Scores affected by short-term teaching and learning
Results designed for external user	Results meaningful to teacher and student

the student and teacher would surely be the prime target of the assessment experience and the results. Approaching assessment from the two perspectives suggests some dramatically different criteria for designing and judging the quality of the assessments.

Experience to Date with the Instructional Goal

In the past, teachers focused on assessment for instruction and test makers on assessment for measurement. But, in fact, neither group was entirely satisfied with the other's assessment efforts. Teachers were not keen on the tests of the test makers because of the lack of (or perceived negative) instructional implications. Test makers were not keen on the assessment teachers often built because of its measurement inadequacies. If new kinds of assessment are to serve instruction, we have to attend carefully to the design of assessment for that purpose, and we have to bring together the skills of the teacher and the test maker directly to that end.

Several years ago, Educational Testing Service (ETS) committed itself to examining its own activities in terms of their impact on student learning. That commitment extended not only to its present testing programs but also to some quite new directions. As a consequence, ETS funded a series of projects to approach assessment entirely from the instructional perspective, quite separate from its ongoing large-scale testing programs. Significant resources have been allocated over several years to thinking about the right-hand column of Table 1. The driving question was, What would assessment be like if it were driven totally by instructional purposes? Three projects provide examples of what we have begun to learn from these efforts.

The first project, ArtsPropel, involved ETS with Harvard Project Zero and the Pittsburgh Public Schools with the support of the Rockefeller Foundation.[6] The instructional domain is the arts—creative writing, music, and the visual arts. This is not an area in which testing has a strong history. The task was to work with the Pittsburgh teachers in their schools to think about instruction assessment to assist that instruction.

The teachers worked with student portfolios which are, of course, not at all new to the arts. Now, however, the portfolios are being used not just as a collection of finished products but as a focal point of student and teacher work, to identify and coach students in the standards of good work and in how to improve. Students study their own progress, and work with other students and teachers to recognize and understand their own accomplishments and to set goals for improvement.

A second example is critical thinking in middle school language arts. In this area ETS staff worked with teachers, first in Brookline, Massachusctts, and subsequently in several schools in New Jersey, to design a computer-based activity to support critical thinking. The result is a computer module called Inside Story, organized around the study of a myth in which students take an active role as journalists in collecting and analyzing information about the myth and then in writing about it. Teachers using the several-week module often continue the approach through the year and, in some schools, are working to generalize it to other subject areas as well. Teachers also sought to prepare themselves better to use the new materials and approaches. As an outgrowth of this interest, ETS staff and teachers from around the country have been involved in experimental workshops on how to use critical thinking approaches in the classroom.

A third project involves the middle school science curriculum.[7] Starting with a broad look at a curriculum for middle school science, the project then focused on some smaller sections of the broader curricular approach. Again with teachers, researchers tried to understand how assessment could assist in the kind of learning teachers want students to accomplish at that stage. How teachers learn about student background experiences on which they can build the instruction was explored, and researchers tried to model the rather informal questioning of students that produced this information.

There was concern as well with how to describe the progress students need to make and have made. Such a description of desired progress seems to be part of the goal setting students and teachers do as well as an explanation to students of the learning progress they are making. Much effort went into integrating small pieces of instruction (and assessment) into larger "chunks" and then into much larger culminating activities. These large integrating tasks help organize the instruction as well as serve as assessment of major instructional goals.

These few examples are representative of the types of activities going on today around the country as educators try to understand assessment and its role in instruction. We have already learned a number of important things. One is that, for both students' and teachers' sakes, the summary assessment tasks must represent important educational goals. When they do, they must be complex tasks, and they can and should be taught directly. They should also be activities that point students toward the learning goals; they should explain to students what they need to learn. Such assessments should be developed by and with the experts at producing learning. Teachers have, of course, always been involved in setting specifications and in writing tests; however, the active educational role involved in development of these instructional assessment tasks is quite different. Task development is a pedagogical effort.

Another feature of assessment design for instructional purposes is that there are few instances in which scores per se are needed. Although there are often activities or materials that could be scored, a score is often not needed for the instructional purpose. Further, when the right tasks are designed, students become very actively engaged. When that happens, students are assessing their own work as they move through the activity, sometimes with the help of the teacher or other students and sometimes alone. Group work is often a part of these tasks, so students are often helping other students, producing some additional instructional benefits.

Bringing Different Types of Assessment Together

What does it mean to find a large disjunction between assessment that seems to serve classroom practice and assessment for measurement purposes? How can we ever hope to bring two such divergent enterprises together?

When we try to bring the measurement and instructional approaches together by using assessment as a lever for instructional change, we should give serious thought to what we are doing. We should remember that it has been tried before with the basic skills assessments of 1970s and 1980s, and educators have generally been disappointed with the results. Some dramatically new approaches will be necessary.

We seem to be moving toward the needed new approaches in the 1990s.

Today's attention to performance testing or "authentic" assessment is likely in part a reaction to this previous era. We now recognize that the tasks we use for measurement must represent the most important instructional goals in an integrated way. In science, our goals are

- that students understand the general principles of science and their relation to their own lives,
- that they observe carefully and learn to frame questions about the world around them,
- that they attack the process of learning in searching and analytical ways,
- that they are creative in their questioning as well as in their problem-solving,
- that they are competent in applying knowledge from multiple areas to solutions of issues and problems and in seeing connections among the things they are learning, and
- that their learning is an active process involving probing and questioning and wondering.

These are all critically important educational goals, and we have to have assessment tasks that represent the complexity and integrated nature of these goals. We also must explain these goals more effectively than we have done to date to the public and repair some of the damage of our simplistic earlier efforts.

The two different purposes of assessment (measurement and instruction) cannot be dealt with totally separately in spite of their differences. There are two approaches to reconciling the two and both are well under way today. On the one hand, we can work on making tests used for the measurement purpose more instructionally compatible in the test tasks used. On the other hand, we can continue to learn how to construct assessments for instructional purposes and try to use that learning for all assessments. The examples cited above illustrate that the latter learning is well under way. There is also great progress being made on the former approach as well.

As an example of the change in large-scale tests, consider the changes announced recently for the SAT.[8] They include emphasis on critical reading, student-produced responses in math, students' short essays in writing, and more attention to subject areas. Eventually, I expect to see more open-ended questions in several areas, including the subject areas in the SAT. Such changes are not a solution to the differences of the two types of assessment, but they are a constructive step toward bringing new and instructionally important educational tasks into these tests that have such a large impact on the educational process.

Another example of this positive trend is the National Assessment of Educational Progress (NAEP). There has been a recent increase in performance-type measures, even at a time of rather severe budgetary constraints. The NAEP began with large portions devoted to performance measures, and as it became larger and the financial resources became more limited, those performance elements were decreased. Now, the 1992 assessment will include considerably more and better performance-type tasks.

Another example of change is the assessment being built to replace ETS's National Teacher Examination. The new assessment will include classroom performance assessment of candidates for the teaching force by trained observers. Measuring new teachers' classroom performance is a complex and difficult task. It takes professional judgment, and it focuses attention on some of the most important parts of teaching, parts that would be omitted with past approaches. In addition, it attends to some of the major educational goals of teacher preparation — an important instructional link.

Each of these new assessment efforts is more expensive, more time-consuming for the test takers, and more difficult to judge objectively than the tests they replace. But they are examples of beginning to bring measurement and instructional assessment closer together. We are making progress both in understanding how to design assessment for instructional purposes and beginning to use some of that understanding to maximize the positive educational impact of assessment used primarily for measurement. The most encouraging part is the extent to which many people, from test makers to researchers to teachers, are working together on these issues that are central to developing and maintaining quality education in this country.

References

1. Mullis, I.V.S., Owen, E.H., and Phillips, G.W., *Accelerating Academic Achievement,* and Policy Information Report: *The Education Reform Decade* (Princeton, NJ: Educational Testing Service, 1990).
2. Dorr-Bremme, D.W., and Herman, J. L., *Assessing Student Achievement: A Profile of Classroom Practices,* Monograph No. 11, CSE Monograph Series in Evaluation (Los Angeles: UCLA Center for the Study of Evaluation, 1986).
3. Messick, S., "Validity," in R. L. Linn, ed., *Educational Measurement,* 3rd Edition (New York: Macmillan, 1989).
4. Frederiksen, J.R., and Collins, A., "A Systems Approach to Educational Testing," *Educational Researcher* **18** (1989), 27–32.
5. Cole, N.S., "A Realist's Appraisal of the Prospects for Unifying Instruction and Assessment," *Assessment in the Service of Learning: Proceedings of the 1987 ETS Invitational Conference* (Princeton, NJ: Educational Testing Service, 1988).
6. Camp, R., Gitomer, D., Wolf, D., in R.E. Bennett and W.W. Ward, eds., *Construction Versus Choice in Cognitive Measurement,* to be published by Lawrence Erlbaum. See also Wolf, P., "Portfolio Assessment:

Sampling Student Work," *Educational Leadership,* April 1989, 35–39.

7. Gong, B., and Venezky, R.L., "Instructional Assessments: Lever For Improvement In Science Education," paper presented at the American Educational Research Association Annual Meeting, April 1991, Chicago, Illinois.

8. "The New SAT: Debating Its Implications." *The College Board Review,* **158** (1990-91), 22.

National Initiatives for Assessing Science Education

Lisa Hudson

Since the first intelligence test for screening school children was developed in 1905, educational testing has grown into a billion-dollar industry. Today, mandated testing alone consumes $700–$900 million and 20 million school days annually.[1] But, as assessment grows in prominence, its limitations become more obvious. Of particular concern is the degree to which assessment helps or hinders instructional practice. Like the doctor who treats the symptoms of a disease without attacking the cause, educators are finding that an over-reliance on assessment not only has failed to solve educational problems, but also may even have generated undesirable side effects, such as a narrowing of curricula.

The 1990 AAAS Forum, "Assessment in the Service of Instruction," presented efforts that are being made to restore assessment to its proper role as a positive diagnostic and monitoring tool rather than merely as a treatment. However, this restoration requires a clear understanding of what assessments exist, how they are used, and what policy implications are involved in their use. This chapter examines national initiatives for assessing science education in these terms.

Definition and Purposes of National Science Education Assessments

In this chapter, two types of national science education assessment initiatives are defined: first, the National Assessment of Educational Progress (NAEP) science assessment, and, second, science tests distributed by commercial publishing companies as part of their grades K–12 assessment program packages. Not included are the standardized tests used for college entry and placement decisions or tests that accompany publishers' science textbooks.

The NAEP and the commercially distributed tests serve two very different purposes. The NAEP was designed to serve solely as a monitoring instrument at the national level; it provides an assessment of the relative performance of the *nation's* educational system in different subject areas, for different types of students, and at different times. By contrast, commercially developed tests assess *individual* student performance relative to that of students in the country as a whole. This information can be used to evaluate either individual students ("Does Jane know as much science as the typical student at her grade level?") or school, district, or state instructional programs.

These purposes are predicated on one very important assumption: that the goals of instruction are reflected in the knowledge and skills measured by the tests. This is where national initiatives run into trouble, for two general reasons. First, these assessments focus on covering the breadth of science topics addressed most commonly by current science curricula or viewed most generally as important educational outcomes. Thus, it is inevitable that there will be an imperfect match between many schools' curricula and the material covered by the test. Further, when a test attempts to cover the full range of the science curricula and to remain generally applicable, the result is that no one topic can be covered in depth. This makes these tests more appropriate for schools using a curriculum that covers more topics in less depth rather than one that covers fewer topics in more depth.

The second problem is that certain skills and abilities are ill-suited to measurement by multiple-choice items, the predominant question format on national tests. Multiple-choice items are best suited to assessing the recall of factual information and the solution of simple, one-step problems.[2] Other abilities acknowledged to be important components of science instruction (such as the ability to solve multistep problems or to develop experimental hypotheses and procedures) cannot be assessed adequately with multiple-choice items (at least in their typical current form). This leaves these assessments open to a number of related charges. One claim is that they are not "authentic" because they do not reflect accurately the nature and process of science. A further charge is that, by reducing science to a series of multiple-choice items keyed to recalling factual information, these tests portray science as a boring endeavor involving little more than rote memorization. Finally, there are the assertions that multiple-choice tests are not supportive of good educational practice and that they encourage the teaching of science as an accumulation of facts and concepts rather than as an investigative discipline involving exploration, curiosity, inductive and deductive reasoning, and collaboration.

So, why are multiple-choice tests used? First, because educational decisions must be made and priorities set. This is virtually impossible without information on how the system is performing. Student achievement is the most crucial educational performance indicator; national assessments, using multiple-choice items, provide the most readily available and efficient means of obtaining student achievement data. This efficiency cannot be overestimated. Standardized, multiple-choice tests provide data that are reliable and valid (for limited purposes), at a reasonable cost, and in minimum time (for both taking and scoring the tests). In short, although admittedly not ideal, such tests are a practical and affordable means of monitoring educational progress throughout the United States (using the NAEP) and of evaluating students and programs in a larger context (using commercially developed tests).

Nonetheless, growing dissatisfaction with the shortcomings of the standardized, multiple-choice format has led to new efforts to develop alternative forms of assessment that are more authentic and supportive of desired educational practices. To estimate how likely it is these more innovative assessment strategies will be incorporated into national assessments, it is useful to review the history and current use of national assessments.

The National Assessment of Educational Progress

The NAEP currently assesses the educational achievements of students in grades 4, 8, and 12, using a nationally representative sample of students at each grade level.[3] Students are tested in science and nine other subject areas on a rotating basis. Science assessments were conducted in 1970, 1973, 1982, 1986, and 1990, and are currently on a four-year cycle. Trend data from these assessments show that student performance in science declined through the 1970s and early 1980s, but increased slightly in 1986. (Data from the 1990 assessment will be available later in 1991.)

The NAEP was established in 1969 in response to a congressional mandate for periodic assessment of the educational performance of American students; the "national assessment" was to serve as a barometer of the nation's educational progress, reporting on achievement status and trends at regular intervals. When it was first developed, there were serious concerns that the NAEP would lead to a national curriculum, to "comparisons with an invidious national standard," or to "unwanted" comparisons among states or local school districts.[4] These concerns resulted in a number of decisions designed to maintain the NAEP as a census instrument only. Consequently, one feature of the NAEP is that it is neither a norm-referenced nor a criterion-referenced test. Performance criteria were deliberately not set and administration procedures were designed to support group but not individual comparisons. It was only by placing such restrictions on the nature of the National Assessment that it was possible to guarantee passage of its enabling legislation. Today, the NAEP is administered by the Educational Testing Service (ETS), under contract to the U.S. Department of Education's National Center for Education Statistics. The policies that govern the NAEP are set by an independent National Assessment Governing Board (NAGB).

The NAEP science education assessments use predominantly multiple-choice items, although a small number of open-ended items usually are included. The objectives from which these items are developed are generated by a panel of expert science educators and science specialists. The members of the panel use the frameworks established by previous panelists to determine what scientific knowledge and abilities students should have at each grade level. The objectives, and therefore the items, cut across science content areas and include the three cognitive areas of knowledge, use, and integration. Judging from the publicly released items, which are reportedly fairly representative of the total set of items,[5] factual and conceptual knowledge, and the direct application of that knowledge, dominate NAEP's assessment of science instruction. Items that could be characterized as measures of "process skills" or "higher-order thinking skills" appear to be few in number. Fortunately for those who argue for increasing the focus on these skills in science education, the NAEP, as was intended, has minimal, if any, effect on instructional practice. However, this situation is likely to change.

NAEP's assessment results are still used primarily as they were originally intended — that is, to monitor the status of educational progress in the country as a whole. In this role, the NAEP has been particularly influential on Capitol Hill, where testimony on the results of NAEP assessments often has been used to support educational legislation. The National Science Teachers Association (NSTA) also publishes articles discussing NAEP findings, and press releases encourage dissemination through the news media. This publicity ensures that both science teachers and the general public are kept informed of national trends in science achievement.

Over the years, however, the NAEP has been called upon increasingly to serve other purposes. Originally, the NAEP was designed to include very little in the way of background data; this was done intentionally to keep the NAEP narrowly defined as a monitoring tool, "not as a tool to explain problems, to support particular solutions, or to predict future trends." [6] However, the National Assessment's research potential quickly overwhelmed concerns about its proper role. This led to two new NAEP features. First, the background section has expanded to include a number of student and school characteristics that can be used to "explain" test results. Second, since 1979, data tapes have been made available to the public from each National Assessment. These data tapes have become an important research tool, supporting many types of educational analyses. For example, the RAND Corporation is currently reanalyzing data from the 1986 science assessment to provide new, national-level indicators of students' achievement.

On occasion, states and local school districts have used NAEP items for their own assessment purposes. Typically, the items they select are not test items, but the background items about students and schools. In the area of science, schools sometimes use these data to determine their relative status on characteristics such as the proportion of students enrolled in science courses, the availability of science equipment, and the time devoted to science instruction. These data can be employed to inform parents and taxpayers about the quality of science instruction in the schools or to document the need for increased funding or policy changes.

The Changing Face of the NAEP

The past few years have seen the most dramatic and potentially far-reaching changes in the NAEP, as pressures have grown to make the National Assessment more useful to national- and state-level policymakers. Taken together, these most recent changes in the administrative policies and procedures underlying the National Assessment could have a profound impact on the influence that the NAEP exerts in science classrooms. One change has resulted from increasing demands for student achievement data that can be used to monitor the educational performance of individual states. To meet this need, the 1990 NAEP asked states, on a voluntary basis, to supplement their 8th grade mathematics sample, for the purpose of obtaining a state-representative sample. Thirty-seven states (and three territories) elected to do so. These state comparisons will be expanded to include 4th grade mathematics and reading in the 1992 assessment. It seems safe to assume that science will be added to the list of subjects to be compared by the states when it is assessed again in 1994.

A second change was announced recently by the NAGB, which has decided to create an alternative to the existing proficiency levels for reporting NAEP results.[7] The alternative is to establish a set of national performance standards that will classify student achievement at each grade level as basic, proficient, or advanced. Again, the 1990 mathematics assessment is where the performance standards will be implemented first, with expansion to all subject areas planned in the 1992 assessment. Because state-level data also will be available beginning in 1990, it will be possible for participating states to receive — and compare — data that report the proportion of students in each state meeting each performance standard.

These state-level NAEP scores will provide the first and only valid means for comparing student achievement among states. Although states do not have to obtain state-level test results, pressure to do so is likely to come from the same forces that led to the institution of the state-sample option. For a number of years, the Council of Chief State School Officers (CCSSO) has been working on the development of a system of state-level indicators of educational status; the CCSSO has earmarked the NAEP as the instrument it will use to provide achievement data for this indicator system.[8]

This use of the NAEP has been reinforced further by the joint goals established by President Bush and the National Governors' Association (NGA), which include making the United States first in science and mathematics achievement by the year 2000. The NGA also has acknowledged that the NAEP is the most likely vehicle for measuring progress toward this goal.

Clearly, the NAEP is on its way to becoming an assessment program with tremendous power. Inevitably, pressures will be placed on school districts — at least those participating in NAEP assessments — to perform well on the NAEP and, thus, potentially to *teach* the NAEP. As a result, at least one group, the International Reading Association, has voiced its opposition to using the NAEP for state comparisons. In addition, at a recent conference on large-scale educational assessment in Boulder, Colorado, an entire symposium was devoted to the issue of whether the NAEP was driving the country toward a national curriculum. Perhaps not surprisingly, there was a decided lack of consensus, with opinions ranging from the view that the country has a national curriculum already to the view that the NAEP is indeed becoming "the nation's curriculum" as well as "the nation's report card."

As the NAEP gains influence, the nature of the items on the science assessment becomes critical. Do these items measure important science education goals? Do they reflect an acceptable depth and breadth of scientific knowledge? Do they assess both basic scientific knowledge and students' ability to use their knowledge of scientific facts and processes? As the NAEP is structured currently, most educators probably would answer "no" to these questions. Fortunately, recent efforts to improve the teaching of scientific inquiry skills have led to a growing interest in innovative assessment techniques, which also have affected the NAEP.

For example, a grant from the National Science Foundation allowed ETS to supplement the 1986 NAEP science assessment with a pilot test of "higher-order thinking skills assessment techniques." The pilot test consisted of 30 tasks, or items, in the following four skill areas: sorting and classifying, observing and formulating hypotheses, interpreting data, and designing and conducting an experiment. Of these 30 items, 21 involved "hands-on" tasks, where students actually manipulate apparatus, 1 involved observing a demonstration, 3 were administered by computer, and 5 were open-ended, written items. Table 1 lists descriptions of some of these tasks. ETS's summary of the pilot test noted the enthusiasm of the advisory panel, students, and participating schools for these items; the inordinate amounts of time required for preparation and administration; and the extensive costs involved.[9] Overall, however, ETS decided that performance assessment is feasible on the NAEP.

Other chapters in this book examine this type of innovative assessment in more detail. Here, it is noted only that before the extra costs of including performance items in a *national* assessment are incurred, several important questions need to be answered. Of prime importance is whether the advantages of such items (for example, encouragement of "hands-on" instruction, more valid assessment of scientific process and problem-solving skills) outweigh their disadvantages (for example, additional costs, potential scoring bias).

Perhaps the first question that should be asked is, Exactly what do these items assess? They clearly have face validity as measures of students' understanding of the procedures and methods used to solve scientific problems. However, it could be argued that the skills they tap are not very different from the general reasoning skills assessed by IQ tests. Many performance items also require that students listen to detailed instructions and that they explain their procedures or findings. This may confound scientific understanding with listening, speaking, or writing abilities. The extent to which the "process skills" called for by these items are (or should be) distinct from other abilities needs further clarification. In other words, the construct validity of performance items needs to be established.

It is also important to examine the extent to which the skills drawn upon by performance items can be assessed validly, using less costly, open-ended, or even multiple-choice items. All of the performance items listed in Table 1, for example, could be converted into paper-and-pencil items. The question is, Does this change in item format lower the validity of the items? Finally, one important rationale for including performance items on science assessments is that only by doing so will assessment encourage "hands-on" or inquiry-based instruction. This may be true. On the other hand, it also may be true that improved paper-and-pencil items that validly assess the skills learned best through "hands-on" experience can be equally effective in encouraging inquiry-based instruction. If so, the use of these items, rather than the more costly performance questions, would free funds for the development of inquiry-based science curricula, the purchase of laboratory equipment, and the training of science teachers—all areas that also need attention if science instruction is to be improved.

In spite of ETS's conclusion that performance assessment is feasible as part of the NAEP, neither time nor funding levels permitted the inclusion of most of these types of items on the 1990 science assessment. That test did include a stronger focus on open-ended assessment; about 75 percent of the items were multiple-choice, with the remainder being divided between two open-response formats — "figural response" and essay.

The figural-response items require students to draw their responses to questions. Students are asked to "draw arrows to indicate direction, make marks to indicate location, interpret data and sketch graphs, draw simple illustrations to indicate how objects are arranged within systems, and

Table 1. Examples of NAEP 1986 Pilot Test Assessment Items

Observing (Grades 3 and 7): Students watch a test administrator's demonstration of centrifugal force and then respond to written questions about what occurred during the demonstration. Students need to make careful observations about what happens as the administrator puts the steel balls in different holes on the Whirlybird arms and then infer the relationship between the position of the steel balls and the speed at which the arm rotates.

Formulating Hypotheses (Grades 3 and 7): Students describe what occurs when a drop of water is placed on each of seven different types of building materials. Then, the students are asked to predict what will happen to a drop of water as it is placed on the surface of an unknown material, which is sealed in a plastic bag so that they can examine, but not test, it. For this exercise, students need to make careful observations, record their findings, and apply what they have learned by hypothesizing what the water will do when placed on an unknown material.

Interpreting Data (Grades 3, 7, and 11): This paper-and-pencil task requires students to evaluate the results of five children in three athletic events (i.e., frisbee toss, weight lift, and 50-yard dash) and decide which of the five children would be the all-around winner. Students need to devise their own approaches to reviewing and interpreting the data, applying it, and explaining why they selected a particular "winner."

Classifying (Grades 7 and 11): Students are asked to sort a collection of small-animal vertebrae into three groups and to explain how the bones in the groupings are alike. To complete this task, students need to make careful observations about the similarities and differences among the bones and to choose their categories according to sets of common characteristics.

Conducting Experiments (Grades 7 and 11): Students are given a sample of three different materials and an open box. The samples differ in size, shape, and weight. The students are asked to determine whether the box would weigh the most (and the least) if it were filled completely with material A, B, or C. The focus is on which of a variety of possible approaches the student uses to solve the problem.

Source: Educational Testing Service, *Learning by Doing: A Manual for Teaching and Assessing Higher Order Thinking in Science and Mathematics* (Princeton, NJ: Educational Testing Service, 1987), 8, 12, 16, 21, 28.

demonstrate their understanding of events in physical, life, and earth science."[10]

A potential advantage of the figural-response item is that it can reveal students' misconceptions about scientific phenomena; awareness of these misconceptions can then help to improve instruction (assuming teachers are trained to use these data). However, multiple-choice items can be developed that are keyed to such misconceptions. Thus, the issue is the same as that for performance assessment: What are the trade-offs in costs, validity, and instructional usefulness between figural-response and better designed, multiple-choice items?

The 1990 essay problems require that students write short essays "to demonstrate their ability to solve problems, interpret information or data, draw conclusions on the basis of available information, generate researchable questions, evaluate the best experimental procedures, organize a series of logical steps, and design an experiment." From this description, it appears that these items should go far beyond the typical NAEP multiple-choice questions in assessing students' scientific problem solving ability and process skills. Again, however, not enough is known about how essay items compare either to performance items or to more innovative, multiple-choice items designed to assess these skills. Of particular concern with essay items is their dependence on students' writing ability.

In sum, the NAEP is an assessment program that traditionally has served a limited function. It was designed primarily to assist federal policymakers. In this role, the NAEP has been far removed from instructional practice, providing little information that teachers can use to improve their science instruction. At this point, the NAEP probably has had a diffuse but positive effect on science instruction because the regular assessment of students' achievement helps to keep public attention focused on science education. However, the NAEP is undergoing some radical reforms, reflecting the change in political climate since it was first proposed in the Congress. These changes in the National Assessment will increase greatly the potential that the NAEP will be used for accountability as well as for monitoring purposes, with possible repercussions to the influence that the NAEP has on science instruction.

Commercially Developed Science Assessments

It is interesting to compare the role and structure of the NAEP with those of commercially developed science tests, which often are used for accountability purposes. There are three main publishers of science tests, each publishing two different test instruments. The most widely used commercial tests are the Comprehensive Test of Basic Skills (CTBS) and the California Achievement Test (CAT), published by CTB/McGraw-Hill. Tied for second place in popularity are the Iowa Test of Basic Skills (ITBS) and the Iowa Test of Educational Development (ITED), published by Riverside Publishing, and the Metropolitan and Stanford Achievement Tests, published by Harcourt Brace Jovanovich.[11]

It is difficult to discover how many students take these tests; publishers resist giving out this information, presumably to keep the competition guessing. A CTB/McGraw-Hill representative did report that over half of the schools in the country use either the CTBS or the CAT; one author of the ITBS estimates that about two to three million students take the ITBS.

Obviously, most students take at least one of these tests at some point in their school careers.

All of these test instruments are standardized, norm-referenced, multiple-choice tests. As this suggests, test publishers have, in general, a rather pragmatic view of the assessment of science education. First, they acknowledge that such assessment, particularly at the elementary school level, is complicated because the field of science is so diverse and there is little consensus on what should be taught or when. As a result, the items on these assessments tend to cover a wide range of topics in a rather haphazard style. Test publishers also show a businesslike skepticism of performance assessment. After all, unlike the NAEP, commercial test publishers cannot shift the costs of performance assessment directly onto taxpayers; they would have to charge schools for this service.

On the other hand, growing dissatisfaction with existing commercial tests is exerting its own financial pressure on publishers. It is difficult to predict the ultimate results of these opposing economic pressures. The most optimistic possibility is that the competitive forces underlying commercial testing will lead to cost-effective methods for performance assessment. This will require that schools continue to function as discriminating consumers — a likely event as long as accountability pressures do not intervene and force the use of *any* national science test as a means of easily satisfying the public's or policymakers' desire for comparative science achievement data.

Given that they are all constrained by the same economic and psychometric limitations, the science tests distributed by commercial publishers are fairly similar in appearance. They have all the typical advantages and disadvantages of multiple-choice tests, with the major disadvantages being failure to test much beyond students' basic recall and application of scientific facts and concepts. Also, all of the tests except the Iowa Test of Educational Development are based on curricula in use. This would be ideal if everyone were happy with the curricula. However, the number of recent reports calling for a restructuring of science curricula attests to the extent of dissatisfaction with the curricular status quo.[12-14]

Thus, schools using curriculum-based tests find themselves between the proverbial rock and a hard place. They can continue with a curriculum that deemphasizes inquiry learning and maximize their probability of scoring high on national tests, or they can adopt a more innovative, inquiry-based curriculum and risk lowering their test scores. A third option, for states and districts that have sufficient freedom and resources, is to elect not to use any national assessment instrument and to substitute a locally developed assessment program. All of these pressures and possibilities can be found in current uses of commercially developed tests.

Uses of Commercially Developed Tests

The importance placed on nationally normed, standardized tests was illustrated by West Virginia physician John Cannell. In a survey of over 3,500 school systems, Cannell found that virtually none had standardized test scores that were below average.[15] He concluded that schools employ a variety of methods — ranging from the use of outdated scoring norms to outright cheating — to reassure themselves and the public that they are performing well. In Cannell's view, the main use of these tests is to lull educators and the public into a false sense of complacency. In certain ways, Cannell is correct; however, the issue is not as simple as he makes it sound.

Schools use nationally normed tests for a variety of purposes, the main ones being to assist teachers in monitoring student progress, to determine student placements, to monitor and evaluate instructional programs, and to establish accountability. As one might expect, these tests are not uniformly suited for these purposes. For example, as a tool for assisting instruction, these tests tend to be of little or no value. Teachers frequently note that nationally standardized tests do not match their curriculum or do not test the types of skills in which they are most interested. In some cases, this mismatch results from teachers' use of inquiry-based, instructional strategies, which they feel are not assessed adequately by any existing commercial tests; in other cases, it is simply a consequence of publishers' attempts to satisfy everyone, resulting in a test that satisfies no one.

Student placement is another function that these tests serve, typically in conjunction with other data such as grades and teacher recommendations. The advantage of using standardized tests in this context is that they add a more objective measure to the decisionmaking process, making the process more defensible to students and parents.[16] The disadvantage is that, frequently, these tests are used to make placement decisions for which they are not appropriate sources of information, specifically for labeling students as "gifted and talented." This designation should result from an assessment of students' aptitude, not their achievement. Too often, however, standardized tests (and other achievement-based data) are misused as measures of aptitude for "gifted and talented" placement decisions.

Standardized test results also are used to evaluate districts' science programs, to reveal their relative strengths and weaknesses. To the extent that the test matches the curriculum, this is a valid and potentially useful instructional role for these tests. As long as the tests are not used for accountability purposes as well, schools can use them primarily as formative evaluation measures; that is, as assessments that contribute to efforts to improve programs. When accountability becomes an additional testing purpose, however, the evaluation tends to take on a different tone as pressures to score well on the test may overshadow other instructional decisions.

It is important to understand the dynamics of these accountability pressures because they tend to affect individual schools and districts differently. First, pressures usually come from two sources. The first source is the group that establishes the accountability system — typically, state or district policymakers. These groups have the power to hold schools accountable through the allocation of funds and the tightening or relaxation of regulations. The second source, and the one that consistently is most enamored of test scores, is the tax-paying public. The public's power comes from its use of test results to make school-related decisions ranging from where to buy a house to whether to vote for increased taxes for schools.

There are also a number of reasons standardized test results are generally the preferred tool for determining whether schools are performing satisfactorily; that is, for accountability. One reason is that they measure the educational outcome of most interest to most people — student achievement. Second, they provide an easily interpreted, objective, comparative standard (the performance of students across the country). Third, they provide data that are relatively inexpensive and easy to collect and that can be collected every year. This last factor is especially crucial because annual assessments can fuel the desire for quick improvements.

Interestingly, even given the role of standardized tests in an accountability system, the effects that these tests can exert on school curricula vary widely. Two factors seem to be critical in determining the degree to which testing dominates curricula. The first factor is the nature of the accountability system; specifically, the degree to which it focuses exclusively on standardized test scores and the nature of the rewards and sanctions that ensue from performance on the tests. The greater the reliance on standardized tests and the greater the consequences their results will have for schools (or school administrators), the more likely it is that obtaining high scores will overshadow other curricular goals. Also, if sanctions result from *decreases* in test scores from one year to the next, curricular changes that move the curriculum away from the content of the tests will be discouraged.

The second factor, which interacts with the first, is the performance of the school itself. Schools and school districts whose achievement are above average typically experience less accountability pressure and are, therefore, more free to experiment with new instructional practices or curricula. In schools or school districts whose levels are below average, which typically translates into schools serving at-risk students, accountability pressures tend to be greater and pressures for improving test performance often dominate and direct all other educational decisions. Consider the following two examples.

South Carolina is a poor state whose students, historically, have performed abysmally on indicators of educational status (such as graduation rates, Scholastic Aptitude Test scores, per-pupil spending). In the 1980s, the state initiated a concerted effort to improve its educational performance. This effort includes the mandated use of both state-developed and national tests; every student is tested each year with one test or the other. Needless to say, local communities take the test results very seriously; often, communities will withhold funds from districts that do not score well, while providing more funds to districts that do score well. The state also has tied its funding to performance on standardized tests, awarding "bonuses" to schools that score well. The message to teachers is clear: Teach to the tests.[17]

District Heights Elementary School provides an extreme example of the "teach-to-the-test" phenomenon. This school is located in a predominantly poor and minority school district where the superintendent has requested that school performance, as well as his own performance, be judged on the basis of CAT reading scores. As a result, teachers instruct students primarily on the skills tested and often teach test-taking skills explicitly. Some of the practices used to raise scores on reading tests in District Heights are special classes on test-taking given in the second grade, the year before students first take the CAT; use of the multiple-choice format in at least one lesson each day; distribution of 100 pages of practice work sheets (as part of a take-home packet) to students in the grades to be tested; and the completion of practice tests during the full day preceding the test. These teaching strategies have resulted in improved test scores, but other evidence of student performance, such as the low grade-levels of reading texts, suggests that the test scores are, under these circumstances, invalid measures of academic achievement.[18]

There are at least three morals to these stories. The first is merely a reiteration of the argument above: Test-based accountability keeps higher scoring schools free to develop new curricula and assessment methods that focus on higher order and process skills, while lower scoring districts, which are disproportionately poor and minority, are held hostage instructionally by pressures to raise test scores. In this way, the use of tests for accountability purposes may exacerbate "true" achievement differences among social classes, while "tested" achievement differences may, in fact, decrease.

The situation at District Heights Elementary School also illustrates the double dilemma that science education often faces with respect to accountability. Based on how accountability is structured at present, it may appear advantageous for science to be left out of these accountability systems, as it is in District Heights. This has the beneficial effect of freeing science instruction from the confining hold of test content and structure. The disadvantage is that excluding science from an accountability system may result in science instruction being ignored in favor of those subjects that are assessed.

Finally, the South Carolina scenario illustrates another way in which accountability can potentially increase the inequities inherent in the educational system. In theory, educational accountability should provide performance-based rewards and sanctions that assist schools in improving performance. Financial incentives for schools or school districts that show exceptional progress are one inducement that works toward raising educational standards because they provide schools with increased revenues for expanding services or programs. In the case of schools or districts making unsatisfactory progress, the response should be not to withhold funds, but to evaluate whether the district has been allocated enough funds to deliver services effectively in the first place. If funding levels are determined to be adequate, the district should be provided assistance in developing an improvement plan; if satisfactory progress is not made after implementation of this plan, more direct intervention may be needed. In any case, all interventions should function to improve the education provided to students, rather than to punish the schools. After all, the goal should not be to "weed out" certain schools, but to allow each school to blossom.

Other National Initiatives

While the above discussion characterizes the current status of science education assessment, a number of studies are in progress that attempt to improve the ways in which science instruction is assessed. Most of this work is funded by the National Science Foundation (NSF), through its Directorate for Education and Human Resources (formerly the Directorate for Science and Engineering Education).

The NSF has responded to calls for improving assessment and its link with curricula by initiating a number of studies examining the role and nature of science assessment.[19] One study supports ETS in developing performance assessment measures for the NAEP; the 1985–86 pilot test of hands-on items[20] and the 1990 figural-response items are two outcomes of that work. A second study supports the state of Connecticut in developing science performance items for a statewide assessment, while a third study examines alternative methods for ascertaining students' science process skills. Finally, other studies are examining critical issues in science assessment, such as the relationship between assessment and science curricula, the impact of assessment on teaching and learning, and methods of incorporating assessment into curricular reform efforts.

The ETS is funding its own study of methods of integrating science assessment with instruction at the middle school level. This study, "Learning Progress Systems in Middle School Science," is developing approximately 24 instructional assessment modules involving computer simulations, laboratory experiences, and discussion activities, in addition to paper-and-pencil tasks. Each module contains five hours of instructional assessment activities and is designed to provide teachers with tools to monitor and assess students' scientific content and process knowledge.

National Assessments in the Service of Instruction

The projects described above have one major feature in common—they all view assessment primarily as an instructional tool; that is, as a tool to help teachers rather than policymakers. As tools to assist teachers with instructional practices, national-level tests always will defer to those developed locally because local tests are necessarily better tailored to each school's curriculum. Nonetheless, national tests are likely to continue to play a role in educational assessment because they provide comparative data that are useful to policymakers as well as to educators. Thus, the issue is, How can the instructional usefulness of these tests be increased?

The NAEP

Because the NAEP is administered on a sampling basis and most test items are confidential, there is not much that can be done to make it instructionally useful for individual teachers. The NAEP could be made more useful for teachers and policymakers in general, however, if the items on the assessment were classified in an instructionally meaningful way and achievement results were reported for each classification area. For example, items assessing "process skills" could form one classification group; items assessing "recall of life-science facts and concepts," could form another. This type of analysis and reporting structure would permit a more meaningful assessment of the strengths and weaknesses of the science curriculum.

As the NAEP becomes a part of state indicator systems, it is essential to ensure that the National Assessment does not drive curricula away from instructional topics and curricular goals that are valued components of science education. There are at least two ways to prevent this. First, educators and policymakers must insist on the inclusion of these topics and goals in the assessment. Similarly, research is needed to address several issues related to the format and validity of test items assessing inquiry-based skills; this research could help to ensure the cost-effectiveness and construct validity of assessments that measure these skills. Second, the NAEP's role in state indicator systems must be maintained strictly as a monitoring role and must not be expanded to include accountability. Given current administration procedures and schedules, this may not be an immediate threat. Nevertheless, the NAEP is likely to be quite appealing as an accountability tool. NAEP's history already is a lesson in how one single-purpose, federally funded assessment can expand well beyond its original role.

Commercially Developed Tests

Because teachers have access to the items on these tests, commercially developed tests have greater potential than the NAEP for increased instructional usefulness. The advantages of classifying items into instructionally relevant categories on commercial tests are even greater than they are on the NAEP. Such a classification system would allow teachers and schools to assess the relative strengths and weaknesses of their instructional programs and of individual students. This system also could be used to develop tailor-made standardized tests. Each school system could choose a selection of item categories from among those available to create an individualized test that better matched its curriculum. If desired, schools or school districts could be compared on the item categories that they have in common. In fact, schools' achievement and curricula could be compared simultaneously in this type of system, thus providing a more complex but more meaningful assessment.

The keying of item responses to student misconceptions is another way that standardized tests can be made more instructionally useful. Providing teachers with analyses of misconception-related errors (such as, "item 5 shows that 67 percent of the students believe the sun orbits the earth") would transform these tests into valuable guides for improving instructional practices. It should be noted that the critical feature for making this revision work is a reporting system that teachers can understand and use; more than just a reworking of the test items is required.

Finally, it is important to ensure that these tests do not dictate or distort instructional practices. The development of tests based on analyses of commonly used curricula makes most commercially developed tests highly conservative instruments; schools must change their curricula before the tests change their items. Without accountability pressures, this is perhaps only a minor inconvenience, but with accountability pressures, it becomes an impossible situation. Thus, these tests should be used for accountability purposes *only* when it is agreed that the items in the test reflect the *desired* curriculum for the schools — whether that curriculum is already in place or not.

These are just a few ways in which national assessment initiatives can be made more relevant instructionally. The likelihood that any one of these suggested improvements will come about depends upon the interplay of a complex set of political, economic, and educational forces. Perhaps the most crucial force at present is the accountability movement, which simultaneously is increasing the use of standardized assessment and revealing the weaknesses inherent in existing assessment practices. The growing frustration with standardized assessment resulting from accountability issues, combined with current efforts to increase science achievement, provide a window of opportunity for science assessment reform that may not exist again for decades. Moreover, as national assessments become increasingly popular, their potential effects on educational practice also become greater. The science assessment practices and instruments that are selected now are likely to have more far-reaching effects than any that have been used in the past. It is indeed time to place assessment in the service of instruction.

Acknowledgments

The author gratefully acknowledges the information and advice provided by the following individuals: Patricia Brougan, Audrey Champagne, H. D. Hoover, Clyde Hudson, Daniel Koretz, and Steven Million. Many teachers and district testing directors also shared their experiences and views generously; their contributions are appreciated as well.

References and Notes

1. National Commission on Testing and Public Policy, *From Gatekeeper to Gateway: Transforming Testing in America* (Chestnut Hill, MA: Boston College, 1990).
2. Shavelson, R. J., and Carey, N., "Outcomes, Achievement, Participation, and Attitudes," in R. J. Shavelson, L. M. McDonnell, and J. Oakes, eds., *Indicators for Monitoring Mathematics and Science Education,* Report No. R–3742–NSF/RC (Santa Monica, CA: RAND Corporation, 1989).
3. Originally, NAEP surveyed students at ages 9, 13, and 17. This was changed to a combination of ages and grade levels, using grades 3, 7, and 11. Beginning with the 1990 assessment, students in grades 4, 8, and 12 are being surveyed.
4. Sweet, R. J., Jr., *Director's Report to the Congress on the National Assessment of Educational Progress,* unpublished manuscript, National Institute of Education, Washington, DC, 1982, 5.
5. In order to compare student performance across years, the NAEP uses many of the same items in consecutive assessments. Because items from past assessments may be used in future assessments, most NAEP items are confidential. A few of them have been publicly released; 30 of these are reproduced in Mullis, I. V. S., and Jenkins, L. B., *The Science Report Card: Elements of Risk and Recovery* (Princeton, NJ: Educational Testing Service, 1988).
6. Sweet, *Director's Report,* 5. See reference 4.
7. Traditionally, the results of science tests have been examined by five proficiency levels, ranging from "knows everyday science facts" to "integrates specialized science information." These categories were developed by first dividing assessment results into five evenly spaced performance levels; specialists then examined the items that differen-

tiated performance at each level and described the types of skills that characterized performance at one level versus another. Thus, the proficiency levels are not a classification of test items, but a general portrayal of the items that distinguish performance between levels.

8. There is some irony in this fact because the CCSSO was one of the groups that argued most strongly for restricting the scope of the original National Assessment. The persistence of the federally sponsored "Wall Chart," with all of its limitations, seems to have instigated the CCSSO's change of heart.

9. Blumberg, F., Epstein, M., MacDonald, W., and Mullis, I., *National Assessment of Educational Progress: A Pilot Study of Higher-Order Thinking Skills Assessment Techniques in Science and Mathematics* (Princeton, NJ: Educational Testing Service, 1986).

10. Educational Testing Service, "Which Assessment Contains All of These Innovations? The 1990 National Assessment" (undated brochure available from Educational Testing Service, Princeton, NJ), 3.

11. American Textronics also publishes a science education assessment instrument. This newer instrument is used mainly by Catholic schools; it has not (yet) been able to break into the public school testing market. Because of its limited use, this assessment is not discussed here.

12. American Association for the Advancement of Science, *Science for All Americans: A Project 2061 Report on Literacy Goals in Science, Mathematics, and Technology* (Washington, DC: American Association for the Advancement of Science, 1989); reprinted as, Rutherford, F. J., and Ahlgren, A., *Science for All Americans* (NY: Oxford University Press, 1990).

13. Bybee, R. W., Buchwald, C. E., Crissman, S., Heil, D. R., Kuerbis, P. J., Matsumoto, C., and McInerney, J. D., *Science and Technology Education for the Elementary Years: Frameworks for Curriculum and Instruction* (Andover, MA: The NETWORK, 1989).

14. National Science Teachers Association, *Science-Technology-Society: Science Education for the 1980s,* NSTA position statement (Washington, DC: National Science Teachers Association, 1982).

15. Goldstein, A., "Finding a new gauge of knowledge," *The Washington Post* (May 20, 1990), A20.

16. This obviously assumes that the tests are not biased against any group of students. The issue of test bias is politically charged and is more complicated than many parties on either side of the issue frequently acknowledge; see Tomlinson, T. M., "The Troubled Years: An Interpretive Analysis of Public Schooling Since 1950," *Phi Delta Kappan,* **62** (5), (1981), 376. Here, we note only that the best way to minimize all forms of test bias is, as others have noted, to make tests publicly available.

17. Through legislation requiring the assessment of higher-order thinking

skills, and a newly elected state superintendent of education committed to further reform, South Carolina shows signs of moving away from this heavy-handed reliance on standardized basic skills assessment.

18. This description of District Heights Elementary School, District Heights, Maryland, was abstracted from Goldstcin, A., "Thc Sccrcts behind the Scores," *The Washington Post*, (May 20, 1990) A1, A20.

19. See, for example, American Association for the Advancement of Science, 1989. See reference 12.

20. Blumberg et al., *National Assessment of Educational Progress.* See reference 9.

9

State Initiatives in Assessing Science Education

Alan Davis and Jane Armstrong

Science is a relative newcomer to the state educational testing scene. In 1978, 41 states had statewide testing programs, but only nine of them included science.[1] During the past 12 years, the number of states testing the effectiveness of science instruction and learning has grown to 28. In this chapter, we examine the phenomenon of the growing state role in assessing science education. What do states expect to accomplish by such assessment? Are there significant differences in the directions that various states are taking? What are the likely consequences of current assessment practices and trends for science instruction?

We approach these questions first from both the historical context of the politicization of testing and the adoption of testing by states as a policy tool. We then describe the results of a survey of state testing directors, comparing the experiences of states that monitor science achievement by using a nationally normed test with those states that have developed their own tests to promote the implementation of state curricular guidelines. Drawing on recent studies of the effects of assessment on instruction, we believe that many state tests have not realized their goal of improving instruction; however, we offer descriptions of three state initiatives that exhibit strong promise.

Uses of Assessment

Testing by states started essentially as a response to public and political pressures for more information on how well schools were educating students in various districts. Two concerns in particular sparked its proliferation in the early 1960s: concern for equity and concern that American students lacked adequate preparation.[2,3] In most states, interest in testing at the state level was driven initially by the desire of legislatures to monitor one or both of these concerns. Several states began to use readily available, nationally normed tests to collect information to determine how their students compared with students nationally and to monitor trends.

The monitoring function of state testing was not greatly intrusive on most schools. No decisions about individual teachers, pupils, or schools were linked to test scores. However, the indirect effects of testing for monitoring were important.[4] State testing conditioned the public to look beyond the bounds of their local school system in assessing the adequacy of education and to accept test scores as the main measure of the success of schools.[5]

Accountability

The disappointing news that the tests often provided, coupled with constitutional reminders that the states ultimately were responsible for educational quality, brought forth a call for accountability in the 1970s.[6,7] An annual ritual developed of reporting test scores in local newspapers by school, thus bringing pressure to bear on teachers and administrators where scores were low.[8] Testing as information gathering evolved into testing as an accountability pressure and a policy lever.

During this period, several states increased local accountability by developing performance tests that had to be taken by all students. The results of these tests carried with them consequences associated with different levels of performance for students, local agencies, or both. By the summer of 1978, 29 states were pursuing actively some form of minimum competency testing;[9] by the summer of 1984, that number had grown to 40.[10] Nineteen states were implementing tests for high school graduation. Several legislatures passed provisions requiring local school districts to meet certain minimum levels of achievement or face the possibility that the state would declare them "academically bankrupt" and begin to exert direct control over them.

Teachers responded to the increased importance of test scores by teaching what was tested at the expense of everything else.[11] Because reading, language arts, and mathematics were tested, these subjects came to dominate the curriculum—usually at the expense of science, social studies, and fine arts. The result is a situation in which testing begets more testing, as educators perceive that if something is not tested, it is not considered important.[12] The explosion of science testing in recent years must be understood in this context. In some cases, the push for science testing has been spearheaded by science educators as a means of drawing attention to their subject again.[13]

National calls for states to strengthen science education have added momentum to the increase in testing. The nation's governors have developed national goals for achieving scientific and mathematical literacy.[14] Reports have demonstrated an increase in the technical and science-related skills required by higher paying jobs,[15] coupled with the poor performance of American students compared to that of students internationally[16] and to previous cohorts of American students.[17] National proposals for reform have called for comprehensive changes in science curricula and instruction.[18,19]

Improving Curricula and Instruction

Some states have begun to incorporate tests into coordinated strategies to affect curricula and instruction. Bill Honig, Superintendent of Public Instruction in California and a leading advocate of direct state leadership in education, described the ideal strategy in these terms:

> It would require educators to define specifically the kinds of subject area content and skills students must learn if the nation is to compete internationally. It would necessitate determining the best ways to teach each subject to a diverse student population. It would mean developing better testing, undertaking heavy investment to bring teachers up to speed, and giving educators the necessary technological tools to improve their productivity. Finally, this strategy would provide schools and districts with planning and implementation grants to translate these ideas into reality and make structural changes to move from a rule-driven system to one that is performance-based. [20]

Honig explains how California has used testing in this process:

> We changed our state tests to reflect the altered curriculum. We now test in science and history, evaluate writing samples, and assess for higher levels of understanding in reading and math. We also instituted an accountability system that set specific targets and gave each school and district information on how it was doing. We publicize the results annually.[21]

Most states have undertaken at least some of these initiatives and a few, like California and New York, have involved nearly all of them in coordinated and concerted efforts to improve the quality of education statewide. These efforts reflect a faith that the test scores are valid indicators of achievement and that attempts by teachers and administrators to obtain higher scores will have a generally positive effect on curricula and instruction.

The four broadly defined intended uses of state testing described according to their historical development — monitoring, accountability, certification of individual student mastery, and a policy tool to lead to improved curricula and instruction — remain useful constructs in comparing state testing programs today.

Characterizing State Science Assessment Initiatives

To determine why and how states were testing science education, we surveyed state assessment directors in May 1990. A questionnaire was sent to the directors in advance and followed up with a telephone interview. The questions emphasized science assessments in the middle grades — the type of assessment, its history and development, test content, reporting, and use. The middle grades were emphasized because they were considered most likely to reflect the national trends in science education. We conducted follow-up interviews in greater depth with testing directors in California, New York, and Connecticut and analyzed documents from those states. Several states sent us copies of their science tests. We also drew on our previous case studies of state initiatives to improve science education.[22]

We found it useful to distinguish between states that relied upon a commercially available, normed test and states that had developed tests of their own. In general, the first group emphasized purposes of monitoring and accountability, while the second reported improving curricula and instruction as the primary purposes for assessment, often with accountability as a secondary concern. A third category included a few states that have added science to the competencies tested for promotion or graduation. These states often use two tests: a commercial, norm-referenced test and a locally developed competency test. Each of these categories of state practices will be considered below.

Commercial Norm-referenced tests for Monitoring and Accountability

Sixteen states assess science education by administering a commercial, nationally normed science test (Table 1). Two of these states administer an additional science test, which is needed for promotion or course credit (Arkansas and Mississippi); in one state (Louisiana), an additional test is required for graduation.

Most of the states that currently administer nationally normed science tests share important characteristics: they have relatively low education budgets, relatively small populations, and half of them are in the South. The following case history, adapted from Armstrong et al., is a composite study describing the history and uses of the norm-referenced science assessment in a representative state:[23]

> In 1974, State A began administering a commercial, norm-referenced test every year to all students in grades 4, 8, and 11 in reading, language arts, and mathematics. In 1983, in response to concerns by educators and the business community that science should be given more visibility, the state started giving the science subtest in addition.

The content of the science test had little to do with its selection. Rather, the content of the reading and mathematics sections of the test and the publisher's competitive bid largely determined its choice.

No promotion standards, district funding determinations, or graduation decisions rest upon the results of the test, but the annual reporting of scores in local newspapers receives a great deal of attention. The scores of schools in affluent suburbs are almost always high; the scores of urban and poor rural schools are almost always low. Low-scoring schools experience pressure to improve their results. However, most of the attention is directed to mathematics, language, and reading scores. Rarely are the science scores a topic of conversation.

There is little support for the science test in the educational community. Members of the assessment staff in the state education agency are apologetic about the test because it assesses mostly factual knowledge and simple concepts. Science educators, including the state science coordinator, are still more critical of the test because they say it does not fit well with the types of science curricula they would like to see in classrooms (that is, fewer topics taught in more depth; more emphasis on processes of investigation and on technology and society) and because they believe it rewards reliance on textbooks and lecture modes of instruction. Yet, there is little evidence that the science test has much effect on instruction. In the nine grades where the test is not given, teachers have no idea what it contains. In the grades tested, teachers are familiar with the reading and mathematics tests. Many teachers report pressure to improve scores in those subjects, but few pay much attention to the science test.

The general lack of enthusiasm for the content of nationally normed science tests reported in the preceding case history was confirmed in our survey of testing directors. Far fewer than half of them reported that the test they administer is a good match with either the curricula they believe are actually implemented in schools, or with the curricula they would like to see implemented. Their opinion reflects the problem of designing tests for a mass market. The tests are developed to reflect the content of major textbooks which, in turn, reflect pressures for "coverage" rather than depth.[24] The result is a cautious curriculum, which mirrors current practice, rather than an exemplary product.[25]

Only six nationally normed tests are used by states to assess science (Stanford Achievement Test [SAT], Comprehensive Test of Basic Skills [CTBS], Iowa Test of Basic Skills [ITBS], California Achievement Test [CAT], Metropolitan Achievement Test [MAT6], and Test of Achievement and Proficiency [TAP]). They represent three publishers (CTBS-McGraw Hill, Riverside, and Psychological Corporation). Gong, Lahart, and Court-ney[26] analyzed the content of the middle grade levels of these tests with

Table 1
Results of a 1990 Survey of State Science Assessment Practices [1]

State	Commercial, Norm-Referenced Test/Grades Tested	Minimum Competency Test or Graduation Test/Grades Tested	Other State-Developed Tests/Grades Tested	Type of State-Developed Tests	Tied to State Curricular Initiative?	Response Format: MC = Multiple Choice OE = Open Ended JP = Judged Performance	Pupil Sample	Frequency of Test	Number of Forms
Alabama	SAT*/4,8				No	MC	All pupils	Annual	1
Arkansas	MAT6†/4,7,10	Minimum Performance Test/6,8		CRT‡	Yes	MC	All pupils	Annual	MAT6-1 MPT§-3
California			CA Assessment Program/6,8,12	State-normed and CRT	Yes	MC, OE, JP	All pupils	Annual	10–15
Connecticut			CT Assessment of Education Program/4,8,11	Domain-referenced	Yes	MC Field Testing OE, JP	Students sampled	Every 5 years	20–36
Delaware	SAT/11				Yes	MC	All pupils	Annual	2
Florida			FL Student Assessment Program/3,5,8	CRT	Yes	MC	Students sampled	Annual	1–3
Georgia	ITBS¶/2,4,7,9				Yes	MC	All pupils	Annual	1
Idaho	ITBS/6,8,11			CRT	No	MC	All pupils	Annual	1
Illinois			IL Goal Assessment Program/3,6,8,11	State-normed	Yes	MC, JP	All pupils	Bian-nual	5

Table 1 (continued)
Results of a 1990 Survey of State Science Assessment Practices[1]

State	Commercial, Norm-Referenced Test/Grades Tested	Minimum Competency Test or Graduation Test/Grades Tested	Other State-Developed Tests/Grades Tested	Type of State-Developed Tests	Tied to State Curricular Initiative?	Response Format: MC = Multiple Choice OE = Open Ended JP = Judged Performance	Pupil Sample	Frequency of Test	Number of Forms
Indiana			IN Statewide Testing of Education Program/3,6,8,11	CRT	Yes	MC	All pupils	Annual	1
Louisiana	CAT#/4,6,9	Graduation Exit Exam/11		CRT	Yes	MC	All pupils	Annual	1
Maine			ME Education Assessment/4,8,11	Domain-referenced	No	MC, one OE	All pupils	Annual	12
Massachusetts			MA Education Assessment Program/4,8,12	State-normed	No	MC, OE, JP	All pupils	Bian-nual	12–20
Michigan			MI Education Assessment Program/5,8,11	CRT	Yes	MC	All pupils	3–5 years	1
Minnesota			MN Education Assessment Program/4,8,11	State-normed CRT	Yes	MC	Sample	4 years	4

Note: In a spring 1990 survey, 28 out of 50 states were identified as periodically collecting statewide measures of student achievement in science.
*SAT = Stanford Achievement Test; †MAT6 = Metropolitan Achievement Test; ‡CRT = Criterion-referenced Test; §MPT = Minimum Performance Test; ¶ITBS = Iowa Test of Basic Skills; #CAT = California Achievement Test; **CTBS = Comprehensive Test of Basic Skills.

Table 1 (continued)
Results of a 1990 Survey of State Science Assessment Practices[1]

State	Commercial, Norm-Referenced Test/Grades Tested	Minimum Competency Test or Graduation Test/Grades Tested	Other State-Developed Tests/Grades Tested	Type of State-Developed Tests	Tied to State Curricular Initiative?	Response Format: MC = Multiple Choice OE = Open Ended JP = Judged Performance	Pupil Sample	Frequency of Test	Number of Forms
Mississippi	SAT/4,6,8	End of course biology test		CRT	No	MC	All pupils	Annual	1
Missouri	ITBS/2–8	Test of Achievement Proficiency/ 9,10 (optional)	MO Mastery Achievement Test/3–10	CRT	No	MC	Sample	Annual	1
New Hampshire	CAT/4,8,10				No	MC	All pupils	Annual	1
New Mexico		High School Competencies Exam/10–12	NM Achievement Assessment/3,5,8	Combination ITBS items and state-developed CRT items	Yes	MC	All pupils	Annual	1
New York		NY Regents Exam	NY Program Evaluation in Science/3,6,8	CRT	Yes	MC, JP	All pupils	Annual	1
North Carolina		End of course test	NC Science Achievement Test/3,6,8	State-normed CRT	Yes	MC	All pupils	Annual	2
North Dakota	CTBS**/ 3,6,8,11				No	MC	All pupils	Annual	2

Table 1 (continued)
Results of a 1990 Survey of State Science Assessment Practices[1]

State	Commercial, Norm-Referenced Test/Grades Tested	Minimum Competency Test or Graduation Test/Grades Tested	Other State-Developed Tests/Grades Tested	Type of State-Developed Tests	Tied to State Curricular Initiative?	Response Format: MC = Multiple Choice OE = Open Ended JP = Judged Performance	Pupil Sample	Frequency of Test	Number of Forms
Oklahoma	ITBS/3,5,7,9,11				Yes	MC	All pupils	Annual	1
South Carolina	SAT/4,5,7,9,11		SC Basic Skills Assessment Program/3,5,8	CRT	Yes	MC	All pupils	Annual	1
South Dakota	SAT/4,8,11				No	MC	All pupils	Annual	1
Tennessee	CTBS/2–8,10				No	MC	All pupils	Annual	3
Virginia	ITBS/4,8,11				Yes	MC	All pupils	Annual	1
West Virginia	CTBS/3,6,9,11				Yes	MC	All pupils	Annual	1

Note: In a spring 1990 survey, 28 out of 50 states were identified as periodically collecting statewide measures of student achievement in science.
*SAT = Stanford Achievement Test; †MAT6 = Metropolitan Achievement Test; ‡CRT = Criterion-referenced Test; §MPT = Minimum Performance Test; ¶ITBS = Iowa Test of Basic Skills; #CAT = California Achievement Test; **CTBS = Comprehensive Test of Basic Skills.

respect to their match with the curricular recommendations of the American Association for the Advancement of Science (AAAS) Project 2061's *Science for All Americans*[27] and Bloom's taxonomy of cognitive skills.[28] They found that the tests sampled a broad content domain, with emphasis on recall and explicit information skills, in contrast to the AAAS's call for greater depth on fewer topics, with special attention to conceptual understanding and skills of problem solving and investigation. These researchers concluded that the tests as a whole "foster fragmentation of shallow knowledge... [and] provide no reliable assessment of process or thinking skills."[29]

In light of such criticism, why do states use nationally normed tests to assess the effectiveness of science education? Largely because they are inexpensive and simple to use. The notion that legislators insist upon national comparison data is refuted by the fact that nearly every large state that assesses science has developed its own test.[30] In states that lack the resources for this undertaking, the commercial test is viewed as a means of ensuring accountability and providing at least symbolic support for the teaching of science.

Although evidence is mounting that multiple-choice measures have negative implications for instruction when they are perceived as the goal of instruction,[31] testing directors report that science tests do not capture media attention in the way that tests of reading and mathematics do. Consequently, local schools pay little attention to them. Interviews with teachers have supported this view.[32] In short, it is likely that the use of commercial tests for science assessment will have little impact on science instruction (good or bad) under current levels of accountability.

State-developed Science Tests for Accountability and Curricular Improvements

Most of the states that have begun to assess the results of science instruction during the past five years have developed their own tests based upon state curricular guidelines or objectives. In these states, assessment generally is viewed as a policy lever to urge the implementation of state recommendations in curricula and instruction. Although there is considerable variation within this category of states, several common features are captured in the following composite case history.

Since 1982, State B has been developing its own criterion-referenced tests to accompany new state objectives in each subject area. Reading, language arts, and mathematics objectives were developed first, and then tests followed in each of those areas. In 1986, science objectives were developed by a large team of science teachers, curriculum coordinators, a professor of science education, and two applied scientists. The assessment division of the state education agency developed multiple-choice tests for grades 3, 6, and 8, based on the state's objectives. Information sessions for administrators and training sessions for teachers on the new objectives were offered regionally throughout the state.

Many state policymakers view the new science test as an accountability tool; that is, as a means of assuring that science is taught in the schools and as a means of monitoring progress. In this state, no direct consequences for students, schools, or districts ride on the results of the test, but the results are publicized. Officials at the state education agency hope that the test, coupled with the state's objectives, will provide an impetus for improving science education, especially in grades K–8. They believe that tailoring the test to the state's objectives should defuse local objections that what is tested is not what is taught. They also hope that the test will encourage teachers to attend to the state's objectives and spend more time teaching science.

Among science educators, the new science test in State B is controversial. Supporters argue that very little science has been taught in grades K–6 and that the test will help to correct this. They agree that the test includes more items requiring students to go beyond factual knowledge to conceptual understanding, familiarity with issues of technology and society, interpretation of data, and design of experiments. Those who oppose the test argue that it still overemphasizes lower cognitive processes, will narrow instruction to the test items themselves, and does not encourage directly "hands-on" instruction and inquiry. In response to these criticisms, the state agency is piloting six applied performance tasks, which will be added to the test in two years.

Approximately 14 states share important similarities with State B. Most of them are larger, more affluent states. All of them test science with tests developed by the state education agency or by a contractor to state specifications. All but two of them (Maine and Massachusetts) have developed their tests during the past six years as part of a state initiative to improve science education, spearheaded by the development of state science objectives or curricular frameworks. In most of these states, the science test is regarded as a policy lever — that is, an inducement to local education agencies to undertake curricular and instructional changes along the directions recommended by the state.[33]

State-developed tests have several advantages over commercial, norm-referenced tests. First, they are developed to match state curricular guidelines or frameworks, providing a tight alignment between what the state considers important to teach in science courses and what is tested. Second, a much larger number of items is usually developed, representing a larger content domain, which is assessed in multiple-test forms using a matrix sampling technique. Test security and narrowing of the curricula become smaller issues because teachers cannot teach to a single test form.

Nearly every state using its own test as a policy lever has built-in or indirect mechanisms to exert accountability pressure. The exceptions to this rule are a handful of states (Connecticut, Minnesota, and Florida) that sample students, so that results are available only for districts or for the state as a whole and cannot have negative consequences for individuals or schools. Apart from these few states, accountability pressure is associated first and foremost with the release of scores to local newspapers. Three states (Indiana, Mississippi, and North Carolina) apply additional pressure by using mathematics and reading test scores in school accreditation decisions.

For reasons of test security, we were unable to review the content of state-developed science tests. However, our survey of state assessment directors indicates that recent tests put more emphasis on technology and society, investigative processes, and science concepts than earlier tests did, and target higher levels of cognitive functioning. Gong et al. compared the item specifications for current, state-developed tests with commercial science tests and concluded that both contained too many factual knowledge questions distributed over too broad a range of topics.[34]

Tests with Important Consequences for Individuals

Six states include science in state tests, with direct consequences for individual students. Science tests are used in grade promotion decisions in Arkansas and in course credit decisions for certain high school courses in Mississippi and North Carolina. Science is a part of the tests required for graduation in New Mexico and Louisiana. New York gives Regents Examinations in science disciplines, which are used to award diplomas and to make selection decisions by the state university system. California is preparing similar tests in its Golden State Examination (GSE) program. With the exception of the California tests, all of these examinations are composed entirely of multiple-choice items (although the Regents Examinations require documentation of satisfactory laboratory performance as a prerequisite).

When the results of tests have consequences for teachers and students, teachers adjust their instruction in ways that they believe will result in higher scores. One negative consequence of this situation is that teachers under pressure will tend to focus their instruction narrowly on the topics tested and will omit other topics.[35,36]

Reliance on the multiple-choice format in state assessment also has raised concerns. Darling-Hammond and Wise found that when teachers are pressured to raise test scores, they may rely on instruction that resembles the format of the test.[37] If the test items are in a multiple-choice format, teachers may conclude that students will benefit from drill and practice on discrete items of knowledge or skills, rather than hands-on investigation.

The multiple-choice format also places constraints on the range of thought processes that can be examined. It is argued widely that such tests cannot provide problems that require students to define solution paths, come up with alternatives, devise a test of alternative strategies, or organize relevant information and present a coherent argument.[38–40]

Both of these concerns apply to the "high-stakes" tests (where results have important consequences for individuals) used by many states. Where states continue to rely exclusively on the use of multiple-choice items and a limited number of forms of the test are used, there is great likelihood that the tests will promote a narrowing of both the curricula and instruction to the tests themselves. Some states have addressed this problem by making annual changes in the form of the test, but high-stakes tests continue to rely predominantly on the multiple-choice format.

Performance Assessment

Several states have introduced innovations in science assessment designed to improve science instruction or science programs. Virginia and New York have expanded their assessments to include program reviews as well as student outcomes. Both states make optional questionnaires available to schools and technical assistance from the state is provided for their scoring and interpretation. New York is assessing students' attitudes toward science as well as cognitive outcomes; other states, including Connecticut, have similar plans.

Some states are beginning to respond to concerns about the limitations of multiple-choice tests by adding open-ended questions and direct performance assessment to state science tests. California, New York, Illinois, and Massachusetts included direct performance measures in their 1990 tests. Connecticut will move to exclusive reliance on performance measures in 1992. Several other states, among them Indiana, Michigan, Missouri, Minnesota, and Florida, are revising their tests to incorporate direct performance measures in future assessments. This change represents the most important current development in state science assessment and warrants closer examination.

Examples of performance assessments from California, New York, and Connecticut are presented below. All three states have moved decisively to assess science education through performance tasks and all of them intend to use assessment to improve instruction. The three states vary with respect to their use of accountability pressure as a policy tool, with California endorsing moderate pressure and Connecticut rejecting it in favor of persuasive modeling.

California. California developed a new state science framework in 1990, which forms the basis for new science assessments in grades 6 and 12 only. The new framework emphasizes themes that cut across the traditional disciplinary divisions of earth sciences, physical sciences, and life sciences. Drawing explicitly on Project 2061 of the AAAS,[41] six themes have been addressed: energy, evolution, patterns of change, scale and structure, stability, and systems and interactions. The thematic organization is proposed to counter a tendency to "reduce and compartmentalize science content and focus on isolated facts and concepts."[42]

The new framework also goes farther than the 1984 science framework addendum in making explicit a call for active learning, defined by Bill Honig in his introduction as "instructional activities where students take charge of learning," and "regularly make new associations between new ideas and their previous conceptions of how the world works."[43]

The California Assessment Program (CAP) has long been aligned in content with the state's curricular frameworks. Now, particular attention is being paid to aligning assessment with instruction in both content and format, so that students' performance and investigation will "play the same central role in assessment that they do in instruction."[44]

California assessed science education in grade 6 in the spring of 1991 and will assess grade 12 in the spring of 1992 as part of the CAP. Both tests are expected to include multiple-choice items, open-ended questions, and performance tasks. CAP also is developing Golden State Examinations (GSEs) in chemistry and biology. These examinations also will contain multiple-choice items, open-ended questions, and performance tasks. High schools throughout the state will field-test the GSEs, and both the GSE biology and chemistry examinations should be ready for statewide use in 1991.

In spring 1989, open-ended science questions were administered to a total of 8,000 6th grade students. The questions were designed to engage students in creating hypotheses, designing scientific investigations, and writing about social and ethical issues in science. Students responded to the questions in one of three ways: by interpreting and entering data on a chart, by drawing a picture to explain an answer, or by writing a short analytical paragraph. Each open-ended question asked students to compose a written response, which required 10 to 15 minutes to complete.

Five performance tasks for grade 6 were field-tested in the spring of 1990 and implemented fully in the spring of 1991. Approximately 1,000 schools participated in the field test, which included controlled studies of the effects of different grouping, time, language of administration, and delivery of instructions. The testing format consisted of five "stations,"[45] with one task to be accomplished at each station. Each task took approximately 10 minutes and the students rotated through the stations, completing all five performance tasks in one testing session.

The performance tasks emphasized scientific skills and processes embedded in the content areas of the physical, earth, and life sciences. The performance tasks field-tested in 1990 asked students to: (i) build a circuit out of the materials provided, predict the conductivity of various materials, test their conductivity, and record the results; (ii) create a classification system for a collection of leaves and explain the adjustments necessary when a "mystery leaf " is introduced into the group; (iii) perform a number of tests on a collection of rocks, record the results of the tests, and classify the rocks based on information provided; (iv) use the limited equipment furnished to estimate and measure a particular volume of water; and (v) perform a chemical test on samples of lake water to determine why fish are dying. The 1991 test, includes 35 tasks, which are randomly sampled and assigned to students.

In addition to the open-ended questions and performance tasks, 200 multiple-choice items were developed to measure understanding of scientific concepts in six categories: physical sciences, earth sciences, life sciences, scientific processes, safety and manipulative skills, and technology and society. Items were sampled to create 20 forms of the test, each containing 10 items.

According to the California State Department of Education,[46] teachers, students, and administrators throughout the state showed great enthusiasm for the "hands-on" component of the science test. One teacher wrote, "High value! Enjoyed it. Kids and teachers loved it. I will 'steal this test' to use as lessons and as teacher inservice."

New York. New York mandated science instruction in elementary schools in 1958, but concerns mounted in the early 1980s that this policy was not borne out in practice and that the instruction that did take place was often dominated by passive reading of textbooks. The Regents Action Plan of 1984 addressed the issue by mandating hands-on science instruction in all elementary school grades and by calling for statewide assessment of science education in grade 4. The assessment was intended not only to place greater emphasis on the importance of elementary school science instruction, but also to encourage the use of hands-on instruction and to provide direct guidance for the improvement of local programs.

The test was developed to improve local science instruction along the lines of the state syllabus, which emphasizes skills of group and individual problem-solving, investigation, science concepts, and students' attitudes toward science. There are two complementary components: a performance test and a multiple-choice test. The performance test administered in the spring of 1989 and again in 1990 asked students to: (i) measure objects to determine their mass, length, temperature, and volume; (ii) compare the absorbency of four different paper products and make a generalization about absorbency; (iii) create a classification system for five types of seeds; (iv) predict which of five materials will conduct electricity, test the predictions by using each material to complete an electrical circuit, and make a generalization about conductivity based upon this evidence; and (v) infer the characteristics of two objects sealed in a box.

The local school districts are responsible for judging their students' achievements on the performance test and for reporting the results to the state education agency. Training is provided by the state to one test administrator in each school. Distributions of scores for each school, not individual student scores, are reported to the state.

The multiple-choice test consists of 45 questions that address factual knowledge and concepts (29 items) and skills of inquiry and data interpretation (16 items). The test also is scored locally and the score distribution is reported to the state.

A significant feature of the New York assessment is its emphasis on pressuring schools to make use of their assessment results and on providing the support and technical assistance to do so. Results for each school become part of its comprehensive assessment report, which must be made public each year. As a part of the assessment, districts also may elect to survey students, parents, teachers, and administrators about their perceptions of the local instructional program in science in order to provide additional data for local program improvements. To facilitate local use of the test and survey data, each school is provided with an extensive set of worksheets in a *Guide to Program Evaluation K–4*, which includes step-by-step interpretation of each set of results. Each district is urged to have a science committee that conducts a program analysis based upon the results of the assessment and uses them to construct recommendations and an action plan to guide their implementation. Technical assistance is available from 93 regional elementary science mentors housed in boards of cooperative educational services and in large cities throughout the state.

Connecticut. Connecticut is preparing to assess science instruction in grades 9 through 12 in 1991–92. If the assessment is implemented along the lines currently planned, it will represent a new conception of large-scale assessment.

The assessment is based on *The Connecticut Common Core of Learning,* a set of state objectives for education. In science, the objectives emphasize problem-solving and investigation — developing hypotheses, designing experiments, drawing inferences, observing natural phenomena, and working with measuring and sensing devices — as well as understanding and applying basic concepts in scientific disciplines.[47] Intellectual curiosity and interpersonal relations are addressed as well.

The goals of the assessment are to model and cultivate these objectives and to measure their attainment. Reasoning that good assessment and good instruction should call upon students to think actively and collaboratively for an extended period of time, the assessment will consist exclusively of group tasks, lasting up to a week each. Students will be scored on the content of their solutions, the processes they used to arrive at them, the interpersonal and communication skills demonstrated by the group, and the manifestation of attitudes such as intellectual curiosity.[48] Here is a sample task for a high school physics class:

> Scientists often use small models to gain understanding so they can confidently analyze larger systems. You will be modeling the impact of a small car crashing into a large truck by using a dart gun and a small toy car. You will also use the following equipment: stop watch, meter stick, and velcro strips. Your goal is to find the velocity of the dart by shooting the dart horizontally into the back of the car and analyzing the coasting car and dart system.[49]

As a group, students calculate the velocity, then discuss and test at least two other methods for finding the velocity. Finally, students apply individually the concepts of the task to a problem involving a Corvette hitting a truck.

The assessment is not intended to place direct accountability pressure on individuals, schools, or districts. Accordingly, matrix sampling of schools, teachers, classes, and tasks is anticipated. Thirty schools will be chosen statewide, and science teachers at them will be sampled. One class period will be identified for each teacher selected and will be matched with one assigned task and two optional tasks.

Summary

We have greeted the threefold increase in state assessment of science education since 1978 with ambivalence. The use of promotion tests and norm-referenced tests has been found to have dangerous side effects on curricula and instruction. States that have developed tests to match their curricular framework have minimized some of these side effects and the tests can be successful in promoting the implementation of high-quality science curricula. Early indications suggest that performance tests may have very beneficial effects on classroom instruction and on students' integration of skills.

If states choose to test, then the principle that must govern the design of assessments must be this: the greater the consequences of the test to teachers, the more their instruction will resemble the test itself. This means that if a state chooses to use a test as a lever, it must attend closely to the design, quality, and instructional implications of the test.

Of the various state scenarios presented in this chapter, the use of state tests for promotion and graduation (as they are designed currently) has the most negative implications for instruction because they present a small number of multiple-choice items as the de facto goal of science education.

The use of commercial, norm-referenced, multiple-choice tests in science for general purposes of monitoring and accountability is also hard to justify. If the objective of these tests is to serve as a place-holder for science in the curriculum, they will serve that objective only if the consequences are commensurate with those perceived for reading and mathematics. This has not happened, but, if it should, the tests would fall short as a model for instruction.

Among states that have developed their own tests, California and Connecticut offer intriguing alternatives. California's test represents an enlightened attempt to couple accountability pressure with strong design. The content of the test and the use of applied performance measures suggest a test that will be "taught to." Yet, teachers cannot "teach to" it narrowly because the number of items and tasks is too great, and they are assigned to more than a dozen test forms per grade. Public reporting of scores focuses attention on achievement, and training and assistance are available to help schools improve their performance. Connecticut promotes a similar model of science instruction, but relies on persuasion and demonstration, rather than on pressure, to bring about results. Only time will tell what impact this approach will have eventually on instruction in Connecticut classrooms, but early outcomes are very encouraging.

Unfortunately, neither of these models is likely to have much appeal in those states that need to improve science education — states with limited resources to devote to a statewide effort to train teachers to administer and score a wide range of performance tasks. Nonetheless, it is our belief that such states will be served better in the long run by a limited effort in this direction than by reliance on existing commercial tests or current multiple-choice, minimum competency measures. States should consider developing consortia along the lines of the Connecticut Multi-State Assessment Collaborative Team (CoMPACT), where test development resources are pooled and tests are shared.

References and Notes

1. Kauffman, J. D., *State Assessment Programs: Current Status and a Look Ahead* (Boulder: Scholastic Testing Service, 1978).
2. Airasian, P. W., "State-mandated Testing and Education Reform: Context and Consequences," *American Journal of Education*, 95 (1987), 393–412.
3. Madaus, G., "Test Scores as Administrative Mechanisms in Educational Policy," *Phi Delta Kappan*, 66(9) (1985), 611–617.
4. Airasian, *State-mandated Testing*. See reference 2.
5. Gallup, A. M., "The 17th Annual Gallup Poll of the Public's Attitudes Toward the Public Schools," *Phi Delta Kappan*, 67 (1985), 35–47.
6. Baker, E., "Mandated Tests: Educational Reform or Quality Indicator?, in B. R. Gifford, ed., *Test Policy and Test Performance: Education, Language, and Culture* (Boston: Kluwer Academic Publisher, 1989), 3–24.
7. Anderson, B., and Pipho, C., "State-mandated Testing and the Fate of

Local Control." *Phi Delta Kappan,* **66**(3) (1984), 209–212.

8. Haertel, E., "Student Achievement Tests as Tools of Educational Policy: Practices and Consequences," in B. R. Gifford, ed., *Test Policy and Test Performance: Education, Language, and Culture* (Boston: Kluwer Academic Publishers, 1989), 25–50.

9. Kauffman, *State Assessment Programs.* See reference 1.

10. Anderson and Pipho, "State-mandated Testing." See reference 7.

11. Darling-Hammond, L., and Wise, A. E., "Beyond Standardization: State Standards and School Improvement," *The Elementary School Journal,* **85** (1985), 315–336.

12. Baker, "Mandated Tests." See reference 6.

13. Armstrong, J., Davis, A., Odden, A., and Gallagher, J., *The Impact of State Policies on Improving Science Curriculum* (Denver: Education Commission of the States, 1988).

14. National Governors' Association, *National Goals Adopted by the National Governors' Association* (Washington, DC: National Governors' Association, 1990).

15. Johnston, W., and Packer, A. H., *Workforce 2000: Work and Workers for the 21st Century* (Indianapolis: Hudson Institute, 1987).

16. Jacobsen, W., *Analyses and Comparison of Science Curricula in Japan and the United States,* Report for the International Association for Evaluation of Education Achievement (New York: Columbia University, 1986).

17 Hueftle, S., Rakow, S., and Welch, W., *Images of Science: A Summary of Results from the 1981–82 National Assessment in Science* (Minneapolis: Minnesota Research and Evaluation Center, 1983).

18. American Association for the Advancement of Science, *Science for All Americans: A Project 2061 Report on Literacy Goals in Science, Mathematics, and Technology* (Washington, DC: American Association for the Advancement of Science, 1989); reprinted as, Rutherford, F. J., and Ahlgren, A., *Science for All Americans* (NY: Oxford University Press, 1990).

19. National Council on Science and Technology Education, *What Science is Most Worth Knowing?* (Washington, DC: American Association for the Advancement of Science, 1987).

20. Honig, B., "Comprehensive Strategy can Improve Schools," *Education Week,* (1990, February 28) 31, 56.

21. Ibid., p. 31.

22. Armstrong et al., "The Impact of State Policies." See reference 13.

23. Ibid.

24. Tyson-Bernstein, H., "America's Textbook Fiasco: A Conspiracy of Good Intentions," *American Educator,* **12** (1988), 22–27, 39.

25. Shepard, L., "Why We Need Better Assessments," *Educational Leadership,* (1989, April) 4–9.

26. Gong, B., Lahart, C., and Courtney, R., *Current State Science Assessments: Is Something Better Than Nothing?*, paper presented at the annual meeting of the American Educational Research Association, Boston, April 1990.

27. American Association for the Advancement of Science, *Science for All Americans: A Project 2061 Report on Literacy Goals in Science, Mathematics, and Technology* (Washington, DC: American Association for the Advancement of Science, 1989); reprinted as, Rutherford, F. J., and Ahlgren, A., *Science for All Americans* (NY: Oxford University Press, 1990).

28. Bloom, B. S., ed.,*Taxonomy of Educational Objectives, Book 1: Cognitive Domain* (New York: Longman, 1954)

29. Ibid., 7.

30. There is a recent trend in several states to use both norm referencing and criterion referencing, which allows for the assessment of state science goals *and* comparisons to national norms.

31. See for example, Shepard, "Why We Need Better Assessments." See reference 25.

32. Armstrong et al., *The Impact of State Policies*. See reference 13.

33. Marsh, D., and Odden, A., "State-initiated Curriculum Reform in Elementary School Mathematics and Science Programs," paper presented at the annual meeting of the American Educational Research Association, Boston, April 1990.

34. Gong et al., *Current State Science Assessment,"* 6. See reference 26.

35. Darling-Hammond and Wise, "Beyond Standardization."See reference 11.

36. McNeil, L., "Contradictions of Control, Part 3: Contradictions of Reform," *Phi Delta Kappan*, **69** (1988), 478–485.

37. Darling-Hammond and Wise, "Beyond Standardization."See reference 11.

38. Shavelson, R., Carey, N., and Webb, N., "Indicators of Science Achievement: Options for a Powerful Policy Instrument," *Phi Delta Kappan*, **71** (1990), 692–697.

39. Shepard, "Why We Need Better Assessments." See reference 25.

40. Raizen, S., Baron, J., Champagne, A., Haertel, E., Mullis, I., and Oakes, J., *Assessment in Elementary School Science Education* (Washington, DC: National Center for Improving Science Education, 1989).

41. American Association for the Advancement of Science, Project 2061's *Science for All Americans*. See reference 18.

42. California State Department of Education, *Science Framework for California Public Schools* (Sacramento: California State Department of Education, 1990), 7.

43. Ibid., vii.

44. Ibid., 9.

45. Stations are worktables set up with science activities and instructions. Students perform each activity and are assessed.

46. Anderson, R., and Comfort, C., *New Directions in CAP Science Assessment* (Sacramento: California State Department of Education, 1990).

47. Connecticut State Department of Education, *The Connecticut Common Core of Learning* (Hartford: Connecticut State Department of Education, 1987).

48. J. Baron, personal communication, April 1990.

49. Greig, J., "Shot from Behind," in J. Baron, *Using Performance Tasks and Collaborative Learning to Assess Higher Order Thinking Skills in Science and Mathematics,* paper presented at the annual meeting of the National Council for Measurement in Education, Boston, April 1990, 39.

10

A National Standing Ovation for the New Performance Testing

Kathleen B. Comfort

> When I took this test, it didn't seem like a test. It was more like an experiment. I had so much fun, it was like having an hour of laboratory experiments. This test made me realize how much I really do know about science — *8th grade student, Fresno, California*[1]

Imagine that you, too, are an 8th grade student. The time of year is early spring: it is state testing week. As you think about the science portion of your state test, you imagine the usual multiple-choice questions and how they will look on your computerized test form. You might even rehearse some of these questions in your head: "What are the parts of a plant?" "What are the parts of a cell?" or "In what year was helium discovered?" You enter the classroom expecting to spend the next hour "bubbling in" answers to multiple-choice questions with a #2 pencil on your computerized test form. But instead of the usual neat rows of bare desks, you find tables set up around the classroom with familiar science materials on them. As you glance around the room, you notice hand lenses, balances, magnets, electrical circuit testers, microscopes, rocks, leaves, bottles of solutions, and even some live sow bugs. Your teacher assigns you to work either with a partner or by yourself. You are asked to design and conduct an investigation to see if sow bugs like wet or dry environments or if a smaller magnet is stronger than a larger one; you might even be asked to test lake water to try to discover why the fish are dying in a mountain lake or, perhaps, to investigate the properties of a black box. You think something may be wrong. This doesn't look like the usual state science test; it looks more like science class. It might even be fun!

Sure enough, students throughout the nation are discovering that these tests are not only fun, but they give students a chance to show what they have really learned. In science, as in other content areas, many states are moving "beyond the bubble," away from standardized multiple-choice tests, toward a more authentic form of assessment called performance assessment. Instead of finding #2 pencils and computer-readable test booklets before them, students from various states and many grade levels are finding a variety of performance assessments. These assessments — tests that ask students to show that they can write, think, solve problems, and do experiments — are reinforcing one very important message: if we want to make effective reforms in science education, we must discover what students actually know, how well they can think, and what they can do.

The Limits of Multiple Choice

In the past, large-scale assessment in most states has taken the form of standardized, multiple-choice tests that ask students to choose the correct response from among a set of alternatives and to "bubble in" their choice.[2] These standardized tests have become the main criteria that many schools use for making decisions that affect instruction and the quality of teaching. Unfortunately, these tests have many negative consequences: they tend to narrow the curriculum, encourage the teaching of disconnected, low-level facts; frustrate teachers and students; confuse the public; and undermine school improvement efforts.

Research shows that students learn what they are taught and that teachers teach what they are held accountable for. The conclusion that achievement testing profoundly affects the quality of U.S. education is inescapable.[3-5] It is not surprising, then, that both the general public and the professional community are dissatisfied with constant use of paper-and-pencil, multiple-choice tests, not only to assess student knowledge and understanding of subject matter but to establish accountability.[6,7] Current research also suggests that to achieve educational reform, assessment must be realigned to match reform. With such realignment, when educators "teach to the test," they will be teaching the "big ideas" advocated by reformers, and the quality of science instruction will be improved.[8-11]

States Move "Beyond the Bubble"

Moving "beyond the bubble" has become the major theme underlying the development of new science assessments in New York, California, Connecticut, Michigan, Texas, Kentucky, and several other states. The nation's governors have developed national goals for achieving scientific and mathematical literacy,[12] and national groups such as the American Association for the Advancement of Science have addressed the critical need for the reform of education in science, mathematics, and technology and provided recommendations to improve scientific literacy in the United States.[13] Six states — California, New York, Connecticut, Michigan, Texas, and Kentucky — have been leaders in responding to the national call for improved education in the sciences, and have developed state science frameworks (or state science objectives) and new, richer, more thought-provoking forms of assessment.

Instead of discriminating among bubbles on a multiple-choice test, students will be required to demonstrate their understanding of the concepts and processes of science and their ability to solve problems and carry out hands-on performance tasks. If the goal of instruction is to teach students the expressive side of communication and problem solving, tests must be patterned after instruction: they must emphasize production, creation, and performance — doing rather than discriminating.[14] The assessment tasks themselves must be intentionally complex, moving away from the unidimensional, factual purity of the past. Only tasks that call for integrating and applying learning can reinforce the highest goals of instruction. The tasks should assess multidimensional skills and present multisensory stimuli and multimodal response formats.[15] A variety of such innovative performance measures are currently under development, in field testing, or being implemented. They include performance tasks, open-ended or free response questions, science modules, science portfolios, and new conceptual-thematic, multiple-choice items. These new forms of assessment are expected to reinforce good curriculum practices, encourage thematic teaching, emphasize learning through hands-on experiences, and, most importantly, go beyond the level of recall and paraphrased recall to include activities in which students can use the concepts learned and relate them to other concepts.

Performance Assessment in the States

The six states cited above are at different stages in the development, field testing, and implementation of performance assessments in science. New York was the first to implement performance assessment statewide, and since 1989 all 4th grade students have been required to take the manipulative skills component of the New York State Elementary Science Program Evaluation Test (ESPET). Connecticut has been developing and field testing performance assessment tasks in both mathematics and science for grades 9–12 since 1989; the new performance tests will be implemented in 1991–92. In Kentucky, science assessment task forces have begun developing models for performance assessment measures in science after surveying educators and business and community leaders regarding graduation requirements for high school seniors. The task forces then developed goals or performance standards for graduation and created ten performance assessment task teams to develop and pilot tasks for grades 4, 8, and 12.

Texas science educators, working with the Texas Education Agency, have developed several new performance tasks for students in grade 9. These tasks will be piloted in fall 1991 and will eventually become part of the Texas Assessment of Academic Skills. The California Assessment Program, in conjunction with the California Science Project, the California

Science Implementation Network, and the California Secondary Scope, Sequence, and Coordination Project, conducted statewide field tests of performance assessment tasks in science at grade 6 (in 1990), and grades 5, 8, and 11 (in 1991). Participation in field tests is voluntary at present, but all schools will be required to implement performance measures as part of the state test in 1993.

New York

> Hey, mister, can we come back tomorrow and take the test again? — *4th grade student, New York public school*[16]

The 4th grade New York State Elementary Science Program Evaluation Test focuses on Levels I (K–2) and II (3–4) science content and skills from the New York Elementary Science Syllabus, and is administered in the spring to all 4th grade students in the state. The test was developed by the New York Bureau of Science Education and Elementary Test Development in conjunction with school administrators, classroom teachers, university staff, and subject supervisors. The purpose of this multifaceted assessment is to provide information to local and state educators on the effectiveness of the science program in each elementary building. Test results are used to evaluate programs, and no individual scores are provided.

Of the seven ESPET components, two are required and five are optional. One of the required components consists of a paper-and-pencil objective test (multiple-choice) of 45 questions, of which 29 items evaluate content and 16 evaluate skills. This untimed test takes about an hour to complete. The second required component is the manipulative skills test, which consists of five stations that can be set up in a variety of locations, including the classroom, school library, or cafeteria. Individual students rotate among the stations performing short investigations and recording responses on a student answer sheet.

At station one, students must determine: (i) the mass of a cup using pennies and a balance; (ii) the height of a cup using a ruler; (iii) the volume of water in a cup if poured to the blue line; and (iv) the temperature of the water in the cup using a thermometer. All four tasks involve measuring the physical properties of objects. Station two calls for students to make a prediction based upon performance of a scientific investigation that they have never performed. Using different examples of paper, marked A, B, and C, the student places a drop of water on each paper. After recording his or her observations the student must then predict what would happen if a drop of water were to be placed on a piece of unknown material (plastic). Station three requires students to create their own classification systems using an assortment of seeds and to state reasons for their sortings and classifications. At station four, students use an electrical device to test a variety of different materials, then write a generalization based upon the materials and their knowledge of electricity. Station five, billed as the most popular test, features a black box. Students must infer the properties of the materials inside the box within seven minutes. Students call it the "magic box" and insist on knowing what is inside.

The five optional components consist of student science attitude survey, student survey, teacher survey, administrator survey, and a parent/guardian survey. Surveys take between 15 and 30 minutes to complete.

Teachers administering the ESPET must attend two workshops organized by the New York Department of Education (NYDE) staff and curriculum development network. The first workshop addresses the administration and general features of the test, and the second provides training in scoring and analyzing test results. Usually one teacher per school is trained, and that teacher is responsible for administering the test to all 4th grade students in the school. Teachers trained in the scoring process score student papers and send results to the NYDE. The NYDE collects a subsample of student papers and scores these papers a second time. Scores from the manipulative tests are aggregated with scores from the other components and results reported to the field. Materials for the hands-on component of the test are packaged in kits containing five sets of each of the five tasks, and schools are required to purchase their own kits of testing materials. The cost of the equipment to test 35 students is approximately $125. The NYDE is responsible for mailing all written materials, including test administration manuals and student answer sheets, to the schools.

Connecticut

> Performance assessment encourages the feedback process. A kid tries something, sees its outcomes, and may try a modified approach . . . I saw sparkles in eyes that had never shown any. — *Gordon Turnbull, Connecticut*[17]

In 1987, the Connecticut State Board of Education approved a set of learning outcomes (standards) for high school graduates. Referred to as the Connecticut Common Core of Learning, these standards cover: (i) attitudes and attributes; (ii) skills and competencies; and (iii) applications and understandings. Classroom teachers, working with Connecticut Department of Education assessment consultants, are currently developing and piloting performance measures in mathematics and science at the high school level that incorporate all of these standards into an integrated assessment system. These new holistic assessment measures require students to demonstrate not only their understanding of content, but their ability to communicate and to work collectively as well. One of the most important points of the Connecticut project is that this new form of assessment has become an empowerment program. Performance assessment empowers both students and teachers and puts control of rating student progress back into the classroom — away from the machines and back into the hands of the teachers.[18]

Connecticut is creating many different formats for performance assessments. Some consist of tasks that take only 10 minutes for students to complete, while others require from 40 minutes to a week to complete. Some tasks are administered individually, others in small cooperative groups. One popular format consists of three parts. In part one, students work individually on a problem, thinking through the task and developing multiple solutions. For part two, students work in small cooperative groups to solve the problem. Each member of the group must contribute a solution and make an oral presentation. Part three requires students to finish individually. A similar task is selected, and individual students are asked to solve this problem based on information they learned from the group experience. Although students have only 45 minutes in which to construct their response in part three, a task of this nature may take from three days to one week to complete. In a science task students may be requested to recognize and solve problems as well as to develop hypotheses and design sets of experiments to test them.

Participation in the piloting of these new performance measures is not mandatory at this time. It is anticipated that in the future about 30 schools will be chosen to administer the new test.

Michigan

Suppose we decided that, instead of generating data that could be collected in an efficient and reliable manner, the primary goal of our assessment program was to help teachers understand their own students better? —*Charles Anderson, Michigan State University*[19]

In Michigan, personnel from the Michigan State Department of Education, Student Assessment and Science Curriculum Units, university staff from Michigan State University and the University of Michigan, and classroom teachers have been developing teaching materials, inservice education programs and new assessments that will be coordinated to the New Directions in Science Education for Michigan: State Objectives for K–12 Science. Drawing on recommendations from Project 2061's *Science for all Americans*,[20] the new Michigan science framework emphasizes objectives to promote scientific literacy for all students.

Michigan is just beginning to develop its first pilot of performance measures for grades 5, 8, and 11. These measures will include open-ended questions in which students will be required to construct their own responses. Data collected by researchers from Michigan State University indicate that these types of questions will help teachers to understand their own students better. Students will respond to open-ended questions with a short written paragraph. The responses will demonstrate to teachers students' common misconceptions regarding scientific concepts. Such shared misconceptions are not random, but are associated with sensible and often deep-seated patterns in students' thinking.[21] By gaining access to their students' thinking, teachers will gain perspective not only in what is being accomplished the classroom, but how students view the world. It is anticipated that teachers will be trained in these new assessment practices and will have a role in evaluating their own students' papers. Michigan is mandated to assess all students at grades 5, 8, and 11.

Texas

Some Texas educators believe that to encourage problem solving and higher-order thinking in science, we must have some kind of performance testing.— *Victoria Young, Texas Education Agency*[22]

Texas, like Michigan, also plans to develop alternative forms of assessment aligned to recommendations from their state-mandated science curriculum. The Texas Education Agency intends to design performance measures that will assess academic skills such as problem solving, critical reading, and higher-level thinking instead of basic skills. In science, these performance measures have been divided into four domains: (i) acquiring and classifying scientific data and information; (ii) communicating and interpreting scientific data and information; (iii) solving problems; and (iv) investigating solutions and application. Moreover, Texas is planning to develop performance assessment tasks that will integrate the life, earth, and physical sciences, as well as tasks that will integrate science and mathematics. Texas is currently mandated to assess annually all students in grades 3, 5, 7, and 9 and to administer an "exit level" test to all graduating seniors. In Texas, information is reported back at the student level. Performance tests are expected to be piloted at grade 9 in fall 1991, and they will eventually become part of the Texas Assessment of Academic Skills.

Kentucky

Some of our performance tasks will have a very practical orientation; other kinds of items will be mental problem-solving exercises. Practical performances might include tasks that address real-life situations like waste management or air pollution. Mental problem-solving exercises might ask students to design a menu for Superman. — *D. Ochs, Kentucky Department of Education*[23]

Kentucky is in an unusual situation since its system for funding schools was recently declared unconstitutional. Therefore, Kentucky has the rare opportunity to restructure its schools without barriers, rules, or regulations. A newly appointed task force will adopt themes, processes, and concepts from Project 2061's *Science for all Americans* and develop a state science framework for Kentucky's schools. Performance assessments will be designed to match the objectives from this framework.

As part of the restructuring process, a science and technology task force was established and charged with designing standards for high school graduation, as well as models for performance assessments in science. This task force is currently focusing on two goals: (i) development of a matrix of conceptual themes and (ii) development of performance tasks. The performance tasks will focus on scientific skills and processes embedded with the content areas of science. The tasks will be designed to focus on the practical applications of science to real-life problems such as waste management or air pollution, as well as on mental problem-solving exercises. Kentucky is planning to test a random sample of students in grades 4, 8, and 12. Test results will be used to evaluate school systems and individual schools. Individual student scores will not be provided.

California

> The value of my participation in this program has been to make me aware of the future of assessment, to open my eyes to the areas lacking in our science programs, and to remind me that the excitement and motivation generated in the children by this type of testing cannot be taken lightly. — *Teacher, Visalia Unified School District, California*[24]

The California Assessment Program (CAP) has begun introducing performance-based assessments intended to indicate directly what students actually know, how well they can think, and what they can do. In 1987 CAP introduced its first performance-based test, the grade 8 writing assessment in which students composed an essay in response to a prescribed topic. This assessment was extended to grade 12 in 1988. A change at grade 12 occurred in 1990 when open-ended mathematics items, for which students must construct their own solutions, were administered, scored, and reported. CAP has also initiated portfolio pilot projects in English/language arts and mathematics, in which students and teachers collect and evaluate an array of student work throughout the school year. The portfolio pilot will be extended to science in fall 1991. Language arts examinations integrating reading and writing were field tested for the first time in spring 1990. Students were asked to read and write a number of short responses, participate in group discussion, and compose an essay to provide evidence of comprehension. In history and social science CAP is developing assessments that will allow students to demonstrate breadth of learning as well as ability to clarify issues, recognize relationships, determine causes and effects, interpret evidence, and present an argument. Assessment modes under development include written tests, portfolios of student work, and integrated performance tasks.[25]

In science, these new assessments will be aligned to the "big ideas" recommended in the Science Framework for California Public Schools, Grades K–12.[26] These big ideas (for example, "Matter and energy interact at the microscopic level") are used to frame an entire K–12 science curriculum including life, earth, and physical sciences; technology; and environmental education. These disciplinary areas are connected, integrated, and interwoven by the themes of science (energy, scale and structure, patterns of change, stability, systems and interactions, and evolution). Scientific processes (observing, communicating, comparing, ordering and categorizing, relating, inferring, and applying), attitudes, and manipulative skills that contribute to this "thinking curriculum" are also incorporated into these new performance measures.

In the spring of 1990, CAP, in conjunction with the California Science Project (CSP) and the California Science Implementation Network (CSIN), conducted the first statewide field test of performance assessment in science at grade 6. Approximately 1,000 schools participated. The testing format consisted of five stations with one task per station. Each task took about 10 minutes, and the students rotated through the stations, completing all five performance tasks in one class period. The tasks challenged students to integrate manipulative and thinking skills with their knowledge of science content. Students engaged in hands-on activities and recorded their observations and conclusions on student response forms. For example, students had to build a circuit with materials provided, predict and test the conductivity of various materials, and record the results. Other tasks required students to observe, classify, sort, infer, detect patterns, formulate hypotheses, and interpret results. In each task, students had to move beyond the activity and develop a conceptual understanding of natural phenomena. A research pilot was also conducted during the field test in which a sample of schools administered two versions of the test: one version utilized the regular station approach; the other was administered by teachers utilizing one of six variants. The variants included time, delivery of instructions, and format (cooperative groups and diads).

In spring 1991, CAP, CSP, CSIN, and the California Secondary Scope, Sequence, and Coordination Project (SSC) conducted a second field test of performance-based assessment at grades 5, 8, and 11 in over 2,000 schools. Students at grade 5 worked with a partner to complete three performance tasks designed to assess a student's ability to: (i) analyze nonliving substances and determine their suitability for living things ("Spaceship U.S.A."); (ii) sort, classify, and give a rationale for their system of classification using a variety of fossils ("Fossils"); and (iii) determine the effect of wind direction on the motion of pinwheels, comparing the amount of work done by pinwheels of different sizes and relating this learning to other situations ("Wind Energy").

Students at grade 8 also worked with a partner to complete three performance tasks. The three tasks were designed to assess a student's ability to: (i) analyze common attributes of fossils and establish relational patterns ("Fossils"); (ii) perform a chromatography experiment and use the information obtained to apply to real-life investigations ("The Mystery Note"); and (iii) manipulate variables and determine the relationships among distance, force, and work ("Happy Trails"). The "Fossil" task at grade 5 spiraled through grade 8, building in complexity. Fifth graders were asked to sort a variety of fossils found in a dig into groups that appear related. They were asked to develop their own classification system and defend it by adding another fossil to one of the groups. By the 8th grade, students were asked to observe a group of fossils and develop a rationale for their development and differences over time. The grade 8 "Happy Trails" task integrated science and mathematics.

Students in grade 11 worked in triads to complete one integrated investigation. The objective of this task was to assess students' ability to use scientific processes and tools, communicate thinking processes, and demonstrate understanding of concepts that are connected and integrated among the sciences. Eleventh graders were asked to investigate a man's death in order to determine whether or not a crime had been committed. Various information sets had been provided that required the students to use a microscope to investigate properties of different hair samples, conduct a chromatography test to determine who wrote a note found on the body, and conduct soil profiles and pH tests on samples of soil found on the victim's shoe.

Overall, the performance component of the CAP science field test was enthusiastically received by teachers and students alike. Nearly all teachers reported that performance tasks provided students with a meaningful and exciting learning experience, and, in addition, yielded valuable assessment information unobtainable through traditional testing methods. Advocates of hands-on teaching and testing felt that this assessment provided them with the validation needed to implement innovative teaching and testing techniques at the local level. Teachers from textbook-driven systems felt especially motivated to seek out and bring more hands-on projects into their classrooms. A typical comment from a teacher:

> I learned so much about my students, especially the so-called "poorer" student. Some of them did quite well on this type of assessment.

Teachers reported that students who participated in the performance assessment exhibited high levels of curiosity and motivation and enjoyed tackling the questions posed by each investigation. Many wanted to repeat the tasks and explore the implications of their results in greater depth. Some typical student comments:

> The hands-on test was fun and terrific. It is a lot better than the ordinary test. It is easy to understand, and you do not have to memorize anything. The hands-on tests I did made me realize how much I really do know about science. I wish every test could be like this.

> The first test on electricity was neat. I had a little trouble building the electrical circuit and kept thinking over and over, "Benjamin Franklin was a genius." I couldn't believe I actually made an electric circuit that worked.

> I thought this kind of test is less stressful than a regular fill-in-the-bubble test. I would rather take hands-on tests than fill-in-the-bubble tests. You still must study, but not remember every answer. This way you do things by brain, not by memorization.

Scoring guides or rubrics were developed for all tasks. Student papers were scored by teachers at regional meetings held throughout the state, and samples of papers and scores are currently being analyzed by CAP. In the past, CAP reported group-level scores for all content areas. CAP is currently undergoing a restructuring process in which some results will be reported at both the individual and group level. Results from these two field tests will not be reported formally, but a report will be developed that will address all aspects of the performance test. All teachers administering the performance components of the test received two days of training: one day in test set-up and administration and a second day in scoring and analyzing results. Teacher training was conducted at 24 County Offices of Education.

CAP also conducted statewide field tests in spring 1990 and 1991 in thematic, conceptual, and integrated multiple-choice items and open ended questions. These field tests were also voluntary. Several districts that administered the performance component administered this component, as well. All test items and performance tasks used on the CAP tests are developed by science educators, including grade-level teachers, science supervisors, university representatives, and scientists on the CAP Science Assessment Advisory Committee.

Summary

If tests determine what teachers actually teach and what students will study for—and they do—then the road to reform is a straight but steep one: test those capacities and habits we think are essential, and test them in context. Make them replicate, within reason, the challenges at the heart of each academic discipline. Let them be authentic.— *Grant Wiggins*[27]

As is evident by the conscientious and enlightened efforts of many states, we are well on the road to amending our outmoded "empirical" traditions in assessment. By establishing new and more substantive means for measuring true intellectual accomplishment — what students actually know, how well they can think, and what they can do—we are closer to achieving the national goal of scientific literacy for all students.

References

1. Comfort, K., *New Directions in CAP Science Assessment* (Sacramento: California Department of Education, 1990); in revision to include 1991).
2. Anderson, R., *California: The State of Assessment* (Sacramento: California Department of Education, 1990).
3. Shavelson, R., Carey, N., and Webb, N., "Indicators of Science Achievement: Options for a Powerful Policy Instrument," *Phi Delta Kappan,* **71** (1990), 692-697.
4. Resnick, L. B., and Resnick, D., "Assessing the Thinking Curriculum: New Tools for Educational Reform," in B. R. Gifford and M. C. O'Connor, eds., *Future Assessments: Changing Views of Aptitude, Achievement, and Instruction* (Boston: Kluwer Academic Publishers, in press).
5. Walker, D. F., and Schaffarzick, J., "Comparing Curricula," *Review of Educational Research,* **44** (1)(1974), 83-111.
6. Frederiksen, J. R., and Collins, A., "A Systems Approach to Educational Testing," *Educational Researcher,* **18** (9) (1989), 27-32.
7. Shavelson, R., Carey, N., and Webb, N., "Indicators of Science Achievement: Options for a Powerful Policy Instrument," *Phi Delta Kappan,* **71** (1990), 692-697.
8. Wiggins, G., "A True Test: Toward More Authentic and Equitable Assessment," *Phi Delta Kappan,* **70** (1989), 703.
9. Raizen, S., and Kaser, J., "Assessing Science Learning in Elementary School: Why? What? and How?" *Phi Delta Kappan,* **70** (1989), 718.
10. Anderson, C. W., "State Assessment Programs and Improving Instruction," paper presented at Forum 90: Assessment in the Service of Instruction, American Association for the Advancement of Science, Arlington, Virginia, November 1990.
11. Hein, G., *The Assessment of Hands-On Elementary Science Programs* (Grand Forks, North Dakota: Center for Teaching and Learning, University of North Dakota, 1990).
12. National Governors' Association, "National Goals" (Washington,

DC: National Governors' Association, 1990).

13. American Association for the Advancement of Science, *Science for All Americans: A Project 2061 Report on Literacy Goals in Science, Mathematics, and Technology* (Washington, DC: American Association for the Advancement of Science, 1989); reprinted as, Rutherford, F. J., and Ahlgren, A., *Science for All Americans* (NY: Oxford University Press, 1990).

14. Carlson, D., "The Relevance of Data Gained from Assessing State Schools: The California Example," paper presented at the symposium, Large-scale Assessment in an International Perspective, Deidestreim, Germany, 1989.

15. Ibid.

16. Reynolds, D., "Science Assessment in the State of New York," paper presented at Forum 90: Assessment in the Service of Instruction, American Association for the Advancement of Science, Arlington, VA, November 1990.

17. Baron, J., "How Science Is Tested and Taught in Elementary School Science Classrooms: A Study of Classroom Observations and Interviews," paper presented at the annual meeting of the American Educational Research Association, Boston, April 1990.

18. Rindone, D., "Science Assessment in the State of Connecticut," paper presented at Forum 90. See reference 10.

19. Anderson, C., "State Assessment Programs and Improving Instruction," paper presented at Forum 90. See reference 10.

20. American Association for the Advancement of Science, *Science for All Americans*. See reference 13.

21. Anderson, C., "Assessing Scientific Understanding," paper presented at the annual meeting of the American Association for the Advancement of Science, New Orleans, February 1990.

22. Young, V., "Science Assessment in the State of Texas," paper presented at Forum 90. See reference 10.

23. Ochs, D., "Science Assessment in the State of Kentucky," paper presented at Forum 90. See reference 10.

24. Comfort, K., "Moving 'Beyond the Bubble' in Science Assessment," paper presented at Forum 90. See reference 10.

25. Anderson, R., "The New CAP Assessments," *California Curriculum News Report,* **16** (3) (1991), 4-6.

26. California State Department of Education, *Science Framework for California Public Schools* (Sacramento: California State Department of Education, 1990).

27. Wiggins, G., "A True Test." See reference 8.

11

Performance Testing and Science Education in England and Wales

Wynne Harlen

Before passage of the Education Reform Act for England and Wales in 1988,[1] there were no legal requirements for curricula and assessment except that religious education had to be included in the education of all pupils. Although curricular decisions were the legal responsibility of the local education authority, in practice they were left to the individual schools. Science was taught, either as a combined science course or as separate courses in biology, physics, and chemistry, to all children from the ages of 11 to 14. Outside these ages, the extent of science education varied considerably from pupil to pupil and school to school.

The provisions of the Education Reform Act began to be put into operation in schools in 1989. In its first part, the Act created the National Curriculum comprising 10 foundation subjects plus religious education, all of which have to be taught to pupils throughout the compulsory years of schooling; that is, from the ages of 5 to 16 (apart from the foreign language which is not taught before the age of 11). Details of the National Curriculum will be considered later. At this point, it is relevant to note that science (as a whole subject as opposed to separate courses in biology, chemistry, and physics) is included in the curriculum as a "core" subject; that is, one of the foundation subjects accorded particular importance. This status has significance in relation to assessment.

Before examining the most recent developments in curricula and assessment, however, it is relevant to recount some of the recent history of science education in England and Wales. Significant changes have taken place since the curriculum reform movement began in earnest in the 1960s, and they have played an important part in preparing the ground for acceptance of the current legislation. Indeed, it is claimed that the content of the National Curriculum for science was framed to reflect existing good practices in science education and, to that extent, the National Curriculum was welcomed by those concerned to spread these good practices more widely. The introduction of National Assessment at the same time as the National Curriculum could not be said to reflect existing practices, but in its design and implementation, the National Assessment has drawn extensively on the developments in assessment that have taken place during the past 15 years. In tracing these developments, which have been inspired as much by political as by educational motives, it is appropriate to consider the primary and secondary phases separately and then to examine some issues concerning recent activities in both phases.

Science and Its Assessment at the Primary School Level

The term "primary" is used in England, Wales, and Northern Ireland to describe the years of schooling for children from their point of entry to school, at the age of 5 years, to the age of 11 years. Children may not be in schools described as "primary" throughout this time because, in some areas, there are separate infants' schools (5–7 years) and junior schools (7–11 years), and in others, there are first schools (5–8 or 5–9) and middle schools (8–12 or 9–13). In Scotland, primary education extends to the age of 12. Thus, in the United Kingdom, the word "primary" is used where other countries may use the word "elementary." Primary school teachers, by and large, teach all subjects and do so in regular classrooms, not in specialist rooms. Science is taught using everyday, rather than specialized, science equipment.

Prior to the legislation of 1988, which applies to England and Wales only, there had been no central control of curricula in the United Kingdom since the beginning of the century. Until the end of the 1950s, however, primary school curricula were driven by the examination that all pupils had to take at the end of the primary phase (called the 11+ examination) and that tested mainly English and arithmetic. The purpose of this examination was to select pupils for entry to the different types of secondary schools, which children attend from the ages of 11 through 16 or 18. With the introduction of comprehensive secondary schools, this examination died away in all but a few staunchly conservative areas of England, and broadening of the primary curriculum became possible in the 1960s.

The introduction of science was a part of this broadening. It began in the 1960s in a few schools and was stimulated by curriculum development projects that paralleled such projects as Science Curriculum Improvement Study, Elementary Science Study, and the American Association for the Advancement of Science's Science — A Process Approach. The projects in the United Kingdom were the Nuffield Junior Science Project, 1964–66,[2] and the Science 5/13 Project, 1967–75,[3] both essentially process-based and child-centered, in line with the prevailing primary education philosophy of the 1960s.

Both the Nuffield Junior Science Project and Science 5/13 provided teachers' guides but no materials for pupils, and advocated that teachers should pursue the ideas suggested by their pupils' questions so that the activities would be suited to the pupils' interests, locality, and stages of development. Teaching science in the way proposed by these projects was demanding on teachers, most of whom, as in other countries, lacked confidence, knowledge, skill, and motivation in this area of the curriculum. Little or no attention was given to pupil assessment even though the notion of "matching" activities to children's development was implicit in the projects' philosophies. Later projects and publications provided materials that had greater structure and were less dependent on teachers' decisions, many comprising sets of work cards for pupils. One project included specific assistance for teachers to help them assess their pupils' development and record their progress.[4] It also argued that assessment is a part of teaching, an idea not taken seriously until the end of the 1980s.

Despite these efforts in the 1960s and early 1970s, a survey of primary schools by Her Majesty's Inspectors of Schools deplored the low incidence of what they judged to be good practice in science.[5] A later survey of recordkeeping in primary schools revealed a general absence of attempts to assess and record children's work in science.[6]

In the 1980s, various factors combined to renew concern to implement the teaching of science at the primary level. Among them was research showing that children's ideas about the world around them are being built up in their early years whether or not they are taught science.[7,8] Many of these ideas are nonscientific and obstruct learning in science at the secondary school level. They are constructed by children from everyday experiences without the benefit of scientific thinking and systematic testing. From this research, it was inferred that there is a role for science education in primary schools because it will help children, through firsthand exploration, to make sense of the world around them in terms of ideas that are related more closely to scientific ones.

Research also had revealed the decline in interest and in positive attitudes toward science at later ages, which it was felt could be addressed by earlier involvement in scientific activity. Finally, educators were concerned that children's early education was excluding an increasingly important part of daily life.

These concerns led to a firm recommendation by the Department of Education and Science, in the first of a series of policy statements on the curriculum, that "All pupils should be properly introduced to science in the primary school."[9] At the same time, financial support was made available for inservice courses in primary school science and for local education authorities to support teachers through a scheme of "advisory teachers" in science. Primary schools were required to develop policy statements for science and to create a post of "coordinator for science" (a teacher who has responsibility for science in the school, but exerts it by helping other teachers, not by teaching science for them).

In responding to these concerns and requirements, teachers were preoccupied with establishing programs of activities in science and still gave little attention to assessment. To some extent, it is logical to wish to have something to assess before devoting energy to assessing it, but this ignores the role of assessment in teaching. More probably, the unwillingness to be concerned with assessment in science was a reflection of the persistent lack of commitment to science as a basic and important part of the curriculum and the absence of a framework of development, which is a necessary prerequisite for assessment.[10] It seemed to require the force of law to change this situation.

To some extent, the lack of interest in assessment can be regarded as an advantage. It meant that there was no market for the "quick and easy to administer" test based on multiple-choice items and assessing mainly recall, and there was no clamor for standardized measures to be created. Had these existed, they might well have been ignored in the same way as the checklists for observation that have been offered in various local education authority guidelines. Instead, the developments in assessment in primary science have been for research purposes, where the validity of the instruments and procedures, rather than the ease of their use in the classroom, was paramount. I shall return later to the developments in assessment that were taking place during this time and that fed eventually into the creation of a system of national assessment.

Science and Its Assessment at the Secondary School Level

Science has been included in secondary education for most of this century. It is significant that the Association for Science Education had its origin in the Science Masters Association, founded in 1902. This latter association, created for the purpose of promoting the teaching of natural science in secondary schools, was followed by the Association for Women Science Teachers in 1922. The two associations combined into the Association for Science Education in 1963, at the same time as primary teachers were first admitted to membership in the organization. The membership of this association reflects the changing pattern of science education. In 1918, 82 percent of its members were public school masters (that is, they taught at fee-paying private schools), while in 1939, this proportion had fallen to 12 percent,[11] reflecting the increase in the number of science teachers at state-funded secondary schools.

Unlike science education in the primary schools, science curricula in the secondary schools had been dominated by external examinations. Until recently, these were the General Certificate of Education "O" level and

"A" level examinations (taken at the ages of 16 and 18, respectively). While there was no national curriculum before 1988, the syllabi of the various public examination boards prescribed closely what was taught. These syllabi, certainly until the early 1970s, were heavily laden with information and the assessment was related mainly to recall and some application of factual knowledge and principles. The exceptions were those syllabi tied to the new programs devised by the projects funded by the Nuffield Foundation and by the Schools Council. However, despite the prominence given to these programs, the extent of their substantial use, as reported in a survey by Her Majesty's Inspectors of Schools in 1979, was less than 20 percent, and only about 6 percent of schools were actually following the programs.[12] The same survey showed that, while almost all pupils in the first three years of secondary school were taught all three sciences, separately or in an integrated program, in the fourth and fifth years large proportions of pupils did not study all three and girls were studying substantially less science than were boys.[13]

Examinations are important in secondary schools because they are the basis of qualification for entry to higher education. In performing this function, they have tended to emphasize the academic at the expense of the applied aspects of the subject; in science, this has emphasized the acquisition of knowledge rather than the processes of science. The changes toward a more process-based approach in the teaching of science, urged by curriculum development projects in the 1960s and 1970s, were seen by teachers for many years as incompatible with success in examinations. So ingrained was this view that when examinations did begin to change so that pupils were required to use, rather than simply to recall, their knowledge, teachers were slow to take the advantage offered. In particular, the tests that they devised for in-school examinations continued to emphasize factual knowledge.

However, in 1988 radical changes were introduced into the examinations taken at the age of 16. The General Certificate of Education "O" level and the Certificate of Secondary Education (taken by less academic pupils) examinations were replaced by a combined system known as the General Certificate of Secondary Education. The assessment objectives for the General Certificate of Secondary Education science examination, published three years before the examination was introduced, included such statements as:

> Candidates will be expected to show that they have the ability to: (i) observe, measure, and record accurately and systematically; (ii) follow instructions accurately for the safe conduct of experiments; and (iii) draw conclusions from, and evaluate critically, experimental observations and other data, etc.[14]

That the assessment of these abilities could be contemplated was due to the developments in assessment procedures and instruments that had been taking place in the late 1970s and early 1980s. Groups of teachers working with the examination boards had been responsible for some of these developments, but a major impact was certainly made by the government's Assessment of Performance Unit (APU). This covered both primary and secondary school science (in addition to English language and mathematics) and is considered in some detail below.

Contributions of the Assessment of Performance Unit

The Assessment of Performance Unit (APU) was set up in the Department of Education and Science in 1974 with the charge to promote the development of methods of assessing and monitoring the achievement of children at school, and to seek to identify the incidence of underachievement.[15] In fulfilling this charge, the unit first set up working groups of subject specialists in six curricular areas and subsequently contracted out to groups of educational researchers the tasks of creating the assessment instruments and procedures. The assessment instruments for science were developed at Chelsea College (now King's College), London University, and Leeds University. Other contracts, for language and mathematics assessment, were placed with the National Foundation for Educational Research, which also was given responsibility for conducting surveys using the assessment instruments.

The educational and political contexts in which the APU was set up are well documented by Gipps and Goldstein, who carried out an independent evaluation of the APU in 1980–82.[16] Briefly, the educational context was an increasing discontent in the 1970s with the education in state-funded schools. To some extent, this was a reaction to the changes in the curriculum of the primary schools in the 1960s, which followed the end of the 11+ examination. The broadening of primary school curricula led to fears that standards were falling. In fact, there was no evidence to support this, but the absence of the 11+ examination, which had served as a quality control measure, allowed critics of the child-centered approach in primary schools to claim that children were not being prepared adequately for secondary education. In addition, considerable publicity was given to the situation at one Inner London school where parents complained of low standards, ascribed to the progressive methods being implemented by the senior staff of the school (the William Tyndale Junior School in Islington). There also was some research purporting to show that progressive methods produced lower performance than did more traditional methods of teaching. Although a reanalysis of these results showed that these conclusions were incorrect, the damage was done. Local education authority officials began to look with suspicion on child-centered education and some made moves to introduce local testing programs in an attempt to ensure that standards in language and mathematics were maintained.

The economy in the 1970s was also a factor in creating a climate in which people were more sympathetic to monitoring standards. The crisis following the sudden increase in oil prices required economies in all areas, and questions were asked about the value obtained for the £6 billion per year being spent on education. In some ways, the situation was similar to that in the United States during its economic crisis in the early 1970s, when the notion of "accountability" in education came to prominence. In the United States, accountability was linked closely with testing, and this link was maintained in the debate in England. Although the APU was not a direct result of this accountability movement, but was more specifically inspired by a concern for standards, the climate created by the call for accountability was favorable to the introduction of testing.

The APU had support from both the political left and right. It was being planned by the Department of Education and Science when there was a Conservative government (and when Margaret Thatcher was Secretary of State for Education), but it was put into place by a Labour government. Both sides wanted to show concern for standards and to respond to the popular call for greater "consumer control over education."[17] When the government changed again in 1979, no alterations were made in the plans and procedures of the APU.

The work of the APU in science occupied the period 1975–89 and had a profound impact on assessment methods and science curricula. Five annual surveys, each of about 1.5 percent of the population of pupils aged 11, 13, and 16, were carried out from 1980–84. The surveys were conducted in England, Wales, and Northern Ireland and, in most of them, two kinds of practical tests were administered as well as written tests. The administration of the practical tests involved visits by trained testers (teachers released from their schools for a period of two or three weeks) who carried with them all the equipment required for testing.

The need for both practical and written tests stemmed from the decisions of what was to be tested in science. These decisions were a major focus of the work of the assessment teams between the start of the project and the first surveys. There being no national statement of the aims and objectives of science education at that time, the creation of a framework for what should be assessed — in order to survey national performance — required a consensus about the aims of science education. There was wide consultation with representatives of all interested bodies, particularly the Association for Science Education. Eventually, the following set of categories and subcategories for science performance was agreed upon:

(1) *Using symbolic representations*: reading information from graphs, tables, and charts expressing information as graphs, tables, and charts;

(2) *Using apparatus and measuring instruments:* using measuring instruments, estimating physical quantities, and following instructions for practical work;

(3) *Using observations*: making and interpreting observations;

(4) *Interpretation and application*: interpreting information presented and applying scientific concepts;

(5) *Planning investigations:* planning parts of investigations and planning entire investigations; and

(6) *Performing investigations:* carrying out entire investigations.[18]

To some extent, the identification of what was to be assessed in the terms of these process-laden categories was a reflection of the values of those involved (many of whom had been active in the development of new science programs in the previous decade). The categories also defused fears that the introduction of national monitoring would bring about a reversal of the trend away from "science as factual knowledge." It was necessary to agree upon a list of the concepts whose application was to be tested in category 4, but this was not controversial at the secondary level, where the examination syllabi already had defined a core of concepts to be taught. At the primary level, there was less harmony and some opposition to the idea of defining anything that would seem to dictate the content of science. This opposition was not entirely overcome until examples of the test items were made public, and the emphasis upon *application* of knowledge was substantiated, and the level at which the concepts were interpreted was exemplified.

In effect, the APU framework and lists of concepts constituted a national curriculum, though without the force of law. Ten years later, when the National Curriculum for science was being drawn up, the work of the APU was an obvious starting point. The fact that it was a framework for *assessment* meant that it was particularly valuable for the development of a national curriculum in which assessment was an integral part.

Considerations of validity in assessing performance in the APU categories led to decisions that categories 2, 3, and 6 had to be tested in a practical mode because the interest was in the processes of using equipment, making observations, and performing investigations, not the theoretical knowledge of how to do these things. Expensive practical testing had not been anticipated by the officials who made the contracts for the science surveys; to include it, the budget had to be renegotiated. That something of the order of doubling the budget was agreed upon indicates the strength of political will to ensure that the surveys would go forward, taking the profession along with them.

A matrix sampling design was used in the science surveys. Banks of items were created for each of the three age groups and for each subcategory except category 6. All of the questions went through a series of stages in development, which included scrutiny by teachers and validation by groups of science educators as well as field trials on about 200 pupils.[19] The items were labeled to indicate their characteristics and stored in a computer bank. Random samples of questions were drawn from these banks for each survey and arranged in packages that required about 40 minutes for answering by 11-year-olds and about 60 minutes by older pupils. Random samples of pupils were drawn by a two-stage sampling process (a random sample of pupils within a random sample of schools), and packages were assigned to them randomly. Thus, any one pupil was tested for one hour or less, but several packages could be used for assessing each subcategory. In the case of written questions, a total sample of about 60 questions for each subcategory was used in the later surveys.

The reason for using a large number of questions to assess a subcategory was the realization, from the trials and from the first survey results, that even with questions that assess process skills and do not depend on subject matter knowledge, a considerable effect on performance is exerted by the context (or setting) within the question. For example, Figure 1 shows a question used to assess the interpretation subcategory, in which the required information about the meaning and significance of tree rings is presented.

This item is one in a family of similar questions about situations where patterns exist. In each case, information is given and does not have to be recalled. However, it is evident from the trials and the survey findings that pupils respond differently to questions that appear to make the same process demand but concern different subjects. Some of these differences are gender-related where, for example, a context seen by girls as pertaining to male interests deters them from attempting it. In designing items, efforts were made to avoid obvious biases associated with gender and with cultural and social backgrounds, but it is not possible to eliminate them entirely. In any case, individual differences remain quite apart from those that correlate with characteristics reflecting a pupil's background.

Figure 2 shows a written question used to assess "Planning parts of investigations" and Figure 3 is a question used for assessing "Making and interpreting observations."

The APU surveys minimized the influence on results of the process-context interaction by averaging results over a large number of questions with different subject matter. However, this approach to the problem requires questions that are in written form or at least that have a written product because it would not be feasible to test pupils in a large number of different contexts where their performance had to be observed. Thus, although these questions are ingenious, their use outside the national surveys is limited.

One consequence of the requirement that items should be selected

Item from APU Survey of Pupils Aged 11, Assessing "Interpretation"

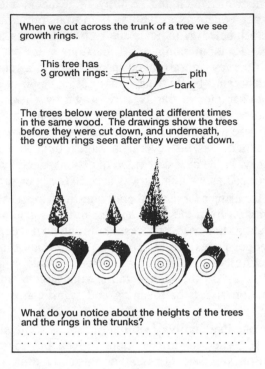

When we cut across the trunk of a tree we see growth rings.

This tree has 3 growth rings: —— pith —— bark

The trees below were planted at different times in the same wood. The drawings show the trees before they were cut down, and underneath, the growth rings seen after they were cut down.

What do you notice about the heights of the trees and the rings in the trunks?
. . . .
. . . .

Marking scheme

Response	Marks
Patterns relating all data e.g. "The taller the tree the more rings there are."	3
Separate statements which mention all the data but not as a relationship e.g. "the tallest tree had the most rings, the next tallest had the next most…etc"	2
Partial reference to a correct pattern e.g. "the biggest has most rings;" "the sizes and rings go together"	1

(DES, 1983a, p. 25)

Figure 1

Item from APU Survey of Pupils Aged 11, Assessing "Planning Parts of Investigations"

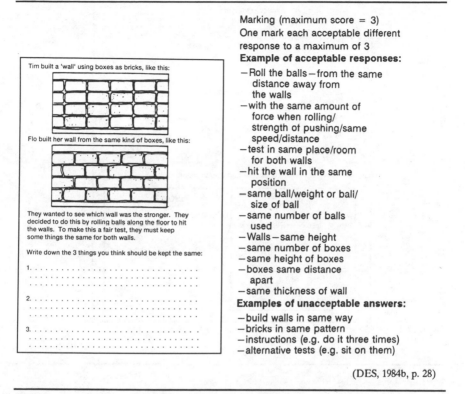

Tim built a 'wall' using boxes as bricks, like this:

Flo built her wall from the same kind of boxes, like this:

They wanted to see which wall was the stronger. They decided to do this by rolling balls along the floor to hit the walls. To make this a fair test, they must keep some things the same for both walls.

Write down the 3 things you think should be kept the same:

1. . . . .
. . . .
. . . .

2. . . . .
. . . .
. . . .

3. . . . .
. . . .

Marking (maximum score = 3)
One mark each acceptable different response to a maximum of 3

Example of acceptable responses:

- Roll the balls - from the same distance away from the walls
- with the same amount of force when rolling/ strength of pushing/same speed/distance
- test in same place/room for both walls
- hit the wall in the same position
- same ball/weight or ball/ size of ball
- same number of balls used
- Walls - same height
- same number of boxes
- same height of boxes
- boxes same distance apart
- same thickness of wall

Examples of unacceptable answers:

- build walls in same way
- bricks in same pattern
- instructions (e.g. do it three times)
- alternative tests (e.g. sit on them)

(DES, 1984b, p. 28)

Figure 2

randomly from a bank is that each item has to be self-contained, presenting all the information required about the content to enable the question to be answered. Pupils were asked to respond to a series of questions, all set in different contexts and requiring reading and willingness to understand what is being asked. The considerable dependence on reading ability for pupils of age 11 was evident in the results from those who took a short "close" reading test as well as a science test as part of the third survey. The results showed correlations of the order of between 0.52 and 0.60 for all tests involving reading and writing, even if a part of the test depended on handling and observing materials.[20] Thus, the validity of written science tests for primary school children has to be regarded as limited.

The two kinds of practical tests used in the surveys were described as "group" and "individual" practical tests. In the group tests, used for categories 2 and 3, a series of eight short practical tasks — constituting a selected package of such tasks from the item bank — was set up, each with

Item from APU Survey of Pupils Aged 11, Assessing "Observation"

Look carefully at the seeds in front of you.
Smell each different type of seed.

Find a way to divide the seeds into two groups.
You can do this in any way you like so long as
you follow the rules.

RULES

1. Something must be the same about all the seeds
   in each group.

2. All the seeds must be in one group or the other.

3. There must be at least two different types of
   seeds in each group.

a) Write down the letters of the seeds in your two groups.

Group 1	Group 2
Letters	Letters . . . .
b) These are all the same because	These are all the same because . . . .

(DES, 1984b, p. 20)

Figure 3

instructions. Eight pupils worked at one of the tasks for a given time, then they moved on to another task until all of the items had been attempted. Some of the tasks required the observer to record actions using a simple checklist, but there could be only one such item in a set of eight. In all cases, the pupils made a written record in their answer booklet and, in some cases, left a physical "product" (such as a weighed quantity of sand), which was labeled and checked later by the test administrator.

Individual practical tests were used to assess category 6, the performance of complete investigations. There was no large bank of investigations and sets of three tests (or two, depending on length) were administered to the pupils drawn in the subsample. These items were considered to be a particularly important part of the survey for several reasons. Their inclusion, despite all the difficulties of arranging for a random sample of pupils across the country to be tested individually by a visiting tester, provided a strong message about the value of practical investigative activity to science education. Further, activity of this kind had not been assessed before, and, indeed, the assumption often had been made that it could not be assessed.

Thus, considerable advances were made in assessment methodology. The recognition that practical investigations could be assessed reliably meant that they could be excluded from the curriculum on these grounds no longer. Also, a great deal was learned about the nature of investigation and the ways in which the processes of science and conceptual understanding are involved in it. A striking finding was the large difference in performance of separate skills in categories 1 through 5, as compared with the use of these skills in category 6 investigations. For example, pupils were capable of using measuring instruments correctly, but failed to make measurements when it was appropriate to do so in the context of an investigation.

There was also a useful outcome from the involvement of teachers as testers in the practical investigations. During the surveys, a test administrator visited each school in the sample and spent up to an hour with each pupil in the selected sample. The child carried out two (sometimes, three) investigations while the administrator made observations using a checklist. A discussion of the results at the end of the activity added further information including, for instance, the willingness of the child to reflect critically on what he or she did.[21]

Those involved in the administration and, to a smaller extent, those who read carefully the accounts of the activities and how they were assessed, noted the following benefits: (i) the importance of designing an activity so that children have opportunities to use various process skills, (ii) the ability to identify the key aspects of performance from the points they were asked to observe, (iii) the value of learning how to formulate questions (used at the end of the investigation) that give access to mental processes not directly observable in behavior, and (iv) the ability to recognize behaviors that indicate that certain process skills are in action. Although, in the course of normal work, teachers are not able to devote the time required for observing individuals in the same detail as in the APU, this work showed convincingly that it is possible to assess the process of performing investigations, and not only the outcome.

At first, it was feared that the APU surveys would have a restricting influence on the curriculum, giving emphasis to those objectives that could be tested readily, but in the event, it has been acknowledged that any influence has been positive rather than negative.[22] The light matrix sampling design[23] was soon recognized as rendering meaningless any notion of "teaching to the test." Moreover, the range of assessment methods used was seen to be far greater and more sophisticated than those used by teachers. Even the written tests were of a wide range; very few were of the conventional, multiple-choice kind, and the surveys demonstrated that short answers to open-ended questions, extended writing, and children's drawings could all be scored reliably.[24]

Certain aspects of the approach to assessment were adapted for use by primary school teachers during the work of an action research project that involved assessing children's performance of process skills and feeding this information to teachers. In order to obtain measures of children's process skills, several tests were created, each dealing with a particular topic related to an overall theme. This theme is introduced into the classroom as a topic for regular activities. The topic includes other activities besides those that lead to an assessment. Within the theme, questions are posed relating to the skills being assessed. Because of the common subject matter, the amount of reading required is reduced, and·children are not mentally leapfrogging from one context to another when passing from one question to the next. This test is described by Schilling et al.[25]

The National Curriculum and Assessment

The idea of a national curriculum, which at one time would have been anathema to the English educational system, gained ground on a broad front during the 1980s. A description has been presented here of how the APU created a consensus view of what the aims of science education should be. Meanwhile, Her Majesty's Inspectorate of Schools had been issuing a series of documents relating to policies in various subjects.[26] For many in the science education community, the acceptability of an imposed curriculum depended entirely on who imposed it and on what form it took. Thus, considerable attention was paid to the proposals made by the Department of Education and Science (DES) and the Welsh Office (WO) (the Welsh counterpart of the DES) in *The National Curriculum 5–16: A Consultation Document*, issued in 1987.[27]

The proposals included the notion of three "core" subjects — English, mathematics, and science — and seven other "foundation" subjects — technology, history, geography, art, music, physical education, and a modern foreign language (the modern foreign language is studied from the ages of 11 through 16 only). (The Welsh language is an additional subject in certain schools in Wales.) Together with religious education, these 10 subjects form the National Curriculum. Consultation took place on a very short time scale and brought about some changes in detail, but the broad composition of the curriculum remained unaltered.

Within a month of publication of the consultative document, working groups were set up by the Secretaries of State for England and Wales to devise curricula for the "core" subjects. The working groups were required to

establish clear objectives — attainment targets — for the knowledge, skills, understanding, and aptitudes which pupils of different abilities and maturity should be expected to have acquired at or near certain ages. To promote these objectives, the Government wishes to establish programmes of study for the subjects, describing the essential content which needs to be covered to enable pupils to reach or surpass the attainment targets. Taken together, the attainment targets and programmes of study will provide the basis for assessing pupils' performance in relation both to expected attainment and to the next steps needed for the pupils' development.[28]

As in the case of the APU, the membership of these working groups was crucial to the kinds of curricula devised and, once again, the membership was weighted with those at the forefront of science education research and curriculum development.

It is clear from the terms of reference that an "objectives" model for curricula was imposed and that assessment was to play an important part in the educational process throughout, not merely at the age of 16 (the end of compulsory schooling). At the same time as the subject working groups were set up, the Secretaries of State also formed a task group to make recommendations on the procedures and possible methods for assessing and reporting pupils' performance. This task group recommended a combination of continuous assessment by teachers (which has become known as Teacher Assessment) and externally devised tests (known as Standard Assessment Tasks) that would serve formative, summative, and evaluative purposes:

> It is possible to build up a comprehensive picture of the overall achievements of a pupil by aggregating, in a structured way, the separate results of a set of assessments designed to serve formative purposes.... It is realistic to envisage, for the purposes of evaluation, ways of aggregating the information on individual pupils into accounts of the success of a school, or local education authority, in facilitating the learning and achievements of those for whom it is responsible...."[29]

Assessment is criterion-referenced, the criteria being statements of attainment that have been identified for each of 10 levels of progression toward each target. In the science curriculum, there are 17* attainment targets; the first, concerned with the processes of science, being of much greater weight than the other 16, which relate to knowledge and understanding. The complete curricular statement comprises the statements of what should be attained at these 10 levels and a description in very broad terms of the learning experiences that should be provided for pupils in four stages, covering ages 5–7, 7–11, 11–14, and 14–16. To give an idea of these components, the parts of the curriculum relating to Attainment Target 14: Sound and Music, are reproduced in the appendix.

National assessment is being introduced at the same time that the National Curricula are being phased in. The first assessment of all 7-year-olds took place in May 1991, using materials that were previously pilot-tested. These materials were not conventional timed tests, but took the form

---

* Currently under review and likely to be reduced to 5.

of classroom activities that have been specially devised to assess the statements of attainment. The administration and marking of the Standard Assessment Tasks was done by teachers whose training for this was the responsibility of the local education authority.

It is envisaged that there will be some form of sampling of the attainment targets or tailoring of tests to pupils because the total number of statements of attainment applicable to 7-year-olds (across levels 1–3 for English, science, and mathematics) is about 250. The problem becomes multiplied to enormous proportions when older children are considered. For example, the population of 14-year-olds is likely to include pupils at all 10 levels.

Meanwhile, teachers also have to carry out continuous assessment of their pupils against the criteria provided by the statements of attainment. It is important that the distinction between Teacher Assessment and Standard Assessment Tasks be grasped; otherwise, there would be a danger of the curricula becoming a series of tasks for assessment purposes, somewhat as in the Standard Assessment Tasks.

Teachers' continuous assessment has the following distinguishing characteristics: (i) carries on throughout the whole of each year; (ii) uses information from regular class activities, not specially devised situations; (iii) can take place in the whole range of contexts of pupils' work; (iv) is an integral part of teaching and learning not taking time away from teaching and learning; (v) requires teachers to plan opportunities for learning that relate to all the attainment targets; and (vi) helps teachers to value, understand, and develop assessment skills as an integral part of their practice.

The Standard Assessment Tasks, by contrast: (i) take place at the end of a key stage only; (ii) provide activities specially designed to address the statements of attainment and are necessarily limited in range of contexts for assessment; (iii) take time away from regular work; and (iv) will be devised externally, with instructions to teachers for administration and marking.

The last point suggests that teachers may not feel a sense of ownership of the information furnished by the Standard Assessment Tasks, which could threaten their reliability. Also, teachers might resent them as a massive intrusion into their time for teaching with few compensating benefits. This is indeed how they have been received by teachers of 7-year-olds who feel that the tests add nothing to their knowledge of the children and only disrupt the work. However, it is the political dimension of national testing that has provided the greatest thrust for objection to the tests. Because of the intention to use the aggregated test results of individual pupils as a measure of the success of a teacher or school, there is widespread reluctance to administer the tests.

Guidance has been given for teacher assessment in the form of written inservice course material and a booklet entitled "Children's Work Assessed" which includes examples of children's work assessed as being at certain levels.[30] This exemplification of how the criteria are to be interpreted in practice is particularly needed in science at the primary school level where the majority of teachers do not feel capable of teaching the subject, quite apart from assessing their pupils' performance. It is clear that continuous assessment of pupils' performance in the context of regular activities requires a sound grasp of the nature of scientific activity and the ability to analyze behavior into component parts.

Implications

These developments in performance assessment in England and Wales have several implications that can be grouped as curricular, practical, and theoretical.

Curricular Implications

The complexity of the relationship between teaching and assessment has been brought into the spotlight. It is hardly conceivable that teaching can be carried out without regard for what is being learned, but the extent to which this is taken into account varies from one view of learning to another. A constructivist view of learning in science has gained considerable support in recent years in the United Kingdom. In fact, the group that devised the National Curriculum in science wrote in its final report:

> In their early experiences of the world, children develop ideas which enable them to make sense of things that happen around them. A child brings these informal ideas to the classroom or laboratory, and the aim of science education is to adapt or modify these original ideas to give them more explanatory power. Viewed from this perspective, it is important that we should take a child's initial ideas seriously so as to ensure that any change or development of these ideas, and the supporting evidence for them, makes sense and, in this way, becomes "owned" by the child.[31]

In order to "take a child's initial ideas seriously," a teacher needs to know about them and must be capable of assessing them. Thus, it is argued that effective teaching depends upon effective assessment. However, it does not follow that assessment by itself will improve teaching. There must be ways of making appropriate curricular responses to information gained from assessment in order to make it worthwhile. The danger in the situation developing in England and Wales at the present time is that teachers will assess because they are required to do so by law, not because they need and can use the information in their teaching. More has to be done to help teachers make genuine use of the information they gain through assessment.

Mention has been made of the value of assessment for advancing teachers' understanding of the objectives of their work. This is particularly useful for science at the primary school level, where the teachers themselves do not have a clear understanding of the meaning of phrases used in the curriculum, such as "formulate testable hypotheses."[32] Involving teachers in structured assessment of their pupils' work can be of benefit in clarifying meaning by exemplification. However, if this is the only help that teachers receive, there is a clear danger of a narrow interpretation of the curricular aims in terms of what is assessed in particular tasks.

Practical Implications

The amount of assessment required of teachers at present by the new regulations is resulting in severe overload. Reports in the press have told of teachers working into the early hours of the morning to prepare and record their work, suffering from stress, and resigning. It must be noted that the curricular changes and assessment were not the only provisions of the 1988 act. Other changes, particularly in the management of schools, have been equally stressful. The situation has been sufficiently near to a crisis that the Secretary of State for Education has reversed some decisions concerning records to be kept on curricular coverage and has reduced the number of topics included in the Standard Assessment Tasks for 7-year-olds to a small sample of the three core subjects.

The degree of detail required in the assessment is the main cause of the overload on teachers. Teachers assessed and kept records previously, but not with all the particulars required in using the statements of attainment. Serious consideration and study now need to be given to whether the detailed information being gathered can, in fact, be used. It seems probable that even the pupils' teachers, who need the most thorough information, are unlikely to be able to make use of all of the knowledge of each child's performance on each statement of attainment. For purposes of reporting to others, the information has to be aggregated and condensed, and the question arises of whether it could have been obtained more economically in a less detailed form in the first place.

Theoretical Implications

As the characteristics of teachers' continuous assessment (mentioned earlier) imply, information is gathered from a range of regular activities. Thus, by its nature and through no inadequacy in gathering it, the information obtained often will be partial, contradictory, and fragmentary. But this is not a disadvantage when the use of the information is formative. Frequently, equivocal findings are the most useful for this purpose; if a child can do something in one context but not in another, this gives useful indications about the kind of support needed, the circumstances that promote progress, and so on. Recognizing such discrepancies and talking them over with the pupil can be the best lead to planning useful learning experiences. Clearly, in such instances, information from the assessment is being used formatively. Uncertain and shifting findings are not pinned down easily for the purpose of indicating that a child is at a certain level. So the value of teacher assessment for formative purposes lies in the very characteristic that makes it unsuitable for use in labeling children.

By contrast, the clear-cut results that will come from the Standard Assessment Tasks will be restricted to specific items and will provide little information that can be used formatively. However, because they will be clear-cut, they will lend themselves more easily to the labeling of children. Thus, the more the reporting of children's performance in terms of being at a particular level is emphasized, the greater the tendency will be to depend on the results of Standard Assessment Tasks or, worse, to turn teachers' assessment into a continuous series of standardized tasks.

One of the harmful effects of assessment on children's learning results from labeling. There is much research to show that a competitive climate, such as labeling produces, does not increase learning outcomes for all.[33] Rather, those who are already successful thrive, and those who are less successful (always the majority) underachieve more and more as they are discouraged repeatedly by the way their efforts are judged. Even for the successful, the emphasis often is on "Has he or she achieved level 5?" rather than on "Has he or she understood this relationship?"

Moreover, because there are multiple criteria at a particular level, there have to be rules for deciding when a level can be said to be "achieved." For example, if three out of four statements of attainment indicate achievement of a certain level, the information becomes of dubious value for helping learning, because it is not known which three statements or, indeed, whether all four apply. Part of the problem here is the level of detail in the curricula, which was determined without regard for the implications for assessment. It is interesting that at about the time the science curriculum was being developed, W. James Popham already was expressing grave reservations about the value of detailed statements of objectives.[34]

Perhaps the greatest theoretical problem with the model adopted for National Assessment in England and Wales concerns using the same information for several purposes. What seemed at first to be a neat solution to the requirement to have information on performance that could be used by teachers, parents, administrators, and politicians has been found to create the kinds of practical and theoretical problems outlined above. It seems necessary to think again about multipurpose assessment and to relate methods more specifically to purposes.

References

1. Department of Education and Science, *Education Reform Act 1988* (London: Her Majesty's Stationery Office, 1988).
2. Wastnedge, R., et al., *Nuffield Junior Science Project, Teachers' Guide No. 1* (London: Collins, 1967).
3. Ennever, L., and Harlen, W., *With Objectives in Mind: Guide to Science 5/13* (London: MacDonald Educational, 1972).
4. Harlen, W., Darwin, A., and Murphy, M., *Match and Mismatch Materials* (Edinburgh: Oliver and Boyd, 1977).
5. Department of Education and Science, *Primary Education in England* (London: Her Majesty's Stationery Office, 1978).
6. Clift, P., Weiner, G., and Wilson, E., *Recordkeeping in Primary Schools* (London: Macmillan Educational, 1981).
7. See, for example, Osborne, R. J., and Freyberg, P., *Learning in Science: The Implications of "Children's Science"* (London: Heinemann Educational Book, 1985).
8. Gilbert, J. K., and Watts, D. M., "Concepts, Misconceptions and Alternative Conceptions: Changing Perspectives in Science Education," *Studies in Science Education*, **10** (1983), 61–98.
9. Department of Education and Science and the Welsh Office, *Science 5–16: A Statement of Policy* (London: Her Majesty's Stationery Office, 1985).
10. Harlen, W., *Teaching and Learning Primary Science* (London: Paul Chapman Publishing, 1985).
11. Layton, D., *Interpreters of Science* (London: John Murray and the Association for Science Education, 1984).
12. Department of Education and Science, *Aspects of Secondary Education in England* (London: Her Majesty's Stationery Office, 1979).
13. Ibid., 167.
14. Association for Science Education, *ASE Science Teachers' Handbook* (London: Hutchinson, 1986), 113.
15. Department of Education and Science, *Educational Disadvantage and the Educational Needs of Immigrants*, CMND Report No. 5720 (London: Her Majesty's Stationery Office, 1974).
16. Gipps, C., and Goldstein, H., *Monitoring Children. An Evaluation of the Assessment of Performance Unit* (London: Heinemann Educational Books, 1983).
17. Ibid., 9.
18. Department of Education and Science, *Science Assessment Framework: Ages 13 & 15, APU Science Report for Teachers No. 2 (London: Department of Education and Science, 1984)*.
19. Johnson, S., *National Assessment: The APU Science Approach* (London: Department of Education and Science, the Welsh Office, and

Department of Education of Northern Ireland, 1989).

20. Department of Education and Science, *Science in Schools: Age 11,* Report No. 4 (London: Department of Education and Science, 1985), 242.

21. Department of Education and Science, *Science in Schools: Age 11,* Report No. 2. (London: Her Majesty's Stationery Office, 1983).

22. Gipps and Goldstein, *Monitoring Children.* See reference 16.

23. Johnson, *National Assessment.* See reference 19.

24. Department of Education and Science, *Science in Schools: Age 11,* Report No. 2. (London: Her Majesty's Stationery Office, 1983).

25. Schilling, M., Harlen, W., Hargreaves, L., with Russell, T., *Assessing Science in the Primary Classroom: Written Tasks* (London: Paul Chapman Publishing, 1990).

26. Department of Education and Science, *Science in Primary Schools: A Discussion Paper by Her Majesty's Inspectorate Science Committee* (London: Department of Education and Science, 1984). See, for example, Department of Education and Science, *Science in Primary Schools;* Department of Education and Science and the Welsh Office, *Science 5–16.*

27. Department of Education and Science and the Welsh Office, *The National Curriculum: A Consultation Document* (London: Department of Education and Science and the Welsh Office, 1987).

28. Department of Education and Science and the Welsh Office, *National Curriculum Science Working Group Interim Report* (London: Department of Education and Science and the Welsh Office, 1987), Appendix A, para. 1.2.

29. Department of Education and Science and the Welsh Office, *National Curriculum: Task Group on Assessment and Testing: A Report* (London: Department of Education and Science and the Welsh Office, 1988), para. 25.

30. Schools Examinations and Assessment Council, *Children's Work Assessed* (London: Schools Examinations and Assessment Council, 1990); Schools Examinations and Assessment Council, *A Guide to Teacher Assessment, Packs A, B, and C* (Oxford: Heinemann Educational Books, 1990).

31. Department of Education and Science and the Welsh Office, *Science For Ages 5 to 16* (London: Department of Education and Science and the Welsh Office, 1988), 7, para. 2.9.

32. Department of Education and Science and the Welsh Office, *Science in the National Curriculum* (London: Department of Education and Science and the Welsh Office, 1989).

33. Crooks, T. J., "The Impact of Classroom Evaluation Practices on Students," *Review of Educational Research,* **58**(4) (1988), 438–481.

34. Popham, W. J., "Two-plus Decades of Educational Objectives," *International Journal of Educational Research,* **II**(1) (1987), 31–41.

APPENDIX
Attainment Target 14: Sound and Music
Program of Study

Pupils aged 5–7
Children should have the opportunity to experience the range of sounds in their immediate environment and to find out about their causes and uses. They should investigate ways of making and experiencing sounds by vocalizing and striking, plucking, shaking, scraping, and blowing, for example, using familiar objects and simple musical instruments from a variety of cultural traditions. Children should explore various ways of sorting these sounds and instruments (p. 67).

Pupils aged 7–11
Children should be made aware of the way sound is heard and that sounds, including musical notes, are made in a variety of ways, and can be pleasant or obtrusive in the environment. They should explore the changes in pitch, loudness, and timbre of a sound, for example, by changing the length, tension, thickness, or material of a vibrating object, and through ways of causing sound, for example, use of different mallets, overblowing (p. 70).

Pupils aged 11–14
Through access to a range of sources of information, pupils should study the way sound is produced and can be transmitted over long distances, how the ear works, common defects of hearing, the effects of loud sounds on the ear, and the control of noise and sound levels in the environment. They should have opportunities to investigate the audible range. Pupils should investigate the effect on sound of the shape and materials of the built environment, such as reverberation times and insulation rates (p. 74).

Pupils aged 14–16
Pupils should explore sound in terms of wave motion and its frequence. They should have opportunities to develop their understanding of the properties and behavior of sound by developing a wave model, for example, through observations of waves in springs and on water. This should be related to pupils' experience of sounds and musical instruments, acoustic and electronic instruments, and recording and synthesizing. They should be given the opportunity to investigate devices such as microphones and loudspeakers which act as transducers and be introduced to the mechanisms underlying various communication systems (telephone, radio) on which a complex society depends. Pupils should investigate the characteristics and effects of vibration, including resonance, in a range of mechanical systems. They should extend this study to include some uses of electronic sound technology in, for example, industry (cleaning and quality control), medicine (pre-natal scanning), and social contexts (musical instruments) (p. 78-9)

Level	Statements of Attainment Pupils Should:
1	• know that sounds can be made in a variety of ways
2	• know that sounds are heard when the sound reaches the ear • be able to explain how musical sounds are produced in simple musical instruments
3	• know that sounds are produced by vibrating objects and can travel through different materials • be able to give a simple explanation of the way in which sound is generated and can travel through different materials
4	• know that it takes time for sound to travel
5	• understand that the frequency of a vibrating source affects the pitch of the sound it produces • understand the relationship between the loudness of a sound and the amplitude of vibration of the source • understand the importance of noise control in the environment
6	• know that when sound waves travel from one point to another they transfer energy through the medium • be able to explain the working of the human ear and some common defects in hearing • be able to describe the working of audio devices, for example, the microphone, loudspeaker, and telephone
7	• know that sound waves can be converted into electrical oscillations, transmitted (as electrical, optical, or radio signals) over long distances and converted into sound waves again
8	• know that a vibrating object has a fundamental characteristic frequency of vibration, and that some systems produce resonant oscillations which can be advantageous or disadvantageous
9	• understand the use of electronic sound technology, for example, in industrial, medical, and social applications
10	• be able to apply a knowledge of wave properties to explain common sound phenomena

Source: Department of Education and Science and the Welsh Office, *Science in the National Curriculum* (London: Author, 1989), pp. 30-31.
* The formulation of the Programs of Study and Attainment Targets is under review. The structure of the curriculum (key stages and 10 levels) remains unaltered.

Part III

Science Assessment in the Service of Instruction

Shirley Malcom

The difficult task in reform is not only figuring out what is wrong with the current system but also identifying (or inventing) replacement components. The limitations of traditional modes of assessment are regularly outlined by reformers both in and outside of school. These include the failure of most standardized testing to assess what is most important for students to learn or to inform practice; the disconnection of testing from real world problems; and the inequity that testing reflects and supports.

Part III of this book reviews some of the alternative modes of assessment which are being explored by researchers and practitioners. In addition to providing an historical perspective on examination systems in Europe Madaus and Kellaghan describe the evolution of assessment there, relating this to the choices being discussed currently by American reformers. Haertel addresses the challenges in measurement using different forms of assessment (e.g., projects, group work, laboratory work, etc.) and the information we gain from each type. The chapter by Baron provides a rich discussion of performance assessment and its essential connection to curriculum and instruction. She also shares experiences from practitioners as performance assessment confronts many of the most often heard criticisms of traditional testing models.

In an engaging discussion Michael Johnson describes the use of existing high stakes testing in place in New York state to document the achievement of students in an out-of-school program aimed at elementary and middle school minority students. Johnson and Johnson describe the "how to" and use of group assessment, a technique of growing importance with an increase in the use of cooperative learning as an effective teaching tool in science and mathematics.

The chapter by Collins describes the purpose, form and context of portfolio assessment for science while Campione discusses the possibilities for science of dynamic assessment techniques which currently are more widely used in special education to look at "potential to learn."

In the final chapter Malcom reconnects the themes of standards raising and equity which introduce the book to the concluding theme: choices in types of assessment. She argues for assessment that promotes, supports and enables equity in science rather than focusing solely on mechanisms to prevent bias in testing.

12

Student Examination Systems in the European Community: Lessons for the United States

George F. Madaus and Thomas Kellaghan

The idea that the United States needs to create a "national" exam or examination system currently has broad appeal, and there seems to be an inexorable movement toward implementation. Such an exam or exam system is seen by advocates as an essential ingredient in creating a world class education system. Advocates argue that: (i) the United States is one of the few industrialized countries in the world without common national examinations; (ii) this absence is a key reason why we are outclassed in international comparisons of achievement; (iii) the national exams must incorporate "authentic" assessment techniques and measure higher-order thinking skills, not just the recall of facts; (iv) national exams will give our teachers clear and meaningful standards and motivate our students to work harder by rewarding success and having real consequences for failure; thereby (v) helping the United States overtake other nations in achievement and emerge the world leader in education.

To date there have been a number of different proposals contained in a series of reform reports, in policy statements, and in legislation which have appeared during the 1980s and 1990s advocating a national exam or system of exams. However, to date proponents of national testing have been vague on the proposed uses of individuals' results, positive effects have invariably been stressed, and the assumptions inherent in the five-point scenario

The Office of Technology Assessment (OTA), in response to requests from the House Committee on Education and Labor and the Senate Committee on Labor and Human Resources, is conducting a comprehensive study of educational testing in the United States. OTA's report, to be completed in the fall of 1991, addresses a wide range of questions surrounding the purposes, design, implementation, and effects of alternative education tests. As noted, this chapter is an abridged version of a longer report that will be available when OTA's report is released. The contents of this chapter have not been reviewed or approved by the Technology Assessment Board, and do not necessarily reflect the opinions or findings of the OTA. The authors wish to thank the OTA for its financial support and Michael Feuer of the OTA for his unfailing assistance.

sketched above have gone unexamined. This paper, excerpted from a longer report prepared for the Science, Education, and Transportation Program of the Office of Technology Assessment of the U.S. Congress, examines the origins and present status, infrastructure, complexity, and uniqueness of the external examination systems in the 12 European Community (EC) countries and then considers the implications of various features of those systems for American education in light of proposals to develop a national exam or exam system for the United States.

The Origins and Development of Public Examinations in Europe

The origins and development of examination systems during the 19th century in Europe are extremely complex. Three major features, however, can be detected in their development. One relates to their selective function, a second relates to the major part played by universities, and a third to the role of examinations in defining student learning — what it is students learn and how they learn. We will describe these features and then we will list some of the advantages and disadvantages that have been ascribed to public external examinations in Europe and elsewhere.

The Selective Function of Examinations

A major feature in the origins and development of public examination systems was their selective function.[1-4] This use in the university context is obvious enough. Students gained access to a university through passing the *Baccalaureat* in France, the *Abitur* in Germany, or the relevant university board examination in England.

However, the use of examinations for selection for the civil service and professions preceded their use for university entrance. A major reason for introducing examinations in Europe in the first place was to replace the old system of patronage and nepotism for making appointments to the civil service which had secured the dominance of the aristocracy. News had been coming through from China since the 16th century by way of missionaries and travellers about the system of examinations in that country that had used written examinations since 2200 B.C. to select personnel for government positions. The Jesuits incorporated examinations into their schools,[5-7] and, at a later date, it was hoped that the introduction of examinations for selection to the civil service would ensure that the most able and talented would be recruited.

Germany (Prussia) led the way with civil service examinations, and by 1748, recruitment to offices in the judiciary and government administrative services was based on examination performance.[8]

Following the Revolution, France adopted the German practice of using selective examinations to identify highly trained professional elites in pursuit of a powerful absolutist state. The first use of such examinations was in 1793 when an official certificate in civic virtue was required of primary-school teachers in the *ecoles centrales*. In 1795, a scholarship entrance examination was established for such schools.[9] In time, examining in France, though developed at a later date than in Germany, became more frequent, covered more subject areas, and was more selective, demanding higher standards and failing more students.[10]

In Britain, professional bodies introduced written qualifying examinations early in the 19th century — the Societies of Apothecaries in 1815 and Solicitors in 1835. At a later date, other nonuniversity examining bodies, such as the City and Guilds of London Institute and the Society of Arts (later the Royal Society of Arts), organized examinations for commercial and technical studies. Competitive examinations for admission to the Civil Service were created under the Northcote–Trevelyan scheme in 1853, and, in the same year, a scheme was designed for selection for the East India Company's Civil Service. In these several spheres, examinations followed Benthamite principles of maximizing aptitude and minimizing expense while at the same time controlling nepotism and patronage. By 1870, almost all civil service appointments were based on examinations, while all but a few branches were open to public competition.[11–13]

Under the influence of British practice, the Civil Service Act of 1883 established competitive examinations in the United States to select personnel for government service. The examinations, however, were abandoned when congress failed to make appropriations to continue them.[14]

University Influence on Examinations

In Germany, the *Abitur,* first introduced in 1788 as a graduation examination for the classical middle school, soon became a qualification examination for university. Students who passed it were automatically admitted to university. The examination was used to upgrade the quality of the universities, rejecting students with poor scholastic ability.[15]

The influence of the university can also be seen in France in the origin of the *Baccalaureat,* established by Napoleon in 1808, which has been traced to the 13th century *determinance* of the Sorbonne University (formalized by royal edict in 1598). The *determinance* consisted of an oral examination to decide whether students were fit to embark on the studies offered in the university.[16] During the 19th century, higher educational institutions administered the *Baccalaureat* examination, which was used both to admit students to the *grandes ecoles* and to government service and other professions.[17] Until recently, the *Baccalaureat* granted access to a university; now, for admission to the most prestigious schools further examinations have to be taken.

In Britain, also, where mass public examining became a feature of education in the 19th century, the universities played an active role in the establishment of examinations. For example, Oxford and Cambridge established systems of "locals" examinations, which were marked by university "boards," to assess secondary school quality, though it was not until 1858 that the examinations were used to examine individual students. It was later again (1877) that certification was provided to students. These examinations survived up to 1918. Other universities (Dublin, Durham) followed the same path and established procedures for examining local secondary school pupils.[18–20]

However, it was London, rather than the older universities (in which examining was mainly oral and in which the examination idea had never really taken strong root), that made the main contribution to the development of the external examination idea. Established as an examining body in 1836, London did not become a teaching institution until 1898. The first London matriculation examination was held in 1838, and was the earliest external school examination in Britain. The examination was conducted entirely by written papers.[21]

Although, as we have seen, competitive examinations for the civil service and for professions preceded their use in the context of university selection, the later development of examinations outside the university was strongly influenced by people who had experienced the use of internal examinations, both in secondary schools and universities. James Booth, who was a member of the Society of Arts and is recognized as having played an important role in the development of examinations, was one such person. Booth had studied at the University of Dublin, where the system of examinations — both written and oral — was more highly developed than at the other old British universities (which placed greater store on residence and meeting other requirements for graduation). The reason for the development of examinations at Dublin was at least partly because a fifth of its students were in effect "external," attending only for examinations each term.[22,23] Booth noted and later promoted the complex procedures he had encountered in Dublin for use at secondary-school level, leading to the establishment of a system at the Society of Arts, which was accepted as an exemplar by other systems. In an 1853 Report of the Society, it was stated that decisive testimony had been obtained in favor of "some system of examination for provincial schools in connection with a central body, which would be empowered to grant certificates of proficiency."[24]

The universities continued to control the development of the secondary-school examination system during the second half of the 19th century in Britain. While recommendations were made, for example, by the Taunton Commission in 1868 and the Bryce Commission in 1895, that a central body be created with overall responsibility for examinations, little was done to implement them. Although the Board of Education (set up in 1899) might have seemed the appropriate body to coordinate, if not administer, the examination system, opposition to central control among teachers and local education authorities was widespread. Besides, the board did not seem keen itself to assume direct control, preferring to exercise an indirect influence. Thus, no attempt was made to wrest the initiative from the universities, which were viewed favorably by the opponents of centralization (particularly, the local authorities).[25]

The Influence of Examination on Student Learning

The third feature of examinations — their role in defining student learning — was in evidence from the earliest days of the French *Baccalaureat*. Originally, the "program" for the examination consisted of a catalog of 500 questions from which a number of questions were randomly selected for the examination.[26] This approach was compatible with the catechetical method of teaching, which had been popular in Europe for several centuries.

Throughout the 19th century, examinations were often developed without any institutional provision for preparing for them. Into the present century, many students prepared for university examinations, not in the university, but in other private educational institutions or in their own homes. Given this situation, it is not surprising that examinations exerted a major influence on what students learned. The influence continued even when formal curricula and teaching facilities were provided.

Foreshadowing contemporary claims in America of strong links between national examinations, school achievement, and increased global competitiveness, the argument was introduced from Professor Liebig of Giessen in Germany (at that time Britain's major industrial competitor and perceived to be ahead of Britain) that, "if no examination is introduced the best schemes will fail, and will produce no effect: introduce the examination, *and all the rest follows of itself.*"[27] Liebig's view from Germany echoed the British view at the time that students and teachers needed to be motivated, that competition would improve motivation and learning, and that examinations were a necessary and cost-effective means of raising educational standards and securing national competitiveness.

Throughout the second half of the 19th century, examinations flourished in Britain. The growth has been attributed to the ethos of the time, which was dominated by the utilitarian values and ideas of Adam Smith and Jeremy Bentham. First, there was the conviction that self-interest was the main motive for study and that since study involved drudgery, it was necessary to reward successful work by distinction in the ways of certificates, prizes, and medals. Second, teaching was also regarded as painful drudgery. Hence, teachers needed to be motivated by pupil success or additional payment (for example, in payment-by-results schemes). Third, it was believed that the products of learning could be measured with some exactness. This led to an emphasis on the reproduction of factual knowledge, which formed the essence of written examinations. Fourth, the state was recognized to have only limited obligations to secure the education of its citizens. When examination systems took off in the middle of the century, there was still no state-funded provision for secondary or technical education in Britain. Fifth, examinations in the educational sphere seemed to meet the important objectives of the Benthamites in industry of uniformity and standardization. Finally, examinations came to be seen as a cheap and effective method of promoting development by demonstrating the value of improved teaching and resources, thus creating incentives to local agencies to raise funds and support education and by focusing attention on the need to provide more relevant curricula in schools.[28]

As Foden (1989) has pointed out, the principles regarding method, motivation, and the effect of examinations which were embodied in the examination movement of the 19th century were almost wholly untested and unvalidated. At the time, there was little substantial critique of examinations as a technique or process, and there was no serious questioning of the utilitarian values of the examination reformers. Booth, however, seems to have anticipated some of the later challenges to examinations when, in 1853, he wrote:

> Objections are sometimes made to examinations, that they cannot always be depended on as true tests of proficiency — that they gave rise to cramming, and to superficial preparation. Now there is no system or plan that was ever devised that does not stand in the shade of some one or other objection. This, however, is no argument against examination as a test; it only proves that the examiner is incompetent to discharge his duties. To be a good examiner requires previous training. A well-trained examiner, who knows the subject in hand, will not only gauge the knowledge but will take true note of the faculties of those who come before him, while he who confines himself to what is set down in text books, who makes no step in deduction, who inquires into mere facts, and not into the bearing of those facts, who does not seek to look at a truth from different points of view, mistakes the duties of his office, and leaves undeveloped the powers of the instrument in his hands. An examination should not consist of strings of leading questions, nor of interrogatories to be answered by a simple yes or no. Neither should the answer be the echo of the question, nor should familiarity with mere tabulated results be sought for. An examination should be something more than the exponent of the strength of a mechanical memory. Examinations of this kind, if they do some good, do more harm. They encourage those principles of association, which rest on verbal similitude.[29]

Booth's description of good examiners and his belief that we can build examinations which can measure what are now called higher-order thinking skills portend many of today's arguments that we need more "authentic tests." As Booth's description of good examining indicates, however, building such tests will not be easy and will be more, rather than less, costly and less, rather than more, efficient. Nor will training teachers to teach higher-order skills be easy or cheap.

Advantages and Disadvantages of External Examinations

Given the importance of external examinations for students' future careers, it is not surprising that they have attracted many apologists and critics. Among the advantages attributed to public examinations is that they are a relatively objective and impartial means of distributing educational benefits. Indeed, as we have seen, one of the reasons for their introduction in the first place was to reduce the effects of patronage and to open higher education and a range of occupations (particularly in the public service) to a wider population of students. Other advantages attributed to public examinations include the degree of national homogeneity in educational standards and practice which they engender, the sense of purpose they give to teachers' efforts, the provision of tangible incentives for students, a diminution of conflict between the roles of teaching and assessment, the provision of an assessment procedure unaffected by personal relationships between teachers and students, their acceptance in the community, and the creation of some measure of social consent among the young, while meeting some definition of comprehensiveness, equal access, and common entitlement or shared experience.[30-38]

At about the time that James Booth was writing in the middle of the last century, criticisms of public examinations and of their effects were also beginning to appear. Since then, there has been an avalanche of observations and analyses, in Europe and elsewhere, which have catalogued the shortcomings of public examination systems. A major criticism of such examinations is that they can have undesirable "back-wash" effects on classwork, not just in examination classes but in lower grades as well, by limiting approaches to learning. They tend to encourage undue attention to material that is covered in the examinations, and since what is examinable is limited, worthwhile educational objectives and experiences may be excluded from the classroom. In effect, the examinations may come to determine the shape of the curriculum rather than the curriculum determining the shape of the examinations, and examination performance comes to be regarded by parents and students as the main, if not the sole, objective of education.

Further criticisms made of external examinations are that they are usually carried out under artificial conditions in a very limited time frame and that they are not suitable for all students and can be extremely stressful for some, causing undue strain and excessive anxiety. They are often viewed by students as unfair since doing poorly, for whatever reason, on an examination at the end of the year can override a year or more of hard work and achievement. They tend to inhibit the development of curricular variety which may be necessary to serve local and student needs. Further, there is often a lack of congruence between course objectives and examination procedures (for example, there may be no examinations for oral or practical objectives); and, given certain kinds of teaching, it would appear that students can do well in examinations without recourse to higher levels of cognitive activity; preparation for examinations often overemphasizes rote memorization by students and drill and practice as teaching methods. There is plenty of evidence that examinations lead to cramming.

Particular criticisms apply to essay-type examinations. Such examinations are inevitably limited in the range of characteristics which they can assess, relying heavily on verbal and logico-mathematical areas. Procedures to reduce unreliability and inconsistency in marking are time-consuming and expensive.

Finally, public examination results are often used to serve a variety of purposes for which they may not have been designed: they may be used to predict future educational and vocational performance as well as to certify the completion of a course of education, though little effort is made to match occupational or educational needs with candidates' talents.[39] In particular, the use of examinations for the dual purpose of certifying the completion of a secondary education and for university admission tilts the examinations towards an academic university domination. Further, examination results may be used by universities and businesses principally to control numbers and to screen out excess applicants with the cut-off score being a function of the ratio of applicants to places. Examinations can also force students out of school before taking the examination or after failing it, or they can result in teachers excluding students who, if they took the examination, would probably perform poorly.[40–88]

The extensive literature on external examinations should serve to underline the need for serious discussion and weighing of the long- and short-term cost benefits associated with the known positive and negative aspects of high-stakes testing before embarking on a national testing program.

The Structure of the Educational Systems of the European Community

The twelve members of the European Community are Belgium, Denmark, France, Germany, Greece, Ireland, Italy, Luxembourg, Netherlands, Portugal, Spain, and the United Kingdom. Our description of education and examination systems refer to the former Federal Republic of Germany and, in the case of the United Kingdom, to England and Wales.[89] Table 1 provides information that allows comparisons among total populations and school populations in European countries and in the United States. In the last column, American states that are equivalent in population to European countries have been identified.

The structures of an educational system and the curricula which students follow are particularly important in the context of considering examinations, since examinations are usually geared to specific curricula. Educational systems in Europe are characterized by the conventional division between primary, secondary (usually divided into lower and upper levels), and third-level education. The primary sector is relatively uncomplicated, offering a free, compulsory, and common education to all students. There have been in the past, and continue to be, differences among countries and even among states in the same country (Germany) in the length of primary schooling. The shortest primary cycle is four years (in Germany and Portugal), while the longest is seven years (in Denmark).

Second-level education has been characterized in the past by its bipartite nature (in some countries, there have been more than two parts), its selective mechanisms, and its relatively low participation rates. In one type of school, a classical academic curriculum in the tradition of the seven liberal arts (*Trivium* and *Quadrivium*) was offered. In this school type (called grammar, *gymnasium,* or *lycee*) students were prepared for third-level education and white-collar occupations. As numbers expanded, the traditional academic curriculum got watered down, subjects were presented at different levels, and practical/commercial-type subjects were introduced for some students. The type of education now offered in such schools includes university-preparation curricula, but in some countries, particularly at the lower level, it also includes more comprehensive programs designed for students who are not likely to go to university. In some national descriptions, the term "general education" is used to describe the activities of these schools.

As an alternative to more academic general education, schools offering technical curricula were established to prepare students for skilled manual occupations. In some countries, curricula were expanded in these schools, as in the case of secondary grammar schools, to accommodate the increasing numbers of students attending. By contrast with grammar schools, however, the schools provided practical, usually short-term, "continuation" education.

After World War II, and particularly during the 1960s, demographic, social, ideological, and economic pressures led to various reviews of educational provision. The division of students at an early age was regarded as inappropriate and curricula were seen to be in need of reform. Since increasing numbers of students were staying at school to receive some secondary education, the idea of developing a system of comprehensive education at the secondary level was considered in most countries. As

Table 1. Total Grade School Population for European Community Countries and the United States

	1985[1] Total population	1984[2] 6-15 First & lower secondary	1984[2] 6-11 Primary	1984[2] 12-15 Lower secondary	1984[2] 16-19 Upper secondary	1988[3] Comparable total population in USA
Belgium	9,858,000	1,288,500	732,600	555,900		MI (9,240,000)
Denmark	5,114,000	696,300	404,400	291,900	327,200	MO (5,141,000)
France	55,170,000	8,039,300	4,578,000	3,461,300	3,426,000	CA (28,314,000) NY (17,909,000) OH (10,855,000)
Germany	61,024,000	6,684,000	3,505,000	3,179,000	4,115,000	CA + NY + OH + MO
Greece	9,935,000	1,435,400	838,500	596,900	618,300	OH (10,855,000)
Ireland	3,540,000	693,400	417,500	275,900	257,300	SC (3,470,000)
Italy	57,141,000	8,388,000	4,773,000	3,615,000	3,806,000	CA + NY + OH
Luxembourg	367,000	42,800	24,100	18,700	21,400	WY (479,000)
Netherlands	14,492,000	2,053,200	1,103,300	949,900	981,200	TX (16,841,000)
Portugal	10,157,000	1,702,800	1,011,300	691,500		OH
Spain	38,602,000	6,511,200	3,883,800	2,627,400	2,660,400	CA + OH
U.K.	56,618,000	7,506,900	4,110,000	3,355,900	3,709,700	CA + NY + OH
USA	239,300,500	34,191,500	19,682,500	14,509,000	14,979,000	

[1] European Community Figures. [2] OECD, *Education in OECD Countries* (Paris: OECD, 1990). [3] Hoffman, M.S., ed., *World Almanac and Book Facts, 1987-88*, (NY: Pharos Books, 1987).

participation rates grew and students remained at school beyond the lower level of secondary education, the classical academic curriculum of the upper level of secondary schools also came under pressure at the upper level, being regarded by many as inappropriate for a student body that was becoming increasingly more heterogeneous in achievement, aptitude, interest, and motivation.

Although all countries have made some moves to comprehensive lower-secondary education (up to age 15 or 16), achievements have been varied[90,91] (see Table 2). Several countries have established lower-secondary school curricula which are largely comprehensive. Denmark and Britain have gone furthest in this with 10 years of comprehensive education, though there are still remnants of the bipartite system in Britain. Greece, Portugal, and Spain, which have had major reforms of education in recent years; and Italy and France have also relatively long periods of compre-

Table 2. Data on Age of Compulsory School Attendance and Structure of the Educational Systems in the European Community

	Compulsory age attendance	Horizontal structure of system	Comprehensive curriculum/ schools lower secondary grades[1]	Differentiated curriculum/ schools grades
Belgium[2,3]	6-16 (16-18 P-T)	6-3-3 or 6-2-2-2	7-10*	11-12
Denmark	7-16	7-3-2 or 7-2-3	8-10	11-12
France	6-16	5-4-3	6-9	10-12
Germany[3]	6-15	4-6-3	5-6*	5-13
Greece	6-15	6-3-3	7-9	10-12
Ireland[2]	6-15	6-3-2/3	7 9*	7-12
Italy	6-14	5-3-5	6-8	9-13
Luxembourg	5-15	6-7		7-13
Netherlands	6-16	6-3-3	7-10*	7-12
Portugal	6-12	4-2-3-2-1	5-9	10-12
Spain	6-15	5-3-3(-1)	6-8	9-13
U.K.	5-16	6-4-2	7-10	11-12

[1] A number of countries are less advanced than others in the development of comprehensive school structures. These countries are marked with an asterisk.

[2] Belgium and Ireland have an additional 2 years preprimary education integrated into the primary-school system. All other countries have provision outside the formal educational system for early childhood education.

[3] Belgium and Germany are federations. There are two states in Belgium with completely independent educational systems. There are 11 states in the former Federal Republic of Germany (16 in the new Germany). Each of the 11 states determines its curriculum under terms agreed by the Council of State Ministers of Education.

hensive education. The remaining countries, with the exception of Germany, can be regarded as aspiring toward, and to some extent in transition toward, a more comprehensive system. However, there still exist systems which have academic-type schools, less demanding schools providing general education, and vocational/technical schools.[92]

There are some comprehensive schools in Germany but, on the whole, the German states have resisted the development of a thorough-going comprehensive system. Both major components of the traditional structure (the classical *gymnasium* and the vocational school) have been sufficiently strong and successful to resist possible incursions from each other. In particular, vocational education, often seen by students as more enticing than the *gymnasium-Abitur*-university route, has been consolidated and improved and is generally regarded as a success of educational policy.[93]

At the lower-secondary level, Germany retains four types of school: the *Realschule* or intermediate school (grades 5 to 10), which is the most popular and offers a technical education; the *Hauptschule* (grades 5 to 9), which is the least demanding; the *Gymnasium* (grades 5 to 13) which offers a classical academic curriculum leading to university, and the *Gesamtschule* or comprehensive school (grades 5 to 9/10 or 13). The situation is even more differentiated than this categorization implies since there are several types of *gymnasium* (classical, modern languages, scientific, economics, and agriculture). The first two grades in all types of schools are conducted in accordance with comprehensive principles and are dedicated to the orientation, observation, and guidance of students. This differentiation can clearly be seen in the scope of the curriculum and how often it is revised. Since 1949 more than 7,000 separate syllabi have been issued for general education alone. With reunification and the addition of six more states, there will be around 2,000 different syllabi in general education and at least twice that number in special and compulsory vocational education. No syllabus is valid in more than one state; no syllabus covers more than one subject; at the secondary level there are 20 different subjects; and no syllabus covers more than one of the different school types or levels.[94]

At the upper-secondary level, all systems offer differentiated schools and/or curricula. In the case of Italy, the system at this level is so complicated that it has been described by Visalberghi as a "jungle."[95] The system is also highly differentiated in Germany. In addition to the types of schools described above, there is a complex and changing vocational sector, in which there are several types of schools (for example, commerce, child care, continuation, and part-time).

Table 3 provides data on the percentages of upper-secondary school students (by gender) following general education and technical/vocational curricula in countries in the European Community. In eight of the countries, a majority of students follow a curriculum of general education. However,

Table 3. Percentages of Upper-Secondary Students in General Education and in Technical/Vocational Education, by Gender, 1985/86.

	Girls (%)		Boys (%)	
	General education	Technical vocational education	General education	Technical vocational education
Belgium[1]	56	44	53	47
Denmark	40	60	26	74
France[2]	65[3]	35	58[3]	42
Germany[2]	51	49	57	43
Greece	83	17	62	38
Ireland	79	21	86	14
Italy[4]	26	74[5]	22	78[5]
Luxembourg	38	62	29	71
Netherlands	49	51	43	57
Portugal[6]	99	1	99.8	0.2
Spain	58	42	53	47
United Kingdom	53	47	57	43

Source: European Communities Commission, *Girls and Boys in Secondary and Higher Education* (Brussels: European Communities Commission, 1990), table 3b.
[1] Lower and upper-secondary education.
[2] 1986/87.
[3] Includes upper-secondary technological education.
[4] 1984/85.
[5] Includes preschool and primary teacher training.
[6] Technical/vocational education was abolished in 1976. New courses were introduced on an experimental basis in 1983/84.

in most countries, a sizeable number of students are enrolled in technical/vocational education courses.

In all systems, with the exception of the United Kingdom (in which the situation is changing), curricula are prescribed for schools by a central authority, usually the Ministry of Education (MOE). The definition of this prescription varies from system to system. In Germany, it is determined by each of the eleven states. In Britain, up to the present, it has, in effect, been determined at the secondary level by the independent examination boards whose syllabi a school has chosen to adopt. In some countries (France, for example), curricula seem to be fairly uniform across schools. In others, particularly Denmark, in which great discretion regarding implementation is left to individual schools, we might expect considerable variation between schools. The trend in several countries is to allow schools a great deal of freedom in the definition of curricula during the compulsory period of schooling.

Table 4. Enrollment Rates for Ages 15-18 in the European Community, Canada, Japan, and USA, for 1987/88

	Age 15	Age 16	Age 17	Age 18
Belgium	95.8	95.5	92.7	72.0
of whom, part-time	2.2	3.6	4.6	4.6
Denmark	97.4	90.4	76.9	68.6
France	95.4	88.2	79.3	63.1
of whom, part-time	0.3	7.9	10.0	5.2
Germany[1]	100.0	94.8	81.7	67.8
of whom, part-time			0.1	
Greece[2]	82.1	76.2	55.2	43.6
Ireland[2]	95.5	83.9	66.4	39.6
Italy				
Luxembourg[3]			83.4	71.1
of whom, part-time			15.8	15.8
Netherlands[4]	98.5	93.4	79.2	59.7
of whom, part-time				
Portugal		32.1	36.9	29.2
Spain	84.2	64.7	55.9	30.4
United Kingdom	99.7	69.3	52.1	33.1
Canada	98.3	92.4	75.7	56.9
Japan[3]	96.6	91.7	89.3	3.2
of whom, part-time	2.6	1.9	1.7	1.4
U.S.A.[2]	98.2	94.6	89.0	60.4

Source: OECD (1990), table 4.2, except figures for Portugal, which are for secondary education in 1983/84 and come from European Communities Commission (1990), table 1c.

[1] Apprenticeship is classified as full-time education.
[2] 1986/87.
[3] Excluding third-level.
[4] Excludes second-level part-time education.

Clearly, the structures of educational systems in Europe vary considerably. There are also many other ways in which the systems differ which might have implications for the nature of examinations and for student achievement, which we cannot consider here in detail. For example, there are differences in enrollment rates at various ages between countries (Table 4). While these rates in all countries are high at age 15, considerable decreases occur in some countries at ages 16 and 17. The largest decrease between ages 15 and 16 is found in the United Kingdom where the participation rate drops from 99.7 to 69.3 percent. By contrast, participation in the United States at age 16 is 94.6 percent.

Examination Systems in Europe Today

Before describing examination systems in Europe, which we will find reflect the complexity of the educational systems in which they function, two general points may be made. First, most countries have had a tradition of external examinations. Second, examination systems in all countries have been evolving as the broader social and educational contexts in which they operate have been changing. This has led in some countries to the virtual abandonment of external examinations. Following our categorization of examination systems, we will briefly consider two further issues: the idea of a single examination for all and the cost of examinations.

Tradition of External Examinations

All countries today have some elements of an external-examination system, though for some it is only at the point of entry to third-level education. At its most general, this means that all have at some point or points in their education system more or less formalized procedures, usually separated from the classroom situation, in which a candidate has to answer questions, usually based on externally devised syllabi, to demonstrate that he/she possesses certain knowledge and skills. Examinations most usually involve written essay-type questions, though in some countries oral examining also features. In some countries (Britain, Netherlands, Portugal, Spain) multiple-choice sections have been introduced to examinations.

On the basis of examination performance, a candidate is usually awarded a certificate or diploma which will contain information on the candidate's performance on each subject in the examination in terms of either letter grades (e.g., A,B,C,D,E), numbers (1,2,3,4,5), or marks. Usually, grades are arrived at by simply summing marks allocated to sections of questions and across questions and papers (if an examination in a subject has more than one paper).[96] The final allocation of grades may also involve some element of norm-referencing in which grade distributions in previous years are taken into account. Marks or grades (converted to points) may be summed across subjects to make decisions about university entry. European countries do little to apply to their examinations psychometric principles of the type developed in the context of testing in the United States. The two major issues of interest are objectivity and comparability. Psychometric issues such as pretesting, item-analysis, IRT, scanning, equating, reliability, and construct validity receive little or no attention. Nor does an extensive technical psychometric community/elite exist in Europe as it does in the United States.

The certificate or diploma, in addition to testifying to the candidate's satisfactory performance in an examination in a particular subject or groups of subjects, may also confer certain rights, such as the right to be considered for (if not actually admitted to) some sector of the social, professional, or educational world. Certificates are "credentials," analogous to stock shares or academic currency.[97] Thus, certification has both educational functions (testifying to and "guaranteeing" students' standard of education) and legitimizing functions at both the level of knowledge (indicating the legitimacy of a new subject, such as computer science) and of the individual (justifying the classification of individuals into social categories and the allocation of educational benefits to certain individuals).

A further feature of educational systems in Europe, which is reflected in examinations, is that syllabi have traditionally focussed on content. At one stage, the study of classical texts was the main feature of syllabi.[98] There have been movements in several countries in curriculum development, largely under the influence of American research, to focus on "objectives," while content has broadened considerably beyond the classics.[99] In the syllabi for the new General Certificate of Secondary Education (GCSE) taken in Britain at 16+, an effort has been made to place greater emphasis on skills.[100] In the technical and trade areas also, examinations may focus on practical skills. However, syllabi in European countries still reflect a major concern with subject-matter content and examinations are organized in terms of traditional subject areas (languages, mathematics, sciences, history, geography, etc.). Within a subject area, syllabuses may be offered at different levels. In some countries, syllabi for the same subject differ in different parts of the country or depending on the examination board offering the syllabi.

Finally, the tradition of examinations in Europe has been to make public the content of examinations and their results. This was so even before examinations were presented to candidates in printed form. For example, after the oral examination of candidates at Dublin University during the first part of the last century, people who had been present wrote up and circulated the questions that had been asked. These questions were then used by students who were preparing for similar examinations in the future, in effect establishing the tradition of allowing examinations to determine curricula. This became easier, of course, when examinations in printed form became available.[101] Broadfoot (1984) has noted about the present French *Baccalaureat* that once examination papers are published, every teacher studies them or makes his/her pupils try them so that the examination questions virtually become the syllabus.[102] Indeed, in France, examinations often make front-page news in newspapers; scholars analyze questions and the cultural implications of examiners' choice of topics.[103] In Germany, in addition to making examination papers available, answer scripts are returned to students who may question the way they were graded with their teachers. If a problem cannot be resolved between student and teacher, the matter is referred to an official of the Ministry of Education.

Changes in Examination Systems

It would be misleading to think of examination systems in Europe as stable or unproblematic. Systems have always been subject to revision. However, the changes over the past three decades have been quite radical in several countries. Some countries have moved in a particular direction, only to retreat at a later date. In the mid-1970s, for example, France abolished external examinations at 16+ with the aims of postponing selection, making assessment more comprehensive, and giving a greater role to teachers in assessing students.[104] However, the examinations were reinstituted in the 1980s, at least partly because the resources to support a school-based system of assessment had not been made available to schools. In Germany, decisions have recently been taken to attempt to restore some of the older standards which people believe have been lost by allowing candidates freedom to select subjects at lower levels of difficulty in the *Abitur*.[105] We can expect further change as social and educational conditions alter and as those responsible for examinations respond to criticisms of systems.

A major force for change in examinations in all countries came from expanding participation rates in education which were accompanied by an increase in variance in the achievements, aptitudes, interests, and needs of students, which in turn necessitated a change in traditional curricula. This situation led many people to the view that the formal and academic nature of traditional examinations (particularly written terminal ones), originally designed for a minority of students, was unsuitable for the assessment of many candidates and curricular areas. Criticisms of external examinations, which have been consistently made during their history, particularly ones about their narrowing effects on students' educational experiences, achieved a new level of significance and relevance when the question of submitting total populations to traditional examination procedures arose. While many commentators judged such procedures to be particularly unsuitable in the case of low-achieving students, perhaps the more striking evidence comes from students themselves, who opt not to take examinations and leave school without any formal certification. In Britain and Ireland, two countries with very formal external examination systems, the number of such students (about 11 percent and 8 percent, respectively) has been a matter of serious concern to policy-makers.

There have been three major approaches to changes in examination and certification systems in recent years. One has been to abolish external examinations and certification completely. The second has been to make greater use of school-based assessment for certification purposes. The third has been to shift the purpose of examinations from selection to certification and guidance.

Most European countries at some time in the past held a national school-certificate examination at the end of primary schooling. Although during their life, criticisms of these examinations were continually made, it was felt that the examinations had certain advantages; in particular, that they helped to clarify for teachers the standards that were expected, they provided a stimulus to pupils, and they were useful for selection to secondary education and in seeking employment.[106] However, in time, various factors — arguments (particularly from teachers) about the limiting effects of the examinations on the curriculum, the fact that schools prevented low-achieving students from presenting themselves for examinations by grade retention,[107] changes in philosophy of education, the raising of the school-leaving age, and the provision of adequate space in secondary schools to accommodate all students — led to abolition of the examinations. Of these reasons, removal of the need to select pupils for secondary education was perhaps the most compelling. No country in the European Community today operates a national external examination at the end of primary schooling. The remnants of the certificate examination exist in Italy, where school-based examinations (set, administered, and scored by pupils' own teachers) are held throughout the country, and in Belgium, where some schools administer an examination at the end of primary schooling. (It seems that these schools feel an examination will help to raise standards.)

While there has been a move towards greater reliance on teacher assessments at the secondary level, examinations and certification have been retained in one form or another at two points in the system in most countries: at the end of lower-secondary and at the end of upper-secondary schooling. External examinations have been abolished and certification is entirely school-based at both levels in four countries (Belgium, Greece, Portugal, Spain). Examinations in all the other countries have some input from teachers. This takes the form either of candidates' own teachers marking examinations that have been set by an outside body, or candidates' teachers contributing assessments which are combined with the results of external examinations. While Britain from the 1960s onwards moved toward a greater dependence on candidates' teachers in the assessment procedure, this position has been reversed in recent years, as central government adopted a more active role in the control of both curriculum and examinations.

A third trend in European examination systems is the shift in emphasis from selection to certification and guidance about future academic study.[108] A shift in function has been possible, especially at lower educational levels, because of the expansion of places in secondary schools. Furthermore, as the numbers taking final school-leaving examinations have increased and as the examinations have become more varied, selection for traditional third-level education is no longer a concern for many students taking terminal school examinations. Increasing numbers of these are now turning to apprenticeships or technical training.[109] A further possible consequence of increasing numbers and diversification of school-leaving examinations, as well as of basing school certification wholly on school-based assessment, has been the introduction of state-controlled examinations for selection to third-level education.

The selective element is evident in other aspects of examinations and, indeed, despite commitment to guidance, most systems retain strong traces of their origins as instruments of selection.[110] In Denmark, for example, a student has to achieve a certain average score (5.5) in the school-based assessment at the end of the comprehensive *Folkeskule* at grade 10 to enter a *gymnasium*. In Germany, the School Leaving Certificate from the *Hauptschule* at grade 9 qualifies for entrance to an apprenticeship or vocational school or transfer to a *Realschule*. Students who leave without this certificate are awarded a nonqualifying certificate, which destines them for unskilled occupations. At the end of secondary schooling, the increase in numbers means that many students completing school are not likely to go to university and so will be interested in certification rather than in selection to a third-level institution. However, for those students with aspirations towards higher education, the competition in many countries (and consequently the importance of examinations for selection) has increased considerably in recent years.

A Categorization of the Systems

As we have noted, formal external examinations have disappeared at the primary-school level in all European countries. Around the end of lower-secondary education, which more or less coincides in most countries with the upper compulsory attendance age, the situation is more complex (Table 5). In six countries (Belgium, Greece, Italy, Luxembourg, Portugal, and Spain), a school-leaving certificate is awarded to students on the basis of school-based assessment (which may involve continuous assessment and/or examinations or simply testify to completion of studies). In the other six countries, examining involves a combination of internal and external procedures. The countries can be placed on a continuum in terms of these practices, from ones in which the external component of assessment is relatively small to ones in which it adopts a major role. In Germany (where there is a variety of certificates at grades 9 and 10, depending on the type of school which the student attends), examinations are set by the Ministry of Education (MOE) but are scored by teachers in the candidates' own school. The external examination results are also supplemented by school-based assessments. In Denmark, examinations are also set by an external agency (MOE) and, in addition to being scored by teachers in the candidates' own school, are also scored by external teachers. In the Netherlands, the Ministry of Education sets an examination which has essay and multiple-choice components. Essay parts are marked with the aid of a marking scheme supplied to schools by students' own teachers and by

Table 5. Examinations/Certificates in European Community Countries

	Primary	End of Compulsory	End of Secondary	Entrance to Third Level
Belgium	Diploma Optional Kantonal school-based exams	Series of diplomas (general, technical) (school based).	Diplomas for general, technical, professional studies (school-based)	Some university departments (engineering) require additional test
Denmark	None	Leaving Certificate set by MOE, marked by own teacher & external teacher.	Atrium set by MOE* and marked by own teacher & external teacher. Vocational/technical alternatives.	
France	None	Brevet de college exam set and marked by 23 academies in basic subjects & teachers' assessment in other curricular areas.	Baccalaureat set and marked by 23 academies (questions selected from centrally approved list). Three types: general, technical, and vocational.	MOE Concours or exam for admission to a grand ecole (after one /two years). For lower status universities Bac. is sufficient.
Germany	None	Series of examinations/diplomas depending on type of school attended set by 11 state MOEs and marked by own teacher.	Abitur set by 11 state MOEs and marked by own teacher. Some weight to school grades.	
Greece	None	Diploma (school-based).	School-leaving diploma (school-based).	Exam set and marked by MOE school-leaving diploma.

Country				
Ireland	None	Two external exams set and marked by MOE; some school assessment in some subjects. (To be amalgamated into one Junior Certificate in 1992.)	Leaving Certificate set and marked by MOE	Some university departments require additional exam, but usually not.
Italy	Primary Certificate under direction of MOE. Set and marked in school.	Middle-school certificate and technical/vocational qualifications. Set by MOE, marked in own school.	Exam/Diploma of General Education; Diploma of Technical Education. Set by MOE. Marked by local exam committees (including teachers from candidates' school).	
Luxembourg	None	School Certificate of completion.	Diploma de fin d'etudes Secondaires set by MOE, marked by school and outside examiners (written).	
Netherlands	None	Exam/certificate based on internal assessment and national (MOE) written exam.	Exam certificate based on internal assessment and national (MOE) written exam.	
Portugal	None	School certificate (based on assessment by teachers).	School certificate (based on assessment by teachers).	Exam set and scored by MOE.
Spain	None	School certificate (based on assessment by teachers).	Bacillerato Unificado Polivant. School certificate (based on assessment by teachers).	One year later: exam set and scored by MOE (additional exams set for some university departments).
United Kingdom	None	GCSE set and marked by 5 regional boards. Incorporates some school-based assessment.	General Certificate of Education, set and marked by 8 Examination Boards.	

* Ministry of Education

teachers from another school. However, the multiple-choice component (which represents about half the written papers) is scored centrally. Oral examinations are administered by candidates' teachers. In France, examinations are set and marked by 23 regional *academies,* each of which produces its own examinations, based on the national curriculum. The examinations cover the subjects of French, the *Brevet* mathematics, and history/geography. For other subjects, school-based assessments are employed.

The last two countries (Ireland and the United Kingdom) show the highest level of external control in the examination system. In Ireland, the administering authority is the MOE; in the United Kingdom, five regional examining groups are responsible for the administration of the examinations. In both countries, the examinations are scored by teachers appointed by the examination authority, who will be unaware of the identity of candidates. While there is provision in both countries, greater in the United Kingdom than in Ireland, for including the results of school-based assessment in the final examination results, the role of the candidates' own teachers is less important than in other European countries at this level.

Diversity between countries, reflecting the position at the end of lower-secondary education, is also to be found at the end of secondary schooling (Table 5). Five models are required to take account of the differences that exist between countries at the upper-secondary level. Four countries, as against the six at lower-secondary level, operate a system of certification based solely on school assessment. The four countries, Belgium, Greece, Portugal, and Spain, which fit this model at the upper-secondary level, also use school-based assessment at the lower-secondary level. In the second model, the examinations are largely school-based but have an element of external control or monitoring. In Denmark and Luxembourg, the examinations are set by the MOE, but are marked by the candidates' own teachers as well as by teachers appointed by the Ministry. In Germany, the examinations are set by the MOE in each of the country's eleven states, but are marked by the candidates' own teachers. The results are sent to the MOE, which identifies ones that it regards as untypical, which are then discussed with the school. In the third model, which applies only in the Netherlands, the system is the same as the lower-secondary model for that country: part of the examination is school-based and part set by the Ministry. The latter part is scored by candidates' own teachers as well as by external teachers. In the fourth model, which operates in Italy, examinations are set by the MOE and are scored by local examination committees which include teachers from candidates' own schools. In the final model, which operates in France, Ireland, and the United Kingdom, examinations are set and scored by an outside agency — the MOE in the case of Ireland and academies and examination boards in the case of France and the United Kingdom respectively. In Ireland and the United Kingdom, there is some provision for including an element of school-based assessment.

In a number of countries, examinations beyond the school-leaving certificate examination are employed to select students for third-level education. This happens in three countries that operate a completely school-based system of assessment and certification at the secondary level (Greece, Portugal, Spain). The examinations are run by the MOE. It will be noted that these three countries have relatively low participation rates at ages 16 and 17 (Table 4) and are among the economically poorest in the European Community. In other countries, individual universities (or schools within universities) may operate their own selection systems, sometimes involving testing but using other criteria as well (for example, school record, work experience), to decide on who they will accept for the limited number of places.

One Examination for All?

A persistent theme in American proposals is the idea that there should be one "national" examination for all students of a particular age or grade level (recommendations about the precise age or grade level vary). At this stage we can ask: how national are European examinations? Is there one examination for all? And, if not, are there problems in comparability?

The term "national" can mean many things. In European systems of examinations, central governments play some role. The role is most direct and influential when the MOE sets, administers, and marks examinations (as in Ireland). It is less direct and influential when the Ministry sets examinations but does not score them (as in Germany). Other types of less direct influence occur when the central Ministry has a general supervisory role in the activities of examining bodies (as in the United Kingdom). Regionalization of the administration of examinations, as occurs in France, Germany, and the United Kingdom, all large countries, dilutes the authority of the central Ministry.

Patterns of authority in examinations may vary within a country according to the type of examination and the grade level of candidates. The administration of technical and vocational examinations, which we did not consider in this paper, can be extremely complex. Sometimes such examinations are organized by the Ministry, but more frequently they are carried out at the level of the school or fall within the province of specific examining groups. There are cases in which students from one country (Ireland, for example) take examinations, particularly in the vocational/technical area, but not exclusively so, which are set and marked in another country (United Kingdom).

Variation relating to grade level occurs when a central body assumes authority for examining at one level, but not at others. In a number of countries, central authorities are more involved in examinations at higher levels than at lower levels (in Greece, Portugal, and Spain, for example).

In other countries, the same organization is responsible for examinations at all levels of the secondary sector (in France and Ireland, for example).

This description may not be of much assistance in deciding what can be considered a "national" examination or who the authority responsible for that examination should be. Different countries have worked out different procedures in line with their school organization and traditions.

What has the European experience to say about having one examination for all students? Obviously, where examinations, even at a single level, are school-based or the responsibility of a number of authorities (regional or otherwise), there is not a common examination. Thus, in France, the area in which a student lives will determine the precise *Brevet* or *Baccalaureat* which he or she will take. In Britain, the choice for GCSE at 16+ and General Certificate of Education (GCE) at 18+ will depend on the choice of the school that the student attends.

Again, in countries that have differentiated secondary education, the possibility of one examination for all does not arise as long as examinations are geared to particular syllabi that are likely to vary from one school type to another. It is impossible to get precise figures on the proportions of age cohorts who take particular examinations. The data in Table 3 on the percentages of students in general education and in technical/vocational education, taken in conjunction with enrollment rates in Table 4, may be taken as a rough indication of the proportion of students taking examinations in the two curricular areas. In Belgium, Germany, Spain, and the Netherlands, one would expect close to half of the students still attending school not to take an academic-type examination.

It could be argued that a similar type of examination, if not the same examination, could be provided for all. Thus, we could say that a common *system* of examinations rather than a common examination operates in Britain and France at 16+.[111] But how common are the experiences of examination candidates in such a system? Given the range of options available, the answer must be that great variation exists in students' experiences.

In Britain, there is a choice of examination board or boards which will usually be made by the school. Then there is the choice of subjects and subject levels — partly a school decision and partly a student one. The choice often creates problems for schools, complicated (it would appear) in Britain by a preference "to do worse on those examinations which carry greater prestige rather than to do better on those that are more useful."[112] Finally, the student, when he or she actually comes to take the examination, may have a choice of questions from which to select. Thus, candidates can achieve the same grade by answering different questions.[113,114]

At the 18+ level, since considerable selection and attrition has taken place, the question of an examination for all does not arise. However, even if we confine our attention to those students taking examinations at the end of academic (rather than vocational) schooling in Germany and France, we find that they have a variety of options and experiences. In Germany, *Abitur* candidates take examinations set by different Ministries, a widely different assortment of subjects, different papers in nominally the same subject, with different weights being given to the results depending on the option chosen. The French *Baccalaureat,* although it retains a large core of general education subjects that all candidates are required to take, albeit with different weights, also offers a considerable variety of options. Four options in 1950 had grown to 38 in 1988.[115]

Given these situations, it is not surprising that questions are frequently heard about the comparability of examinations operating within a single "system."[116–117] In the context of the *Baccalaureat,* Eckstein and Noah (1989) have commented on the strongly demarcated hierarchy of prestige that surrounds the examination, with the mathematical options at the top and the vocational options at the bottom.[118] They conclude that "the French examination authorities have been prepared to yield more and more comparability across candidates in an explicit trade-off to meet what are essentially political demands for 'relevance' and 'access.' "

Faced with eight examining bodies at the GCE level and five at the GCSE level, the greatest efforts to achieve comparability across examinations have been made in Britain. The development of criteria for a range of individual subjects has been a step designed to help improve comparability.[119,120] Further, all the examining bodies operate under the supervision of the School Examinations and Assessment Council, which, among other things, arranges for a team of examiners from one examining board to work at another board for an extended period of time, to observe procedures, the marking of scripts, and standards. The Intergroup Research Committee of the boards, made up of research officers, also carries out a program of continuous research relating to comparability between boards, subjects, and modes of examining. Staff read scripts from other boards, paying particular attention to borderline grades. To supplement these procedures, there is a strong tradition of research on British examinations.[121, 122]

It would seem from the European experience that a single common examination for all can only be provided if the content/skills assessed are relatively basic or if teachers play a major role in assessing students' performance. Attempts in Britain to provide an external examination covering a wide range of achievement do not seem to have been entirely satisfactory. While it was hoped that the General Certificate of Secondary Education would cater to a wider range of students than the GCE/CSE systems which it replaced, this does not seem to have happened. Many syllabi have become less accessible to lower-ability candidates and, in particular, to those candidates who in the past would have been catered to on the basis of teacher-designed syllabi and examinations (CSE Mode 3).[123,124] Ironically, the attempt to provide an examination with a wide-achievement span seems to have been unsuccessful not only in the case of lower-achieving students but is reported (in newspapers and on television) to have lowered the standards of the higher-achieving students who go on to do GCE A-level examinations at 18+. Teachers say that these students are less well-prepared in such areas as science and mathematics, while some university teachers have expressed the opinion that a further year at university will be necessary if students at graduation are to reach the same level as their predecessors.

The Cost of Examinations

We were not able to obtain extensive information about the costs involved in examining in European systems. In countries in which teacher assessments play a major role, costs are largely absorbed in teachers' salaries.

Information on external examinations taken at 16+, from Britain for the GCSE (in which students on average take about five subjects) and from Ireland for the Junior Certificate (in which students on average take about seven subjects), indicates that the cost of examining a student is $107. (In Ireland, candidates pay about 40 percent of the cost.) If the state of Massachusetts were to adopt one of these models to test its 65,000 16-year-old students, the cost would be almost $7 million. At present, it spends $1.2 million to test the reading, writing, and arithmetic achievements of students at three grade levels (3, 6, 9 and 4, 8, 12 in alternate years), using machine scoring for the reading and arithmetic tests. It is clear that the adoption of an external European model of testing would have very substantial financial implications for the Commonwealth.

To test the 3 million 16-year olds in the United States would cost over $320 million using the British or Irish model. Costs would be reduced if multiple-choice tests were used, if students were examined in fewer subjects, if the range of options available to students were reduced, and if examination papers were not released. Costs would be increased very considerably if more "authentic" measures of student achievement were used. It is also likely that labor costs (in scoring examinations, for example) would be higher in the United States than in Britain or Ireland.

In some European countries (Britain, Ireland) and in the Canadian province of Alberta,[125] some students repeat the last year of secondary school in an attempt to raise their already passing scores on examinations to qualify for a university place. This involves considerable expense to taxpayers, which should be considered in making cost estimates of national examinations.

Conclusions

We may take it as axiomatic that an educational system develops in response to the values, needs, and aspirations that characterize a nation. We may also regard it as axiomatic that the components of any system interlock in an idiosyncratic way to contribute to the realization of a nation's goals for education. This is not to say that the performance of educational systems should not be reviewed from time to time to see how effectively they are performing their tasks and, in particular, to determine if they are meeting new needs that continually arise in our fast-changing society. However, it does mean that if one decides to select a feature of another educational system in the search for solutions to one's own educational problems, one should ensure in the first place that one understands the functioning of that feature in the foreign system (including any problems that have been identified in its operation) and, in the second place, that the proposed transplant is likely to be compatible with the host system.

There are many features of an educational system, and of the wider socio-economic and cultural system in which education is carried out, that one could nominate as being likely, in interaction with other features, to contribute to students' achievement in school. For example, some systems (including all European countries) use inspectoral systems to monitor the quality of education in schools. Again, some systems have a longer school year than others (see Table 6), presumably providing more "time on task" for students. Systems vary in the quality of candidates recruited to the

Table 6. Minimum Number of School Days in European Community Countries

Belgium	182
Denmark	200
Germany (North Rhine-Westphasia only)	240
Greece	175
Ireland	180
Italy	215
Luxembourg	180
Netherlands	200
Portugal	?
Spain	?
United Kingdom	?

Source: Stichting Research voor Beleid, The Conditions of Service of Teachers within the Member States of the European Community (Leiden: Stichting Research Voor Beleid, 1988).

teaching profession and in the type and length of training provided. Most European systems recruit better students, on average, to teacher preparation than is the case in the United States. The support that homes provide is also generally accepted as an important factor in student learning.

To isolate national examination systems from their context, to reduce them to a consistent relationship with achievement, educational quality, increased global competitiveness, and to suppose they have the same effect in all contexts is based on a false sense of technological optimism. Indeed, the conviction that the establishment of national examinations, standards, and curricula will usher in a golden age of education is consistent with the great tradition of optimistic technophilia of the last two centuries.[126,127] There is the further danger that focusing on examinations and assessment may lead us not only to overlook other important aspects of the system that need attention (for example, the need to modify instructional techniques for at-risk youth) but actually to aggravate existing problems (for example, difficult examinations and graduation requirements may lead to an increase in drop-outs).[128]

We do not know of any evidence that would tell us whether having or not having external examinations in the United States would have the effects that proponents of national examinations are at present suggesting. In this paper, we described the ethos in which examinations grew in Europe, the complexity of the systems in which they are embedded, and some of the problems that have been noted in their operation. Whether or not examination systems introduced to the United States would work as they do in Europe and, in particular, whether or not they would impact on educational standards are matters that require serious consideration.

We will conclude by considering the relevance of European experience to the proposals for examining that have been made in the two reform waves in the United States since 1983. Although there is considerable complexity and variation in European systems of examinations, there are, as we noted, certain features that characterize several systems. In considering American proposals, we will regard European experience as particularly relevant, if it occurs in several countries; or, even if it is found in only one country, we may still cite it if it seems to speak directly to an American proposal. The British experience seems to fit this latter criterion since some of the proposed American reforms seem to have been inspired by, if not actually modelled on, the British system. In applying any European experience, we should not lose sight of differences between the United States and European countries in the political, social, and educational contexts in which examination systems function.

Purpose of Testing

In general terms, proposals for examinations in the United States are directed toward raising educational standards which, in turn, it is hoped, will improve the economic competitiveness of the nation. It is envisaged that the examinations will have their effect by promoting a common curriculum in schools across the country and by putting pressure on teachers and students to achieve a high standard of performance on the examinations.

When one looks at proposals in more detail, however, one finds a wide variety of purposes being posited for examinations. Some of them relate to decisions within the school — the identification of students for remedial intervention or for advanced or accelerated work (National Commission on Excellence in Education, 1983) or for grade promotion (Education Commission of the States Task Force, 1984). Others relate to decisions that might involve leaving school. The Commission on the Skills of the American Workforce (1990) sees examinations as providing information relevant to whether a student enters a college-preparatory program, studies for a technical certificate, or goes to work. Boyer's (1983) report on secondary education regards examinations as being relevant to decisions at a later stage about whether one goes to work or to third-level education.[129] These proposals all emphasize the use of examinations for guidance; however, they do not make clear why examinations are needed to improve the guidance services already in place in American schools. Neither do they face the issue of how examinations are going to motivate students to work harder if high stakes are not attached to performance on those examinations. While the proposals may not explicitly acknowledge the fact, it would seem that they envisage the use of examination results to make decisions about students, as indeed the Education Commission of the States Task Force (1984) does when it says that examination performance could be used in deciding grade promotion of students, and Public Law 100 297 does when it proposes the use of examinations to identify outstanding students.

Another proposed use of examinations is the certification of students' achievements. This use is proposed by the National Commission on Excellence in Education (1983), the Commission on the Skills of the American Workforce (1990), and Public Law 100-297. However, there is not total agreement among the reports on the use of examinations for certification. Educate America (1991), in proposing the testing of all high-school seniors, specifies that the examinations should be held in the fall of grade 12 so that they would *not* be used for graduation.

Other purposes proposed for testing are accountability of students, schools, and states (Educate America, 1991) and monitoring of standards in schools (National Commission on Excellence in Education, 1983). While the idea that examinations would motivate students to work harder runs through all the proposals, it is most explicit in the Educate America proposal.

The proposals are most compatible with European practice when they emphasize the use of tests for the two interrelated functions of certification and selection. The origin and traditions of examinations in Europe exhibit a major concern with selection, particularly for third-level education and certain jobs. As the educational systems of these countries have come to resemble more that of the United States in its student retention rate and curriculum comprehensivization (up to the age of about 16), the emphasis in examinations has shifted from selection to certification. However, examination results continue to be used for selection both inside and outside the school system. An inevitable consequence of accepting the spirit of European examinations (even if not the details) would seem to involve a greater emphasis on the selection and categorization of students in American schools.

A number of other aspects of the purposes of European examination systems seem relevant to the American proposals. The public examination systems in Europe are not used formally for accountability or monitoring of standards. Rather, examinations are used to make decisions about individual students, not about teachers, schools, or districts (though parents may make judgments about schools on the basis of their examination results). Neither are examinations used to help improve or monitor standards. Efforts in Britain to allow comparability between marks from successive examinations set on the same syllabus are made so that users can be confident that a given grade has the same meaning from year to year, rather than to improve overall standards.

For the most part, quality control and accountability in European educational systems are the function of school inspectorates. All countries also participate in comparative studies of educational achievement and some have national assessment systems similar to the National Assessment of Educational Progress (NAEP), in which individual school performance is not identified. While existing public examination systems do not seem appropriate to serve the functions of monitoring or accountability, efforts are being made in Britain to develop a system that will serve these functions as well as the traditional student certification function of public examinations for students up to the age of 16 years.

While a single assessment system is unlikely to meet efficiently a variety of purposes, neither are the purposes of examinations completely independent of each other. Thus, examinations are likely to have a motivating effect on some students only if performance on them has some real consequence for the students (admission to a third-level institution or selection for a job, for example) or if examination performance is used for some other high-stakes purpose such as teacher accountability.

A final point to note in considering the relevance of the purposes of examinations in Europe for the United States is that European examinations fit into a differentiated (highly so, in some countries) educational structure. All systems have different types of school and curricula at the upper-secondary school level and examination performance at the end of lower secondary (about age 16) is an important factor in determining the type of school and course in which a student will find himself or herself.

Age/Grade of Testing

Apart from one proposal to test students at grades 4 and 8 (President's Education Policy Advisory Committee)[130] and another to test at unspecified major transition points (National Commission on Excellence in Education, 1983), all the American proposals envisage examinations at either age 16 or at some point during the senior-high school years, including the point of graduation.

The proposals that confine testing to age 16+ are compatible with European practice. Formal national external examinations below the age of 15 or 16 years no longer exist in any European country. Even at age 16+, six of the twelve European Community countries do not have an external examination. In two other countries, examinations are set by an external agency but are marked by candidates' own teachers. The trend in most countries, with the exception of Britain and France during the 1980s, has been to reduce the external element in examinations at this stage (age 16+).

Britain also differs from other EC countries in that it proposes to institute national testing of students at ages 7, 11, 14, and 16 with a new form of teacher-administered tests. However, this testing, except at ages 14 and 16, will not form part of the public-examination system.

Responsibility for Testing

Proposals for national examinations in the United States are not clear about who should be assigned responsibility for the examinations. The decision, however, is an important one since the responsible agency will exercise considerable control over teaching and learning in schools. Some indications are provided in some proposals. The responsible body should be external to the school (Commission on the Skills of the American Workforce, 1990) or school district (National Commission on Excellence in Education, 1983) and it should be a national system (but not controlled by the federal government), involving state and local tests (National Commission on Excellence in Education, 1983). However, Public Law 100 297 would be a federal test under the control of the Secretary of Education. A proposal has also been made for an agent to build and administer the examinations in all states (Educate America, 1991), while another recommends the establishment of a National Board of Educational Standards to calibrate to a common national standard examination that would be built under the auspices of state examining boards (National Center on Education and the Economy/Learning Research and Development Center at the

University of Pittsburgh, 1989).[131]

In European countries, with the exception of Britain and Germany, the central government has a major responsibility for curricula and examinations. Until recently in Britain, curricula were a matter for local education authorities and schools, while examinations were controlled by independent examining bodies with loose links to universities (except for the London University Examining Board, which is a part of the university). In the last few years, central government has been adopting an increasingly active role in the specification of curricula and in the control of examinations. In Germany, responsibility for curricula and examinations rests with the eleven state governments.

In these two large countries, as well as in another large country, France, one authority does not assume responsibility for the administration of examinations. Such countries probably provide a better model for the United States than smaller countries in which the Ministry of Education has responsibility for public examinations.

While Ministries of Education in all European countries play a role in examinations, the important role played by other interested parties should be recognized. In the setting of papers and standards, teachers, subject specialists, and university personnel, in addition to Ministry officials (school inspectorate), play a role. Teachers in all countries also mark examinations and assign grades to candidates. At both the 16+ and 18+ levels, in nine countries, whether or not examinations are set by an authority outside the school, candidates' own teachers play a major role in marking examinations. In the other countries, the major role in marking is played by teachers from other schools who are not aware of the identity of candidates.

Areas of Testing

Although there are references to tests of "academic excellence" (Public Law 100-297) and of what students ought to know and be able to do when they leave school, including the knowledge needed to participate in a democratic society (Commission on the Skills of American Workforce, 1990; Educate America, 1991), most of the American proposals for what should be examined are more specific. A number of reports mention basic or general achievement and skills (Education Commission of the States Task Force on Education for Economic Growth, 1984; National Alliance of Business),[132] while several spell these out in terms of traditional subject areas. English, mathematics, science, history, and geography or social studies are proposed by the President's Education Policy Advisory Committee and by Terrence Bell,[133] who also adds computer studies. Educate America (1991) lists the same subject areas but instead of English proposes reading and writing.

European examination systems tend to emphasize broad cultural goals in their examinations rather than preparation for later life, though the latter, of course, is not ignored. Some European systems provide external examinations in a small core of subjects at the 16+ level — two (Danish and mathematics) in Denmark or three (French, mathematics, history/geography) in France. Others allow students to take a larger number of subjects and, with some restrictions, to choose the subjects they will take (Britain, Ireland). In these countries, examinations are offered at different levels. Different types of examinations are provided in countries that have a differentiated educational structure according to the type of school attended by the student (Germany, Netherlands).

No European country offers a single examination for all students at the 18+ level. By this stage, most countries offer one system of examinations for students following academic university-oriented curricula and another system for students following more vocationally-oriented curricula. Within the academically-oriented system, students may be required to take certain core subjects in combination with options they themselves choose. A choice of levels varying in difficulty (higher/lower, honours/pass, higher/ordinary) within subjects will also be available to students. The British systems at 18+ differs from other European systems in its high level of specialization.

It is clear that systems of examination in Europe, especially during the senior-high school years, are much more complex than the systems being proposed for the United States. The European experience can contribute little to the design or implementation of proposals for a single external examination for all students at the upper-secondary school level.

Methods of Testing

Proposals about methods of testing in the American context range from the use of standardized tests (National Commission on Excellence in Education, 1983) to the use of "state of the art" assessment practices (Educate America, 1991), which, in current thinking, would include the performance, portfolio, and project examinations specified by the Commission on the Skills of the American Workforce (1990). Other proposals recommend subject-matter examinations in core curriculum subjects as we noted above, which presumably would use the predominant European mode of having students write extended essays.

There is little that European systems of examining can tell us about the value of such procedures as portfolio and performance assessment. While efforts have been made to develop such procedures in Britain (sometimes called records of achievement), the efforts have been inspired by perceived inadequacies of public examinations to record accurately and in sufficient detail students' achievement records. This work has focused in the first place on lower-achieving "nonacademic" students who were being poorly served by the examination systems, though it is hoped to extend the procedures to all students. With several competing models of assessment being developed or reformed in Britain at the moment (public examinations, Student Assessment Tasks, and profiles of achievement), it is difficult to predict what the final shape of assessment practice will be by the end of the 1990s.

One Test or One System of Testing

American proposals in some cases indicate that a single set of examinations would be used to test all students at a given age or grade level. In other cases, the suggestion is made that a system of examinations, rather than a single examination, is needed.

As we have seen above, the larger European countries (Britain, France, and Germany) have a number of examination authorities that devise and administer their own tests. In France, the examinations of the different authorities are based on a common curriculum; in Britain and Germany, they are based on separate curricula.

In two of the remaining countries that use an external examination at 16+ (Denmark, Netherlands), scripts are marked by candidates' own teachers. In the third (Ireland) students select from a range of subjects offered by the examination authority. Thus, a common examination, as distinct from a common system, exists in only two countries at 16+ (Denmark and the Netherlands).

The situation at 18+ is, as we saw, much more complex. Students' examination experience can vary, depending on the region of the country in which they live (France, Germany), the examination authority they choose (Britain), and the curriculum options they have chosen in the upper-secondary school (which vary in subject matter and in level within subject).

The question of comparability arises when students come to use their certificates either for entry to third level education or in seeking employment. In Britain and Ireland research efforts have been directed toward investigating the comparability of performance of students who take different groupings and/or levels of subjects. On the whole, however, the question of comparability of examination performance within countries does not appear to be a major problem. Rules of thumb are usually devised on the basis of a judgmental process relating to grading and comparability of grades and these are generally accepted by universities and employers.

Effects of Examinations

We have already considered many of the effects that have been attributed to examinations to which high stakes are attached. These include motivational ones ("making" teachers and students work), focusing teachers' and students' activities, cramming, emphasizing memory work, and developing test-taking skills. Here, we will just note that positive motivational effects are likely to operate only if students perceive they have a good chance of achieving the rewards attached to high test performance. For students who are not likely to do well (and thus for whom the stakes are, in effect, irrelevant), the negative effects of examinations have been a matter of serious concern in many European countries. Over the years, efforts have been made in many countries to adapt the examination system to suit these students.

It is important to note that in Europe the impact of examinations on teaching and learning — what is taught and learned and how it is taught and learned — is mediated through the availability of past examination papers. An American proposal (Public Law 100-297) not to release test papers after examinations would diminish the impact of the examinations on teaching and learning in the schools.

Cost

The only United States proposal for a national test that offers a cost estimate is that of Educate America (1991); their figure is $30 per student. As we saw, cost figures in Europe were not readily available except for Ireland and the GCSE in Britain. As we noted, the cost for an essay on demand exam at age 16+ (consisting of between 5 and 7 separate exams in Britain and Ireland respectively) was $107. The labor costs (scoring the exams) would probably be higher in the United States than those incurred in Britain and Ireland. In many countries the exam costs are largely absorbed in teachers' salaries. Costs would surely be higher than the $107 figure if more "authentic" assessment measures of student achievement were used. We also pointed out that repeating the last year of secondary school to improve exam scores involves considerable expense to taxpayers.

Fitting Examinations into the Existing System of Testing

With the exception of Britain, European countries do not have external systems of examinations other than the public examination system. In most countries, little use is made of standardized tests developed outside the school. The formal aspect of internal school assessment mirrors the public examination system. Students take examinations that are similar to public examinations at the end of each school term and may take "mock" public exams some months before the actual public examinations. The United States has an extensive commercial infrastructure for developing, marketing, scoring, and reporting standardized achievement tests. Companies make their money from scoring and reporting rather than from the sale of the reusable test booklets. These tests are widely used at all levels of education.

The place of any national exam or systems of exams within the present system of testing needs considerable thought. For example, to use the essay form (as in Europe) for the national exam while the multiple-choice form continues to be widely used by states or districts could be confusing to teachers and students. In developing a national exam, or a system of national exams, an infrastructure will have to be created for developing and scoring assessment techniques, reporting results, and overseeing the entire exam operation. In Europe teachers are an integral part of the exam infrastructure, as are the MOE inspectors. We would need to consider Europe's experience in this regard, particularly their trust of teachers. Discussion of the infrastructure for an American national examination system also raises serious issues of cost and quality control. In Europe, control is governmental or quasi-governmental through the MOE and the established inspectorates. Cost and oversight issues associated with using commercial companies for development, scoring, and reporting will need to be weighed against developing a new infrastructure for assessment in the United States.

References and Notes

1. Christie, T., and Forrest, G. M., *Defining Public Examination Standards* (Basingstoke, England: Macmillan, 1981).
2. Creswell, M. J., "Describing Examination Performance: Grade Criteria in Public Examinations," *Educational Studies,* **13** 265.
3. Goacher, B., *Selection Post-16: The Role of Examination Results* (London: Methuen, 1984).
4. Ingenkamp, K., *Educational Assessment* (Windsor, England: NFER Publishing Co., 1977).
5. Du Bois, P. H., *A History of Psychological Testing* (Boston: Allyn and Bacon, 1970).
6. Durkheim, E., *The Evolution of Educational Thought,"* trans. by Peter Collins *(London: Routledge and Kegan Paul, 1979).*
7. McGucken, W. J., *The Jesuits and Education* (Milwaukee, WI: Bruce, 1932).
8. Amano, I., *Education and Examination in Modern Japan,* trans. by W. K. Cummings and F. Cummings (Tokyo: University of Tokyo Press, 1990).
9. Broadfoot, P., "From Public Examinations to Profile Assessment: The French Experience," in P. Broadfoot, ed., *Selection, Certification and Control: Social Issues in Educational Assessment* (Lewes, Sussex: Falmer, 1984).

10. Amano, *Education and Examination in Modern Japan*. See reference 8.
11. Foden, F., *The Examiner: James Booth and the Origin of Common Examinations* (Leeds: School of Continuing Education, University of Leeds, 1989).
12. Montgomery, R. J., *Examinations: An Account of Their Evolution as Administrative Devices in England* (London: Longman, 1965).
13. Roach, J., *Public Examinations in England 1850-1900* (Cambridge: Cambridge University Press, 1971).
14. DuBois, *A History of Psychological Testing*. See reference 5.
15. Amano, *Education and Examination in Modern Japan*. See reference 8.
16. Halls, W. D., *Society, Schools and Progress in France* (Oxford: Pergamon, 1965).
17. Amano, *Education and Examination in Modern Japan*. See reference 8.
18. Lawton, D., *Politics of the School Curriculum* (London: Routledge and Kegan Paul, 1980).
19. Montgomery, *Examinations*. See reference 12.
20. Mortimore, J., and Mortimore, P., *Secondary School Examinations* (London: University of London Institute of Education, 1984).
21. Kingdon, M., and Stobart, G., *GCSE Examined* (London: Falmer, 1988).
22. Foden, *The Examiner*. See reference 11.
23. McDowell, R. B., and Webb, D. A., *Trinity College Dublin 1592 1952: An Academic History* (Cambridge: Cambridge University Press, 1982).
24. Foden, *The Examiner*, 75. See reference 11.
25. Mortimore and Mortimore, *Secondary School Examinations*. See reference 20.
26. Prost, A., *Histoire de l'enseignement en France 1800-1967* (Paris: Colin, 1968).
27. Foden, *The Examiner*, 74. See reference 11.
28. Ibid.
29. Ibid., 78.
30. Bowler, R. F., "Payment by Results: A Study in Achievement Accountability," unpublished doctoral dissertation, Boston College, 1983.
31. Commission on Mathematics, *Program for College Preparatory Mathematics, USA* (New York: College Entrance Examination Board, 1959).
32. Consultative Committee on Secondary Education, *Report of the Consultative Committee on Secondary Education with Special Reference to Grammar Schools and Technical High Schools* (London: Her

Majesty's Stationery Office, 1938).

33. Curriculum and Examinations Board, *Assessment and Certification: A Consultative Document* (Dublin: Curriculum and Examinations Board, 1985).

34. Hargreaves, A., *The Crisis of Motivation and Assessment,"* in A. Hargreaves and D. Reynolds, eds., *Educational Policy: Controversies and Critiques* (London: Falmer, 1988).

35. Heyneman, S. P., "Uses of Examinations in Developing Countries: Selection, Research, and Education Sector Management," *International Journal of Educational Development,* **7,** (1987) 251-263.

36. Hotyat, F., "Evaluations in Education," in UNESCO, ed., *Reports in an International Meeting of Experts held at the UNESCO Institute for Education* (Hamburg: UNESCO, 1958).

37. Madaus, G. F., and Macnamara, J., *Public Examinations: A Study of the Irish Leaving Certificate* (Dublin: Educational Research Centre, St Patrick's College, 1970).

38. Morris, G. C., *Educational Objectives of Higher Secondary School Science* (University of Sydney, Australia, 1969).

39. Goacher, *Selection Post-16.* See reference 3.

40. Amano, *Education and Examination in Modern Japan.* See reference 8.

41. Bell, R., and Grant, N., *A Mythology of British Education* (London: Panther, 1974).

42. Bloom, B. S., *Evaluation in Higher Education: A Report of the Seminars on Examination Reform Organized by the University of New Delhi Grants Commission Under the Leadership of Dr. Benjamin Bloom* (New Delhi: University Grants Commission, 1961).

43. Bowler, "Payment by Results." See reference 30.

44. Broadfoot, P., "Social Perspectives on Recent Developments in Assessment and Examination Procedures in France," *Irish Journal of Education* **21,** (1987) 36–52.

45. Calder, P., *Impact of Diploma Examinations on the Teaching Learning Process* (Edmundton Alberta: Alberta Teachers' Association, 1990).

46. Cannell, J. J., *The "Lake Wobegon" Report: How Public Educators Cheat on Standardized Achievement Tests* (Albuquerque, NM: Friends for Education, 1989).

47. Consultative Committee on Secondary Education, *Report.* See reference 32.

48. Cuban, L., "Persistent Instruction: Another Look at Constancy in the Classroom," *Phi Delta Kappan* **68,** 7–11.

49. Cummings, W. K., *Education and Equality in Japan* (Princeton, NJ: Princeton University Press, 1980).

50. Cunningham, A. E., *Eeny, Meeny, Miney, Moe: Testing Policy and*

Priorities in Early Childhood Education (Chestnut Hill MA: National Commission on Testing and Public Policy, Boston College, 1989).

51. Curriculum and Examinations Board, *Assessment and Certification*. See reference 33.

52. Eisemon, T. O., Patel, V. L., and Abagi, J., "Read These Instructions Carefully: Examination Reform and Improving Health Education in Kenya," *International Journal of Educational Development* **7** (1987), 1–12.

53. Fallows, J., "Gradgrind's Heirs," Atlantic 1987 **259**(3), 16–24.

54. Gayen, A. K., Nanda, P. D., Duari, P., Dubey, S. D., and Bhattacharyya, N., *Measurement of Achievement in Mathematics: A Statistical Study of Effectiveness of Board and University Examinations in India, Report 1* (New Delhi: Ministry of Education, 1961).

55. Goacher, *Selection Post-16*. See reference 3.

56. Gordon, P., and Lawton, D., *Curriculum Change in the 19th and 20th Centuries* (New York: Holmes and Meier, 1978).

57. Haertel, E., "Student Achievement Tests as Tools of Educational Policy: Practices and Consequences," in B. Gifford, ed., *Test Policy and Test Performance: Education, Language and Culture* (Boston: Kluwer Academic, 1989).

58. Haladyna, T. M., Nolen, S. B., and Hass, N. S., *Report to the Arizona Legislature: Test Score Pollution* (Phoenix, AZ: Arizona State University West Campus, 1989).

59. Holmes, E. G. A., *What Is and What Might Be: A Study of Education in General and Elementary in Particular* (London: Constable, 1911).

60. Holt, J., *The Underachieving School* (London: Pitman, 1969).

61. Kamii, C., *Achievement Testing in Early Childhood Education: The Games Grown-ups Play* (Washington, DC: National Association for the Education of Young Children, in press).

62. Kellaghan, T., and Greaney, V., *Using Examinations to Improve Education: A Study in Fourteen African Countries* (Washington, DC: World Bank, in press).

63. Kelly, A. V., *The Curriculum: Theory and Practice*, 3rd ed. (London: Chapman, 1989).

64. Kreitzer, A. E., Haney, W., and Madaus, G. F., "Competency Testing and Dropouts," in E. F. Lois Weis and H. G. Petrie, eds., *Dropouts From School: Issues, Dilemmas, and Solutions, Part II* (New York: State University of New York, 1989).

65. Little, A., "The Role of Examinations in the Promotion of the 'Paper Qualification Syndrome'," in International Labour Office, ed., *Paper Qualifications Syndrome (PQS) and Unemployment of School Leavers: A Comparative Sub-regional Study* (Addis Ababa: International Labour Office, 1982).

66. Madaus, G. F., and Greaney, V., "The Irish Experience in Compe-

tency Testing: Implications for American Education," *American Journal of Education* **93**, (1985), 268-294.

67. MDC, *America's Shame, America's Hope* (New York: Charles Stewart Mott Foundation, 1988).

68. Mehrens, W. A., and Kaminski, J., *Using Commercial Test Preparation Materials: Fruitful, Fruitless, or Fraudulent?* (Washington DC: National Council on Measurement in Education, 1988).

69. Meisels, S. J., "High-stakes Testing in Kindergarten," *Journal of Educational Leadership* **46**(7) (1989), 16-22.

70. Morris, "Educational Objectives." See reference 38.

71. Mukerji, S. N., *History of Education in India: Modern Period* (Baroda: Acharya Book Depot, 1966).

72. Murphy, R., "The Birth of the GCSE Examination," in A. Hargreaves and D. Reynolds, eds., *Educational Policy: Controversies and Critiques* (London: Falmer, 1989).

73. National Commission on Testing and Public Policy, *From Gatekeeper to Gateway: Transforming Testing in America* (Chestnut Hill, MA: National Commission on Testing and Public Policy, Boston College, 1990).

74. Srinivasan, J. T., "Annual Terminal Examination in the Jesuit High Schools of Madras, India," unpublished doctoral dissertation, Boston College, 1971.

75. Rafferty, M., "Examinations in Literature. Perceptions from Non-technical Writers of England and Ireland from 1850 to 1984," unpublished doctoral dissertation, Boston College, 1985.

76. Raven, J., *Education, Values and Society: The Objectives of Education and the Nature and Development of Competence* (London: Lewis, 1977).

77. Reynolds, D., "Better Schools? Present and Potential Policies about the Goals, Organization, and Management of Secondary Schools," in A. Hargreaves, and D. Reynolds, eds., *Educational Policy: Controversies and Critiques* (London: Falmer, 1988).

78. Rosenholtz, S. J., "Education Reform Strategies: Will They Increase Teacher Competence?," *Canadian Journal of Education* **95**, (1981) 534-562.

79. Shepard, L. A., "Inflated Test Score Gains: Is It Old Norms or Teaching to the Test," paper presented at a symposium on "Cannell Revisited: Accountability, Test Score Gains, Normative Comparisons and Achievement," American Educational Research Association, 1989, San Francisco.

80. Shepard, L. A., and Smith, M. L., "Synthesis of Research on School Readiness and Kindergarten Retention," *Educational Leadership* **44**(3) (1986), 78-86.

81. Smith, M. L., and Shepard, L. A., "Kindergarten Readiness and

Retention: A Qualitative Study of Teachers' Beliefs and Practices,'' *American Educational Research Journal* **25** (1988), 307-333.

82. Spaulding, F. T., *High School and Life: The Regents Inquiry into the Character and Cost of Public Education* (New York: McGraw Hill, 1938).
83. Stake, R. E., McTaggart, R., and Munski, M., *An Illinois Pair: A Case Study of School Art in Champaign and Decatur* (Urbana IL: CIRCE, University of Illinois, 1985).
84. Stodolsky, S. S., *The Subject Matters: Classroom Activity in Math and Social Studies* (Chicago: University of Chicago Press, 1988).
85. Turner, G., "Assessment in the Comprehensive School: What Criteria Count?," in P. Broadfoot, ed., *Selection, Certification, and Control* (New York: Falmer, 1984).
86. Tyler, R. W., "The Impact of External Testing Programs," in W. G. Findley, ed., *The Impact and Improvement of School Testing Programs, 62nd Yearbook of the National Society for the Study of Education* (Chicago: NSSE, 1963).
87. Wheelock, A., and Dorman, G., *Before It's Too Late: Dropout Pevention in the Middle Grades* (Boston, MA: Massachusetts Advocacy Center, 1989).
88. White, E. E., "Examinations and Promotions," *Education* **8,** (1888), 519–522.
89. European Communities Commission, *The Education Structures in the Member States of the European Communities* (Brussels: European Communities Commission, 1987); Feneville J., *Un Apercu du Systeme Educatif Francais* (Paris: Centre International d'Etudes Pedagogique, 1987); (France) Ministere de l'Education Nationale, *La rentree scolaire* (Paris: Ministere de l'Education Nationale, 1990); (France) Ministere de l'Education Nationale, *Reperes et references statistiques sur les enseignements et la formation* (Paris: Ministere de l'Education Nationale, 1990b); (Great Britain) Department of Education and Science, *Education Statistics for the United Kingdom,* 1989 edition (London: Her Majesty's Stationery Office, 1989); Holmes, B., *International Handbook for Education Systems, Vol. 1. Europe and Canada* (New York: Wiley, 1983); Husen, T., and Postlethwaite, T. N., eds., *The International Encyclopedia of Education: Research and Studies* (Oxford: Pergamon, 1985); OECD, *Reviews of National Policies for Education: Greece* (Paris: OECD, 1982); OECD, *Reviews of National Policies for Education: Educational Reforms in Italy* (Paris: OECD, 1985); OECD, *Reviews of National Policies for Education: Spain* (Paris: OECD, 1986); OECD, *Education in OECD Countries 1987-88* (Paris: OECD, 1990); Solberg, W., and Meijering, P. H., "School Leaving Examinations in the Netherlands," in F. M. Ottobre, ed., *Criteria for Awarding School Leaving Certificates* (Ox-

ford: Pergamon, 1979); Witte, G. B., "Curriculum Research in Spain," in U. Hameyer et al., eds., *Curriculum Research in Europe* (Lisse: Swets and Zeitlinger, 1986); Xochellis, P., and Terzis, N., "Curriculum Research in Greece," in U. Hameyer et al., eds., *Curriculum Research in Europe,* (Lisse: Swets and Zeitlinger, 1986); as well as in personal communications from Vasca Alves, Angela Barone, Patricia Broadfoot, Mrs. Fraincoise Connolly, Mike Creswell, Peter Hoeber, Romain Hulpia, E. Leclerq, Javier Valbuena, Leila Vang Anderson, Monique Vervoort, and Ernest Weis.

90. Wake, R. A., Marbeau, V., and Peterson, A. D. C., *Innovation in Secondary Education in Europe* (Strasbourg: Council of Europe, 1979).

91. European Communities Commission, *The Education Structures in the Member States of the European Communities* (Brussels: European Communities Commission, 1987).

92. In this stage of transition, one cannot always be clear from the type of school what curricula actually are offered. For example, in Ireland, which has traditional grammar schools, vocational schools, and comprehensive schools, there is no restriction on what courses the school may offer.

93. Hearnden, A., "Comparative Studies and Curriculum Change in the United Kingdom and the Federal Republic of Germany," *Oxford Review of Education* **12**, (1986), 187–194.

94. Hopmann, S., "The Multiple Realities of Curriculum Policymaking," paper presented to the symposium on International Perspectives on Curriculum History; The Social System of Curriculum Policy and Reforms, American Educational Research Association annual meeting, Chicago, April 4, 1991.

95. Visalberghi, A., "Italy: System of Education," in T. Husen and T. N. Postlethwaite, eds., *The International Encyclopedia of Education* (Oxford: Pergamon, 1985).

96. Bolger, N., and Kellaghan, T., "Method of Measurement and Gender Differences in Scholastic Achievement," *Journal of Educational Measurement* **27** (1990), 165–174.

97. Solberg, W., "School Leaving Examinations. Why or Why Not?," in F.M. Ottobre, ed., *Criteria for Awarding School Leaving Certificates* (Oxford: Pergamon, 1979)

98. Lundgren, U., "European Tradition of Curriculum Research," in V. Hameyer et. al., ed., *Curriculum Research in Europe* (Lisse: Swets and Zeitlinger, 1986).

99. Hameyer, U., Frey, K., Haft, H., and Kuebart, F., *Curriculum Research in Europe* (Lisse: Swets and Zeitlinger, 1986).

100. Kingdon and Stobart, *GCSE Examined.* See reference 21.

101. McDowell and Webb, *Trinity College Dublin.* See reference 23.

102. Broadfoot, "From Public Examinations to Profile Assessment." See reference 9.
103. Eckstein, M. A., and Noah, H. J., "Forms and Functions of Secondary-school Leaving Examinations," *Comparative Education Review* **33** (1989), 295–316.
104. Broadfoot, "Social Perspectives." See reference 44.
105. Noah, H. J., and Eckstein, M. A., 'Trade-offs in Examination Policies: An International Comparative Perspective,"in P. Broadfoot, R. Murphy, and H. Torrance, eds., *Changing Educational Assessment: International Perspectives and Trends* (London: Routledge, 1990).
106. Ireland Department of Education, *Report of the Council of Education as Presented to the Minister for Education* (Dublin: Stationery Office, 1954).
107. Madaus and Greaney, "The Irish Experience." See reference 6b.
108. See, for example, Broadfoot, "Social Perspectives." See reference 45.
109. Eckstein and Noah, "Forms and Functions." See reference 105.
110. Ibid.
111. Kingdon and Stobart, *GCSE Examined*. See reference 21.
112. Macintosh, H., "The Sacred Cow of Coursework," in C. Gipps, ed., *The GCSE: An Uncommon Examination* (London: University of London Institute of Education, 1986), 22.
113. Creswell, "Describing Examination Performance." See reference 2.
114. Orr, L. and Nuttall, D. L., *Determining Standards in the Proposed Single System of Examining at 16+* (London: Schools Council, 1983).
115. Noah and Eckstein, "Trade-offs in Examination Policies." See reference 105.
116. Murphy R., and Torrance, H., ed., *The Changing Face of Educational Assessment* (Milton Keynes, England: Open University Press, 1988).
117. Noah and Eckstein, "Trade-offs in Examination Policies." See reference 105.
118. Eckstein and Noah, "Forms and Functions." See reference 103.
119. Creswell, "Describing Examination Performances." See reference 2.
120. Gordon and Stobart, (1988) p. 36. See reference 21.
121. Nuttall, D. L., Backhouse, J. K., and Willmott, A. S., *Comparability of Standards Between Subjects* (London: Evans/Methuen Educational, 1974).
122. Nuttall, D. L., and Willmott, S., *British Examinations Techniques of Analysis* (Windsor, England: NFER, 1972).
123. Kingdon and Stobart, *GCSE Examined*. See reference 21.
124. Murphy, "The Birth of the GCSE Examination." See reference 72.
125. Calder, "Impact of Diploma Examinations." See reference 45.
126. Winner, L., *Autonomous Technology: Technic-out-of-control as a*

Theme in Political Thought (Cambridge, MA: MIT Press, 1977).

127. Winner, L., *The Whale and the Reactor: A Search for Limits in an Age of High Technology* (Chicago: University of Chicago Press, 1986).
128. MDC, *America's Shame, America's Hope*. See reference 67.
129. Boyer, E. L., *High School: A Report on Secondary Education in America* (New York: Harper & Row, 1983).
130. President's Education Policy Advisory Committee, 1989.
131. National Center on Education and the Economy, *To Secure Our Future: The Federal Role in Education* (Rochester, NY: National Center on Education and the Economy, 1989).
132. Education Commission of the States Task Force on Education for Economic Growth, 1984, National Alliance of Business.
133. Bell, T. H., "American Education at the Crossroads," *Phi Delta Kappan* **65**, (1984) 531-34.

13

Form and Function in Assessing Science Education

Edward H. Haertel

A seventh-grade student asks why maple leaves turn red and elm leaves turn yellow in the fall, and a teacher answers. Perhaps the student understands, perhaps not. How does the teacher know if this small instructional exchange was successful? The semester is coming to an end, and a high school physics teacher must prepare the final examination. What should be selected from the mass of details covered during the preceding months? How can the teacher discover which students really understood the major principles? The state department of education has decided to include performance testing in science in the annual state education assessment program. What kinds of performance tests are feasible logistically? What should be measured? How should the results be scored? These disparate situations all call for evaluations of student learning, but they differ in purpose, format, scope, significance, and the breadth and nature of the learning objectives addressed.

When such varied measurement problems are juxtaposed, it is clear that they call for quite different solutions. Not only are distinct instructional products and results at issue in each case, but also the practical and psychometric requirements of the measurements are different. The teacher trying to judge whether the answer to a student's question was understood probably would not worry about reliability or objectivity. Eventually, many such small interactions might support generalizations about a student's overall level of understanding, but, in a single specific exchange, there is typically no concern with standardization or replicability and no interest in generalizing to any larger universe of either content or occasions. The most important requirement would be that the measurement is timely and unobtrusive.

The physics teacher faces a different set of challenges. The final examination must be sufficiently reliable so that its answers can be used in good conscience as a basis for assigning grades, but the paramount concern probably will be the validity of its content. The content sampled and the problems posed should represent the major themes of the course, but need not and probably should not exhaust the entire range of facts, principles, applications, and broader understandings that the teacher has tried to impart. Performance on the final examination should indicate something about each student's probable level of mastery of more material than is included on the test itself. Standardization matters to the extent that all students should be treated equitably, but if, in the course of the examination, it appeared that a question was unclear, there would be no reason not to write a clarification on the chalkboard for the benefit of the entire class.

The externally mandated performance test raises still other concerns. Let us suppose that scores are to be reported at the level of individual schools, not individual students. The reliability of each student's score is then quite unimportant.

Average scores at the school level may be sufficiently accurate even if individual-level scores have low reliability. Logistical and cost considerations in both administration and scoring will loom large, however, and standardization across testing sites will be critical.

Unfortunately, the specificity of measurement problems is often overlooked. High reliability is regarded as uniformly important for all measurement problems, and questions of validity are differentiated too rarely according to the particular form of test interpretation intended.

This chapter discusses the relationships between the contexts and purposes of measuring school science education and measurement formats and properties.

External and Classroom Testing

Any consideration of assessment contexts must begin with the broad distinction between *externally mandated* and *classroom* measurement, also referred to as *external* and *internal* measurement[1] or *assessment for measurement* versus *assessment for instruction*.[2] The category of externally mandated assessments includes large-scale testing and assessment programs, administered at the district, state, or national level. These include science subtests in elementary school achievement test batteries, state testing and assessment programs, and advanced placement examinations in chemistry, physics, and biology. The science subtest of the American College Testing program's college entrance examination would be included in this category, also.

Classroom tests are created or chosen by individual teachers or small groups of teachers for use at the classroom or sometimes the school level. They include all of the quizzes, unit tests, midterm examinations, and final examinations used to assign grades, diagnose learning difficulties, monitor progress, and plan instruction. Although measurement specialists have devoted far more attention to externally mandated tests, classroom tests usually matter much more to students and teachers.[3]

Externally mandated and classroom tests tend to differ in many ways. The external tests are administered less often, but on a larger scale than classroom tests. The external tests usually sample a larger domain of content, but are less closely tied to the specific curricular or instructional activities of a particular classroom. Interpretations of external tests tend to be norm-referenced, comparing students, schools, districts, or states to one another. Consequently, reliability and standardization are paramount concerns. The scale of external testing programs usually dictates that they not take too much time to administer (e.g., no more than a single class period) and that they be easily and efficiently amenable to scoring.

Usually, classroom tests are intended to provide much more detailed information about a narrower range of content. They are tied closely to the preceding instruction and may employ a variety of formats. Interpretations tend to rely less on percentiles, grade equivalents, or other derived scores, and more on simple raw scores, often supplemented by narrative comments on specific aspects of students' error patterns or solution methods. Students may be compared to one another, but criterion-referenced interpretations are also common; the absolute level of performance, often expressed as a "percent correct," conveys information about how well students are meeting the teacher's expectations. Classroom tests range in duration from brief quizzes to projects completed over a period of weeks or months. Different formats may be used, including individual or group recitations and performance tests as well as essay, short-answer, or objective written examinations.

Form and Function in External Testing

Most externally mandated tests are "high-stakes"; that is, significant rewards or sanctions may be attached to good or poor performance. Because the tests matter, scoring high is likely to become an end in itself and the test may come to exert a powerful influence on both the form and the content of classroom instruction.[4] Overreliance on standardized tests to evaluate the success of students, schools, or educational systems can lead to neglect of the learning outcomes that are not assessed. Even if, when first introduced, the tests are correlated with a broader range of student attainments than they actually measure, their validity may erode with the passage of time. That is because curricula and instruction are likely to shrink toward more thorough coverage of the objectives evaluated, at the expense of the content and processes that the tests omit. It follows that externally mandated tests should be devised to measure as broad a range of the important products of learning as possible.

Limitations of Objective Test Formats

Several factors limit the range of learning outcomes measured by external tests. One obvious limitation arises from the format of these instruments, which rely almost exclusively on multiple-choice, objective items. The kind of knowledge most easily measured using such items is recognition of facts. With care and creativity, multiple-choice items can be constructed to measure more complex understandings, but fundamentally they are limited to "convergent" thinking processes. There must be a single correct answer (or set of correct answers) to be selected from a list provided. This alone places a basic limit on the ranges of knowledge and skills that multiple-choice questions can measure.

Students' understanding of "controlling variables" is a case in point. One typical kind of "higher-order," multiple-choice item in science sets forth an experimental hypothesis, then asks which of four or five experimental procedures might be used to address it. The correct alternative is the only one that provides a clear comparison between groups differing only with respect to the critical variable. In a limited way, this type of item tests the students' understanding of an important principle of the scientific method, but it cannot probe the students' ability to analyze a situation and decide *what* variables need to be controlled or the ability to formulate hypotheses worthy of testing. Attempts to measure this latter ability, "formulating hypotheses," using multiple-choice items, have proven unsuccessful.[5]

There is increasing recognition of the need for free-response and even performance items to measure some important learning objectives. Items calling for students to write brief essays or descriptions of experiments, or to propose multiple possible explanations for a phenomenon are free-response items that might be used to measure forms of learning that it is nearly impossible to measure by using items calling for no more than a selection among fixed alternatives. Still other important goals of learning might be measured best by "performance tests," in which students create a performance or product to demonstrate their understanding.

Limitations of External Assessments

The inherent limitations of objective item formats are just one reason external tests can measure only a limited range of learning achievements. Other limitations arise from the fact that these tests are brief and self-contained. Higher order learning objectives involve forms of reasoning and problem solving that presuppose a substantial base of knowledge. The educational goal is usually not reasoning in the abstract, but reasoning *applied* to a system, situation, or body of information. Such learning cannot be tested unless the requisite knowledge is in place. In the United States, school curricula vary considerably from state to state, across districts, schools, or even among same-grade classrooms within a single school building. At the elementary school level, especially, no one particular science topic will have been taught necessarily in all or even most of the classrooms where the test is to be used. No specific preparatory activities may be assumed to have occurred prior to the external examination, even though testing dates are fixed months in advance. Thus, test questions usually presuppose only general scientific knowledge, unless more detailed information can be presented as part of the test itself. The information that can be presented in the test is limited because time is short and, in any case, the intent is to measure higher-level thinking, not reading comprehension or the rapid assimilation of new information.

Implications for External Testing

Poorly designed external tests — instruments that measure no more than superficial understandings or factual recall — are worse than no tests at all. They cannot provide valid information about the relative or absolute success of different educational programs or systems, and they jeopardize sound curricula and instruction. There will be costs involved in redesigning these tests, however. More comprehensive, valid tests will be more expensive to build, administer, and score. More instructional time will be lost to testing, nonprint examination materials probably will be required (e.g., computers, laboratory equipment), and more extensive training may have to be provided for test administrators and scorers. To justify these costs, there will be pressure to consolidate redundant testing programs and, perhaps, to force the same tests to provide useful information about individual students. Many of those concerned with their results will need to participate in detailed planning for better external tests. Some general directions can be recommended at this point.

First, science tests should provide more extended stimuli. Rather than dividing the content to be assessed into a large number of separate, self-contained items, fewer problem situations should be presented, in greater detail. Each such situation might pose a series of questions to be answered. One series might focus on an ecological system, for example. After a page or two of description, students might be asked about probable food chains and the likely impacts of introducing new species, drought, disease, or other events. They might draw or interpret graphs, or estimate the number of new species that the system could support. In the earth sciences, students might be presented with information about geological strata in an area and asked to work out the probable sequence of events leading to the present configuration. At the level of individual students, this form of testing might imply less adequate sampling of the content domain (although even that is debatable). At the level of schools or larger aggregates, however, different students might be given different problems to solve. Such a procedure, referred to as "matrix sampling," would permit the presentation of far more problem situations than any one individual student would have time to answer.

Second, redesigned tests probably would use a greater range of item formats and assessment materials. Eventually, descriptions of problem situations might employ videotape or video disk technology, so that scientific reasoning and understanding could be assessed apart from reading ability. Computer simulations might be used as assessment as well as teaching tools. Laboratory work might be included also.[6,7] Students must be asked to *generate* their responses, at least some of the time, and some items must call for more than a few words or numbers by way of an answer.

Third, all or part of the examination might be administered by computer. Computer-based test administration would permit branching and probing on the basis of earlier responses. Students could be given hints or partial information until they solve a problem and could be scored on the basis of how much help they needed. Because gaps in understanding would be "filled in" as students worked through a series of items, later questions would not have to be written so as to be completely independent of the knowledge probed in earlier questions. It is emphasized that this is *not* a proposal for computerized adaptive testing as it is presently conceived. It resembles more nearly the process by which a skillful interviewer might probe different topics in greater or lesser depth, giving as much help as needed to assist interviewees in revealing what they know.

Finally, a *fundamental* requirement for better external testing must be a clearer specification of what is to be measured. There should be an explicit, extensive, genuinely comprehensive domain of specific topics that may be sampled for testing and a clear description, richly illustrated with particular examples, of the kinds of analysis, problem solving, and application of scientific principles that may be called for on the test. Ideally, the only viable and legitimate strategy for improving test scores should be to teach to that entire domain, not by covering everything superficially, but by giving students a solid grounding in fundamental principles and important factual information, together with repeated practice in assimilating specific topics, reasoning about them, and applying them.

Form and Function in Classroom Testing

The results of external tests may matter most to policymakers and the public, but for the students themselves, it is the classroom tests that are more important.[8] These are the tests that determine their grades and enable them to judge their own progress in understanding. Moreover, students are likely to have a greater sense of control over and participation in classroom testing than in standardized testing. Generally, important classroom tests are announced in advance. Students know what is to be covered and are expected to prepare accordingly. Indeed, one obvious function of classroom testing is to encourage students to study.

It is expedient to organize this discussion of classroom testing according to types of instructional activities in science. Classroom testing for "textbook science" activities is discussed first, followed by testing for "hands-on" science tasks. The review of "hands-on" exercises is subdivided into collaborative work, informal investigations, projects and simulations, and laboratory activities. These distinctions are general, not sharp, categories. Actual classroom tests call upon learning that has occurred across multiple settings and even across different subject areas in the curriculum. Science instruction and assessment can provide excellent opportunities for students, within the limits of their ability and understanding, to engage in extended expository writing, for example, as well as to apply their mathematical skills.

Relationship of Testing to Instruction

Classroom testing is intimately related to classroom teaching. Teachers must always remember, however, that it is possible to "teach to" a classroom test as well as to an external test. The most impressive test question, regardless of its format, may call for no more than rote recall if its answer or solution has been taught directly. Teachers bear the responsibility for establishing and maintaining the meaning of test questions, whatever their format, by controlling the relationship of the test to the antecedent instruction. If students are told exactly what is going to be on the test, neither they nor their teachers can take much genuine satisfaction in their good performance on it. In teaching science, it often may be good practice to avoid working through some problems or examples that will be on the test until after the test has been given, in order that those problems can be used to measure transfer or application of learning.

Measuring Text-based Learning

For better or worse, lectures and textbook reading assignments continue to play a substantial role in most science instruction. To oversimplify matters considerably, these are the forms of science instruction through which students learn facts, concepts, and the solution of textually presented problems. (Other types of problems would include, for example, qualitative analysis of a chemical solution or the construction of a simple machine capable of lifting a heavy weight.) Along with reading assignments, lecture-discussions, and recitations, this category would include written problem sets. The learning products engendered by these types of instruction can be tested efficiently using written examinations.

Written tests are sometimes divided into "objective" items versus "essays," but it is more useful to array written assessment exercises in science along a continuum from "convergent" to "divergent." Convergent problems and questions admit of a single correct answer. Multiple-choice and other selected-response (objective) item formats fall at the convergent end of the continuum. Divergent problems and questions do not have any single correct answer. In fact, as one approaches the divergent end of the continuum, one finds some problems where students are asked to come up with as many different answers as possible. Various kinds of short-answer, essay, or other test questions may be arrayed along the convergent-divergent continuum according to the degree of latitude they permit in arriving at correct responses.

Written classroom tests may call on students to write explanations, draw diagrams, or work problems. A set of findings might be presented for which students are to propose as many plausible explanations as possible,[9] or a hypothesis might be presented and students asked to describe two different experiments that could be conducted to test it. When straightforward problems are given, their scoring criteria should include attention to the processes and procedures employed to find a solution as well as to the answer finally attained in order to counteract students' tendency to see the only goal as "getting the right answer."

"Hands-on" Science: Assessing Collaborative Work

One distinctive feature of "hands-on" science is its collaborative nature. Sometimes, students may work alone, but, at other times, in pairs, small groups, or as an entire class. Cooperative science learning raises two broad issues for assessment. First, how should individual learning be assessed in the group context? Second, given that there are learning outcomes (e.g., cooperation, collaborative planning, communication) specific to group work contexts, how should these be assessed?

With regard to the first question, teachers should not assume automatically that they always need to evaluate learning outcomes at the individual level. In well-designed, cooperative learning activities, teachers share the responsibility for instruction with their students. Members of a group are supposed to learn from one another. At the same time, teachers may need to share the responsibility for monitoring that learning. Strong norms should be established in the classroom to ensure that everyone in a group has the right and the duty to participate and to profit from the activity, and that those who grasp an idea or discover how to proceed with a task first have the responsibility of sharing their insights with their peers. Collectively, the members of the class have far more eyes and ears than the teacher as well as the advantage of experiencing the instructional task firsthand. If students are taught to notice and attend to one another's understanding or lack of understanding, the teacher's burden will be lighter. In addition, of course, students may be asked to produce separate reports of their common experiments or to designate their respective contributions to a common written account.

With regard to the second question, concerning the assessment of group-specific learning outcomes, the answers are less clear. Teachers may observe student groups at work and try to judge the quality of the interaction, making sure that all are included and that no one student is dominating, but that hardly seems sufficient. At higher grade levels, students might be asked occasionally (probably not routinely) to complete anonymous evaluations of cooperative learning activities. They should be admonished not to name classmates, but to indicate how many in the group really contributed to the group's efforts, whether the activity was interesting, valuable, and so on, as well as how it might be improved.

Teachers should not be surprised if students unaccustomed to group work find it difficult at first or if things do not go smoothly. The skills that students need to practice while working in groups are important, but they may be unfamiliar. Like all skills, they may be expected to improve with practice.

Monitoring Informal Investigations

At the elementary school level especially, but continuing throughout school and beyond, informal investigation is an important part of "hands-on" science. As part of their science instruction, children may be led to engage in purposive observation, description, and classification, for example. They may collect and classify seeds or leaves, discover what kinds of materials conduct electricity, or chart the wanderings of a planet through the night sky on successive evenings. In carrying out such activities, they may learn to use simple measuring instruments, practice skills of estimation, or formulate explanations for what they see and then collect evidence to support or refute their explanations.

The teacher's evaluation of such activities is also likely to be informal, relying mostly on unobtrusive observations. Teachers may find it useful to observe systematically individual students, small groups, or even the class as a whole. Because different students are likely to engage in different tasks, or at least to approach common tasks in alternative ways, such observations will vary from one student to another. They need not be altogether impressionistic or haphazard, however. The teacher's observations should be recorded in writing, either immediately or at the end of the day, noting the time, date, and activity. These remarks may be quite brief, even cryptic, but should specify in some way what was seen, not just the teacher's judgment of its quality. If the comments are recorded on index cards, say, they can be filed easily by student name, can serve as a record of progress and attainment to be used in planning further instruction, shared with parents, used in grading, and perhaps even shared with the students themselves.

In carrying out classroom observations, teachers must work to counter their human tendency to notice the exceptional rather than the typical. Although it is not necessary to make exactly the same number of observations of each student, teachers should make certain that no students are neglected altogether.

Measurement of Projects and Simulations

The construction and manipulation of models of different kinds are fundamental to scientific inquiry and can be invaluable in scientific pedagogy. From elementary school children constructing a model of the solar system to high school students using a computer to simulate experiments in genetics, modeling through projects and simulations is playing an increasing role in science instruction. Assessment of these activities is important in order to determine their overall educational value, identify general misconceptions or gaps in understanding at the classroom level, and assess students' individual achievement levels.

This range of scientific learning and these assessment purposes are ideally suited to the performance assessment methods described by Stiggins.[10] Projects or simulations generally result in scorable products. These may be artifacts constructed by one or more students (e.g., models of the solar system), reports prepared by students documenting their investigations and findings, or perhaps even unobtrusive logs of students' sessions at the computer. In the best of circumstances, there will be considerable variety in the work produced by different students in response to a given assignment, but the teacher should specify clearly the criteria to be met.

Performance assessments involve judgmental rating of such products. They are carried out for a definite purpose and are based on explicit criteria.[11] In science, performance tests can measure the ability to apply knowledge and understanding to practical problem solving. Thus, these tests should be designed to offer relevant problem contexts in which children can apply, consolidate, and expand their base of scientific knowledge. Like written tests, sound performance assessments require planning and preparation, but, with a modest investment of effort, they can be made to yield not only valid but also highly reliable achievement scores. Performance criteria should be defined in observable terms, with explicit statements of acceptable versus unacceptable levels. Possible criteria might include demonstrating a systematic approach to the problem posed, accurate and appropriate use of scientific information, and clear and effective communication.

Measurement of Laboratory Activities

Accurate and detailed recordkeeping is a fundamental tenet of sound laboratory work in science; keeping laboratory notebooks or other careful records also is a typical requirement in school science laboratories. The step from laboratory notebooks to assessments is not automatic, however. Any performance assessment should be designed deliberately, following the same guidelines given for projects and simulations.

As in all performance assessments, the rating criteria established are critical.[12] In addition to such obvious dimensions as proper laboratory procedures and accurate recording of information, the logical structure of the students' investigations might be assessed. False starts and missteps are certainly a part of science and may even be indicators of good, not poor, laboratory work, but, in the end, there should be some evidence of planned purposive progress to test hypotheses or answer questions. Thoughtful attention to rating criteria may clarify what knowledge and skills laboratory exercises really call for. If students need do no more than follow a prescribed sequence of steps, it may be difficult to assess their genuine understanding of the reasons for those steps or of the scientific principles involved.

Giving students a chance to formulate their own investigations before (or instead of) prescribing the approach they are to follow might enhance the pedagogical value of a laboratory exercise as well as provide an opportunity to measure understanding more directly. After familiarizing themselves with the equipment to be used and the kinds of operations that are possible, students might be asked to outline how they would use the equipment to solve a given problem. Their written responses could be scored for types of understanding complementary to those required in carrying out a step-by-step procedure.

Although almost any kind of "hands-on" science activity may involve some use of measuring instruments, laboratory work generally is distinguished from less formal, "hands-on" science learning by increased concerns for safety and for the proper use of equipment. Testing knowledge of safety rules and equipment use should be straightforward, but probably should occur in advance of the laboratory exercise itself.

Summary

Measurements of school science education may serve different audiences and inform different learning outcomes, subject to different practical constraints. To summarize — after distinguishing between external and classroom tests, this chapter has described possible approaches and improvements to each type of measurement. External tests should use more extended stimuli as well as a greater range of assessment materials and item formats, including some formats in which students must generate, not merely select, their answers. The potential of computer-based tests should be explored, but the greatest potential of the computer for measuring school achievement probably lies beyond current models for computerized adaptive testing. Finally, careful attention must be paid to the relationship between external tests and the content domains they represent, especially at the elementary and middle school levels.

With regard to classroom testing, separate recommendations were presented for textbook-based and "hands-on" instruction. In all kinds of science education assessment, teachers must maintain a proper relationship between testing and instruction. For textbook instruction, written examinations seem most efficient, but they should employ a variety of item formats, possibly asking students to draw diagrams, write explanations, or suggest multiple answers to the same question, in addition to asking for responses to more conventional, objective items.

Systematic classroom observation and performance assessment[13] should be the mainstays in determining the results of "hands-on" science education. Suggestions were offered for improving teachers' classroom observations and for increasing the reliability and validity of performance assessments. In collaborative learning activities, students may take some responsibility for monitoring one another's understanding, sharing their insights, and cooperating to reach a common goal. Designing performance assessments creates opportunities for teachers to reflect on the kinds of knowledge and understanding they really require of their students. Such reflection can lead to better teaching as well as to better testing.

References

1. Nitko, A. J., "Designing Tests That Are Integrated with Instruction," in R. L. Linn, ed., *Educational Measurement*, 3rd ed. (New York: American Council on Education/Macmillan Publishing, 1989), 447–474.

2. Cole, N. S., "A Realist's Appraisal of the Prospects for Unifying Instruction and Assessment," in Educational Testing Service, ed., *Assessment in the Service of Learning,* proceedings of the 1987 ETS Invitational Conference (Princeton, NJ: Educational Testing Service, 1988), 103–117.

3. Stiggins, R. J., and Bridgeford, N. J., "The Ecology of Classroom Assessment," *Journal of Educational Measurement* **22** (1985), 271–286.

4. Haertel, E. H., and Calfee, R. C., "School Achievement: Thinking About What to Test,"*Journal of Educational Measurement* **20** (1983), 119–132.

5. Frederiksen, N., "The Real Test Bias,"*American Psychologist* **39** (1984), 193–202.

6. Tamir, P., "An Inquiry Oriented Laboratory Examination,"*Journal of Educational Measurement* **11** *(1974), 25–33.*

7. Tamir, P., Nussinovitz, R., and Friedler, Y., "The Design and Use of a

Practical Tests Assessment Inventory," *Journal of Biological Education* **16** (1982), 42–50.

8. Haertel, E. H., Ferrara, S. F., Korpi, M., and Prescott, B., *Testing in Secondary Schools: Student Perspectives,* presented at the annual meeting of the American Educational Research Association, New Orleans, April 1984.

9. Frederiksen, "The Real Test Bias." See reference 5.

10. Stiggins, R. J., "Design and Development of Performance Assessments,"*Educational Measurement: Issues and Practice* **6,** 3 (1987), 32–42.

11. Ibid.

12. Ibid.

13. Ibid.

14

Performance Assessment: Blurring the Edges of Assessment, Curriculum, and Instruction

Joan Boykoff Baron

What does performance assessment look like? If one were to visit any one of approximately 100 high school science and mathematics classrooms located in 21 states during the 1990–91 school year, performance assessments like these probably would be encountered:[1]

> In a biology class, students will be working in groups of three or four to design and conduct an experiment to determine the optimal salinity of water to be used to ship brine shrimp to a friend.[2]

> In a chemistry class, three or four students will be collaborating to design and conduct an experiment to determine which of two liquids is the regular soda pop and which is the diet version.

> In an earth science class, students will be meeting in teams of three or four to decide whether a particular site would be appropriate for a nuclear power plant.

> In a physics class, students will be assigned to groups of three or four to design and conduct an experiment to calculate the distance that a small toy car can jump between ramps when released from a given height on an inclined plane.

Similar activities would be taking place in algebra, geometry, general mathematics, and advanced mathematics classes. Such tasks are part of a growing national trend to develop performance assessments in science and mathematics.

This chapter describes both the national context of performance assessment and one state's experience in designing and evaluating performance assessments in high school science and mathematics classrooms. It explores some of the reasons educators are creating performance assessments, how they are being structured, what their theoretical and research-based underpinnings are, what teachers see as their advantages, and some of the impediments to their general acceptance.

A Brief History of Recent Performance Assessment in the United States

Performance assessment is not a new idea. It has been and continues to be used successfully as the predominant form of assessment in much of the world. The United States has been almost alone in its enthusiasm for multiple-choice tests. But even in the United States, there have been several forays into performance assessment over the past 20 years. In the 1970s, the National Assessment of Educational Progress (NAEP) included many performance assessment tasks. In mathematics and science, students were asked open-ended questions to which they generated responses. In art and music, students were asked to draw and sing. In the mid-1980s, NAEP conducted a feasibility study in which it adapted and used several exercises developed by the Assessment of Performance Unit in Great Britain. The NAEP report entitled *Learning by Doing*[3] showed that performance assessments are feasible. Unfortunately, despite their demonstrated feasibility, many of these performance tasks were excluded from the most recent rounds of NAEP assessments because of budgetary constraints. During the 1980s in the United States, performance assessments were kept alive by several state assessment programs. The most popular manifestation was in the assessment of writing; more than half of the states experimented with requiring actual writing samples as part of their statewide testing programs.

Science was another discipline that experimented with performance assessments. In 1988, New York State implemented a 45-minute science performance test for all of its fourth-grade students. Pupils moved from station to station to do a series of short tasks that assessed their abilities to use scientific apparatus and elements of the scientific process. Throughout the 1980s, Connecticut incorporated performance assessments into its statewide testing programs in art and music, business and office education, English language arts, foreign languages, industrial arts, mathematics, and science. In the late 1980s and early 1990s, other states, California and Vermont, for example, have been piloting a variety of approaches that incorporate performance assessments into science, mathematics, language arts, and social studies. Judging from the size of audiences at recent national meetings on performance assessment, interest is growing.

Why Performance Assessments?

In the United States today, most students, teachers, and policymakers are comfortable with multiple-choice tests. They are seen as efficient, economical, easy to administer, and objective. Psychometricians have invested many years of research in demonstrating their validity and reliability. It is legitimate to ask: Why change? Several reasons offered for considering alternative forms of assessment are reviewed below.

Tests as Magnets for Instruction

The first reason for considering alternative forms of assessment is that the tests themselves have recently become magnets for instruction. We are living in an "era of accountability." We are all aware that, in classrooms throughout this country, teachers are held accountable for their students' progress. Often, students' progress is assessed by statewide tests. Thus, these tests have become powerful magnets for the kind of instruction that takes place. Where multiple-choice tests are used, they foster instruction that is broad rather than deep, fragmented rather than holistic, and convergent rather than divergent. Students are encouraged to memorize information, generally in the form in which it is presented. They are not encouraged to take risks in their thinking. Multiple-choice questions have only one right answer and the machines that score them are not equipped to read any comments that students might want to write about alternative interpretations, ambiguous items, or subtle or unusual ways to approach the questions. Multiple-choice tests do an excellent job of assessing large numbers of discrete bits of knowledge, but they fall far short when it comes to assessing many other aspects of what society values in the 1990s.

Assessing What Society Values

Whether one reads a science journal or *The Wall Street Journal*, one sees pleas for educating students who have both mastery of specific subject matter and the more general abilities to think, solve problems, communicate, and collaborate.[4] The latter goals require a fresh approach to assessment. Multiple-choice tests cannot gauge the complex problem-solving skills used to answer loosely structured questions. They are not able to evaluate divergent thinking. They cannot rate groups of students working together. They are not equipped to consider speaking, listening, and the delivery of oral presentations. In short, multiple-choice tests were not built to assess well-functioning communities of inquiry.[5]

Assessment that Serves Instruction

Educators in other countries are starting to regard their assessments as "bits of curriculum." Examples include the new national curriculum in Great Britain,[6] innovative mathematics curriculum and teaching programs in

Australia,[7-8] and the national mathematics project in the Netherlands.[9] In this view, assessment is integrated naturally into the curriculum, and the assessment itself models good instruction. Thus a performance assessment becomes the culminating activity of a unit and provides opportunities for students to synthesize their knowledge, make connections, and deepen their understanding of major concepts. The assessment creates situations that are intended to foster the development of deeper levels of understanding. This new view of performance assessment blurs the edges among assessment, curriculum, and instruction.

Creating Publicly Stated Performance Criteria and Mutually Acceptable Standards of Judgment

As both students and teachers gain access to students' thinking, each will be better able to judge its quality. This is another benefit of performance assessment over multiple-choice testing. Performance assessments require clear scoring standards. What does it mean to have an adequate understanding of a scientific concept? What does it mean to understand the sources of error that may have affected the results of a study? By situating these rich performance assessments and scoring criteria in the classroom, both teachers and students will be in a position to talk about the quality of students' thinking and the understanding that is being demonstrated. Opportunities to gain access to students' thinking and to discuss its quality using shared standards have dramatic potential for teaching and learning.

Creating Assessment Models for Teachers

Teachers currently rely heavily on multiple-choice tests, perhaps because they want to prepare students for "high-stakes" state or national tests with similar formats or because multiple-choice tests are so readily available from textbook publishers. A recent study of classroom assessment practices found that, when elementary school teachers were asked to assign grades in science to their students' report cards, most depended to a great extent on multiple-choice test scores as the basis for those grades.[10] What was particularly noteworthy about this study was that, in a series of follow-up interviews, teachers often acknowledged that the students who "knew the most about science" were not necessarily those who got the highest test scores or the highest grades on their report cards. In other words, teachers were aware that students who were "interested in science, read books about it, and knew lots of things that other students did not know" often did not get the highest scores on multiple-choice tests. The time seems ripe for developing assessment alternatives and for providing teachers with new models for classroom use.

Using Performance Assessments for Policy Decisions

Experiences in Connecticut suggest that when policymakers have access to student performance data on actual tasks and to multiple-choice test scores in the same content area, they tend to place higher value on the performance data. This finding seems to hold true whether performance assessments yield higher or lower results than multiple-choice tests. Recent data from Connecticut provide examples of each pattern. In one case, students performed poorly on multiple-choice tests about how to repair small engines but were able to take apart, repair, and reassemble the engines. In the second case, students taking several foreign languages could read more proficiently on multiple-choice tests than they could speak in an oral interview. Policymakers—like members of the general public—placed greater value on the results obtained by the performance assessments. Such results indicate that policymakers will pay attention to performance assessments and act upon them. It is time to inform public policy by developing a wide array of performance assessments that provide valid and reliable information about what students can do in realistic settings.

How Are Performance Assessments Structured?

Performance assessments, like automobiles, are available in many models. A fully equipped, full-size luxury car and a basic, stripped-down, semi-compact car can both "take you places." The same is true of performance assessments. A 5-minute experience in which fourth graders are asked individually to make measurements using thermometers and scales may seem to have little resemblance to a several-day investigation in which a group of high school chemistry students work together to design and carry out an investigation. But, like the cars, both of them can "take you places." Each task requires students to actively solve a problem. Each calls for students to produce solutions rather than merely to recognize them. Each requires students to construct responses rather than simply recall them verbatim from the text. Exactly how much new construction and how much new thinking students will have to do depend entirely on the nature of the performance assessments used and the kinds of curricula and instruction that precede them in the classroom.

The Structure of Connecticut's Performance Assessments

The performance assessment tasks developed for Connecticut's Common Core of Learning Assessment project blend individual work, at the beginning and end, with group work, in the middle, of each task. At the beginning, each student provides information about his or her prior knowledge of the relevant scientific concepts and processes. Each student is asked for an initial impression, an estimate of the solution, a preliminary design for a study, and/or a list of questions he or she would like to ask about the concepts being assessed. There are at least four important reasons for beginning by determining each student's prior knowledge. First, it helps each student do some preliminary thinking before entering the group discussion. Second, it increases the likelihood that each group can begin its deliberations with different perspectives represented. Third, it makes more obvious what each student brings to the task. The students' conceptions (naive or canonical) can be used as a springboard for group discussion and by the teacher to assess where students are beginning.[11] Finally, it provides the students with records of their early thinking, which can be reflected upon from time to time. (Teachers can use the students' initial thoughts as informal baseline data against which to review changes in their thinking that occur during the group work.)

In the middle section of the task, by far the longest phase, students work as a team to produce a group product. Students plan together and work together. In more complex tasks, the work is divided among the members of the group for part of the time. Throughout the task, interdependence is fostered by having each student feel responsible for telling "the whole story," from the development of the group's initial design to its final conclusions. Also, at various intervals, students are asked to monitor their success both as a group and as individuals working as part of a group. Through a variety of accompanying assessment tools, some written (checklists, optional journals, logs, and portfolios) and some oral and visual (videotapes of discussions, oral presentations), students have frequent opportunities to provide evidence of their deepening understanding and related reflections. To warrant several hours of group time, tasks must meet one of two criteria: they must provide a forum in which students can work together and talk together in ways that intensify their understanding of essential scientific (or mathematical) concepts and processes, and/or their structure must allow students to divide a large amount of work among the group's members and report their findings to the group.[12]

Following the team effort, a related task is administered to students individually to see what each student learned from the group experience. These tasks consist of applications or "near-transfer" activities that attempt to assess the same content and processes as those in the group task. They provide each student an opportunity to apply the learning, synthesis, and integration that occurred in the group experience to a new context. They also provide teachers and policymakers with a summation of what each student knows and can do at the end of a series of meaningful learning and assessment opportunities. Characteristics of enriched performance assessment tasks are shown in Table 1.

Table 1. Characteristics of Enriched Performance Assessment Tasks

Enriched performance assessment tasks have the following characteristics:

They are grounded in real-world contexts.

They involve sustained work and often take several days of combined in-class and out-of-class time.

They are based on the most essential aspects of the content of the discipline(s) being assessed; that is, they deal with "big ideas" and major concepts (energy, form and function, change) rather than peripheral or tangential topics.[13-14]

They are broad in scope, frequently integrating several scientific principles and concepts.

They blend essential content with essential processes, often requiring the use of scientific methodology and the manipulation of scientific tools and apparatus.

They present nonroutine, open-ended, and sometimes loosely structured problems that require students both to define the problem and to determine a strategy for solving it. Optimal problems afford both multiple solutions and multiple solution paths. [15-17]

They encourage group discussion and "brainstorming" in which a problem is considered from multiple perspectives.

They require students to determine what data are needed; to collect, report, and portray the data; and to analyze the data to discern sources of error.

They call upon students to make, explain, and defend their assumptions, predictions, and estimates.

They stimulate students to make connections and generalizations that will increase their understanding of the important concepts and processes.

They are accompanied by explicitly stated scoring criteria related to content, process, group skills, communication skills, and to a variety of motivational dispositions and "habits of mind."[18]

They spur students to monitor themselves and to think about their progress (as individuals, as members of a group, and as a complete group) in order to determine how they might improve both their investigational and group process skills.

They require students to use a variety of skills (such as reading, listening, viewing) for acquiring information and for communicating their strategies, data, conclusions, and reflections (through, for example, speaking, writing, and graphic displays).

Learning and Motivational Principles
That Undergird Performance Assessment

Recent theory and research in cognitive and motivational psychology have provided a foundation for the development of performance assessment. Studies of effective learning environments have yielded many sound learning principles that can be incorporated into the design of a task. They indicate that students perform best when then can: (i) experience an active rather than a passive learning mode; (ii) tie new learning to what they know and believe already; (iii) tell a "whole story" with many interconnecting parts;[19-20] (iv) talk with others about their understandings;[21] (v) monitor their own progress;[22-23] (vi) have a clear statement of expectations; and (vii) realize that their knowledge can be transferred to new situations more easily if they have had experience in learning how and under what conditions it is appropriate to do so.

Research in motivational psychology[24-28] contributes additional findings that can be included in task design. These findings suggest that the best work will be elicited from students when they: (i) tackle problems that are embedded in real-world contexts that have apparent relevance; (ii) have some choice in and control over their learning; (iii) believe they have the requisite knowledge and skills to carry out the task, that is, when they have strong beliefs in their own effectiveness in a given domain; (iv) are given tasks that are intrinsically motivating and have personal meaning; (v) realize that their individual contributions are recognized and valued; (vi) take responsibility for their learning; (vii) undertake tasks that are challenging and engaging; and (viii) attempt tasks that allow for self-regulatory behaviors such as managing their time and resources in order to achieve specific goals.

How Do Theory and Research on Cooperative and Collaborative Learning Support Performance Assessment?

At least three separate lines of social policy, theory, and research have led to encouraging students to work collaboratively on assessment tasks. An important policy impetus comes from society at large, which of late has been calling for students who can work productively in group settings. This call is reiterated in both professional and service-oriented work settings. Whether in fast-food restaurants, industrial settings, or research laboratories, people do not work in isolation. They depend on each other for ideas, implementation, follow-through, and well-being.

A second line of support emanates from the writings of Vygotsky.[29] In a chapter entitled "The Interaction Between Learning and Development," he emphasized the social and interpersonal aspects of learning. Stated simply, people learn from others around them. In fact, a person's potential as a learner is inextricably bound to the surrounding interpersonal learning environment. Vygotsky defined the "zone of proximal development," as "the distance between the actual developmental level as determined by independent problem solving and the level of potential development as determined through problem solving under adult guidance or in collaboration with more capable peers."[30] He went on to claim that "an essential feature of learning is that it creates the zone of proximal development; that is, learning awakens a variety of internal developmental processes that are able to operate only when the child is interacting with people in his [or her] environment and in cooperation with his [or her] peers."[31]

Tasks that call upon students to work together to "make meaning" and to deepen their understanding rest on the work of Vygotsky and a group of researchers who have extended his ideas by developing dynamic assessment models that recognize the power of interpersonal settings to structure and support learning.[32–35] Thus, two or more students, each with an incomplete level of understanding, can use the resources of the group to develop a more thorough understanding than any single member may have brought to the task.

A third line of theory and research has evolved under the title "cooperative learning."[36–43] There is general agreement that cooperative learning has a positive effect on student achievement, but there is much less agreement about what characterizes the optimal set of variables and incentives. For example, Johnson, Maruyama, Johnson, Nelson, and Skon conducted a meta-analysis of 122 studies and concluded that cooperative goal structures produce greater achievement than either competitive or individual goal structures.[44] The conditions that foster productive groups are: (i) clearly perceived, positive interdependence; (ii) task structures that ensure that the efforts of all members of the group are needed; (iii) considerable face-to-face interaction; (iv) a sense of personal responsibility to achieve the group's goals; (v) clear individual accountability; and (vi) students who have the necessary collaborative skills.[45]

Slavin's analysis of 46 studies highlighted the need to differentiate between the task structure (whether individuals worked alone or together) and the incentive structure. He concluded that cooperative learning is most effective when there are group rewards and individual accountability.[46]

Although agreement is lacking on the reasons cooperative learning has been so successful, several researchers have built explanatory bridges between the work in cooperative learning and that in motivation research. Sharan and Shaulov note that "working toward a common goal with peers and involvement in deciding one's own course of work are generally considered to be critical factors for arousing intrinsic motivation in a task and a sense of personal responsibility for its completion."[47–51] In their own work, Sharan and Shaulov studied the effects of motivation to learn on achievement and found that, for some of their subjects, perseverance with the task was the most powerful predictor of achievement.

The emphasis on cooperative and collaborative learning strategies in today's workplace, the strong theoretical basis for peer-centered learning groups provided by Vygotsky, and the accumulated wisdom from those involved in cooperative learning research and practice all support the need to develop assessment opportunities that require groups of students to work together in solving complex, loosely structured problems.

Connecticut's Experience with Performance Assessment

The first two years of the Connecticut Common Core of Learning Assessment project were, in a sense, an experiment. It tested whether complex performance tasks requiring groups of students to work together over several days could be properly used as assessments of what students know and can do. Early indications are that they can be so used, and teachers' comments on issues related to performance assessment are given below. The second phase of the Connecticut project will focus on scoring and examining issues related to validity, reliability, fairness, feasibility, and reporting formats. In order to be useful in the classroom, the feasibility, validity, and fairness criteria must be met. In addition, to be useful to school district and state-level policymakers, the reliability and aggregation criteria also must be satisfied.

Advantages of Performance Assessment over Pencil-and-Paper Tests

Teachers from ten states who participated in the first year of the Connecticut project report that performance tasks have several advantages over written tests. The most commonly stated advantages are that they permit the valid assessment of affective areas (that is, attitudes and dispositions) and process and communication skills. However, a number of other advantages surfaced in teachers' responses. These advantages are stated below, in the teachers' own words:

> I know of no way to assess certain skills development or attitudinal development through the use of paper-and-pencil evaluations alone. If we want to determine whether students have mastered a particular skill, it makes sense to me to place the student in a problem-solving environment and observe whether he or she can use that skill to perform a particular task. *Edward Snyder, New York*

> [Performance assessment] allows for cooperative learning, stresses

[positive] attitudes and thinking skills, and encourages creativity. It keeps *more* of the students *more* actively involved. It is what learning and science are all about. *Sue Battersby, Connecticut*

The method gives students a chance to apply the topics discussed in class to a practical situation, to tie several threads of the curriculum together, to broaden their understanding, and to give them a feeling of the teamwork, "brainstorming," and problem solving involved in chemistry — missing altogether in pencil-and-paper tests. *Mary Johnson, Michigan*

By giving some control over these activities to the students...kids get the opportunity to show what they can do.... They respond in ways we often never guess to look for. They have strengths and abilities that get masked by traditional approaches. *Gordon Turnbull, Connecticut*

The greatest advantage is that we are finally trying to develop thinking skills and positive attitudes toward the nature of science and not just memorization of facts, ideas, and concepts, which kids do not internalize. *David Artt, Vermont*

Students learn how to integrate and verbalize scientific concepts. They use "brainstorming" and they are learning how to listen. Students who are engaged affirmatively in the learning process become active learners. Students get to know each other! New leaders emerge in groups—those who are most creative and process-oriented take on much more important roles. Students feel better about themselves when their opinions are expressed and respected by others in their group. *John Mangini, Connecticut*

Students have an opportunity to practice "doing science" beyond testing content. We can come closer to assessing what they really know and can do. It helps the teacher to consider richer, [more] meaningful experiences for the student. Creativity and divergent thinking can be tapped. Pencil-and-paper tests often do not go beyond recognition. Cooperative group work is a lifelong skill that will encourage retention. *Celia Rainwater, Texas*

You can test items that are difficult to measure with pencil-and-paper examinations—for example, thought processes, analysis, and real understanding of content (which means being able to explain it). *Robyn Ford, Texas*

Success of Performance Tasks

If success is to be measured by student engagement, then the [performance tasks] have been very successful. Students in my school have worked through break periods (voluntarily) and have described these tasks as very enjoyable. *Compton Mahase, New York*

I have enjoyed writing and using the performance tasks in my classroom and the students have requested that we do more of these types of assessments — they hesitate to call them tests because they enjoy doing them so much! *Irene Tlach, Minnesota*

It also integrates and equalizes learning among boys and girls, as well as ethnic groups who normally remain segregated in the classroom. *John Mangini, Connecticut*

The method has been very successful in involving students in the creation of the investigation. They really enjoy having the opportunity to solve the problem on their own. It provides a means for them to use their creativity as well as to see how science is done in the real world. Too often, students view a scientist's job as a lonely one when, in fact, it is not. The students admit they learn more from this approach than from workbook instruction. They are overwhelmingly in favor of developing their own techniques. *Mary Johnson, Michigan*

Some students finally have an opportunity to show their classmates their creativity and enthusiasm. I feel that the number one success has been the better feeling of community and fun in the science classroom. *Bob Segall, Connecticut*

[Performance assessment] encourages the feedback process. A kid tries something, sees its outcomes, and may try a modified approach…. I saw sparkles in eyes that had never shown any. *Gordon Turnbull, Connecticut*

This philosophy has changed the way I view my teaching. The performance tasks are ideal for self-assessment for teachers as well as for student assessment. *Jim Bickel, Minnesota*

Opportunities for Students Who Traditionally Do Not Perform Well on Pencil-and-Paper Examinations

Creative students do well. Minority students also benefit greatly. In a small group, they are no longer a minority and therefore achieve much better results because they have a much more active role. *John Mangini, Connecticut*

My failure rate has been reduced. Students feel like part of a team effort. They now look to each other as a resource for understanding. *Celia Rainwater, Texas*

I have done very open-ended tasks with groups of students. I find my "A" students struggle for ideas and creativity. Students who have trouble completing their written labs come up with creative ideas and do very well in completing the task and drawing a well-rounded conclusion. I certainly see another side of the students. *Irene Tlach, Minnesota*

It shows *great* promise for those students who do not perform well on traditional exams. It allows us to evaluate *process* and gives the teacher a whole new frame of reference. *Sue Battersby, Connecticut*

[Performance tasks show promise] especially for students who have great ideas and thoughts, but who for various reasons cannot put them in writing. This process encourages and does not discourage. *David Artt, Vermont*

Performance Assessment Informs Instruction

This type of assessment has definitely changed my way of teaching. I have taken more time to look at science—the skills, process, and content involved and the way I am presenting them. The assessment has had a positive effect on me, my students, and my state. It has enhanced instruction—led to more "hands-on" activities. Students definitely see the importance of doing the labs, especially when they are tested with a lab. *Irene Tlach, Minnesota*

It has made me focus more on the objectives of the tasks. "Less is more" has to be utilized since some elements of the curriculum were eliminated in order to incorporate more group work. I have a better feel for students' strengths and weaknesses by going around to smaller groups. *Bob Segall, Connecticut*

It helps me to look more closely for integrating threads and forces me to develop better grading and scoring rubrics. *Compton Mahase, New York*

I believe it gives us better insight into the students' process skills. Do they really understand what they are doing? I think good memorizers can "con" us into thinking they know more than they really do. *George Lelievre, Connecticut*

It helps me by realizing the different abilities and interests of my students. When I can focus on these, I can get much more out of them. *David Artt, Vermont*

How Readily Do Teachers and School Systems Accept Performance Assessment?

The jury is still out on this question. The teachers in Connecticut and nine other states were nominated to participate in the Common Core of Learning Assessment project because of their own creativity and willingness to implement performance tasks and are volunteers. When asked this question, a range of responses was received. For the most part, as expressed in the preceding comments, they were very positive about their own experiences.* However, many of them indicated that a large number of their colleagues would be more resistant to change. Some explained that they and their colleagues were oppressed by time shortages. The participating project teachers felt that the time required to develop, prepare, carry out, and evaluate a performance task was formidable. In addition, time would be required to train teachers. Others, especially those who were members of the Coalition of Essential Schools, felt that their colleagues would embrace these activities as valid ways to assess what their schools really valued. At a meeting of the Connecticut Science Supervisors Association in spring 1990, a similar range of sentiment was expressed. When asked to project what proportions of their 258 science teachers would be "very receptive" to using such measures, estimates were about 15 percent. The supervisors estimated that just over 50 percent would be "somewhat receptive" and that approximately 30 percent would be "not at all receptive."

These findings are consistent with those of Blum and Niss who cited three serious obstacles to providing effective problem solving, modeling, and applications in mathematics classrooms.[52] These obstacles came from the viewpoints of the instruction, the learner, and the teacher. Regarding instruction, "many math teachers are afraid of not having enough time to deal with problem solving, etc., in addition to the wealth of compulsory mathematics included in the curriculum." From the learner came complaints that "problem solving, modeling and applications to other disciplines make the mathematics lessons 'more demanding' and less predictable than traditional mathematics lessons. Routine mathematical tasks such as calculations are more popular with many students because they are easier to grasp and problems can often be solved merely by following certain recipes, which makes it easier for students to obtain good marks in tests and examinations."[53]

From the teachers came complaints that "problem solving and references to the world outside mathematics make instruction more open and more demanding for teachers because additional 'nonmathematical' qualifications are necessary and make it more difficult to assess students' achievements. Moreover, many teachers do not feel able to deal with

---

* Fewer than one-fourth of the teachers participating in the first year of the project did not fulfill their responsibility to implement three tasks in their classrooms.

applied examples which are not taken from subjects they have studied themselves. And very often teachers simply either do not know enough examples of problem solving suitable for instruction or they do not have enough time to update examples, to adapt them to the actual class, and to prepare the teaching of them in detail."

Blum and Niss assert that "in the light of arguments put forward in favor of problem solving, modeling, applications, and connections to other subjects, we should continue to make every effort to overcome these obstacles." They feel that this "could be done both by adequate preservice and inservice teacher education, to equip teachers . . . with knowledge, abilities, experiences and in particular with attitudes to cope with the demands required for problem solving,...so that *problem solving and relations to the real world become and remain essential parts* of mathematics instruction at all levels, even in spite of all the difficulties mentioned." [54] Blum and Niss conclude their article with reference to a "bottleneck" hampering widespread integration of problem solving in mathematics instruction. They claim that "very few curricula around the world make substantial problem solving, modeling, and applications abilities the object of systematic assessment and testing." They continue by stating that, "in an examination-based educational system, as most educational systems are, instructional components which are not tested on a par with other components tend to occupy marginal positions only." [55]

With the assistance of the teachers in the first and second phases of the Connecticut project, this bottleneck may be reduced. Sustained performance assessment tasks in both science and mathematics provide viable alternatives for those teachers who are willing to increase the time spent on developing problem solving, collaborative, and communications skills in their students. As the title of this chapter indicates, when assessment occurs in the service of instruction, there will be an intentional and beneficial blurring of traditional pedagogical boundaries.

Acknowledgments

Funding support from the Connecticut State Department of Education and the National Science Foundation (Grant SPA-8954692 Principal Investigator: Joan Boykoff Baron) for the Connecticut Common Core of Learning Assessment project is acknowledged with gratitude. The content of this chapter does not necessarily reflect the views of the funding agencies.

References and Notes

1. During its first two years, the Connecticut Common Core of Learning Assessment project has benefited from the assistance of almost 400 educators in the United States, Australia, Canada, Great Britain, Israel,

and the Netherlands who have helped to shape its work. In July 1989 we began with a group of 50 dedicated high school science and mathematics teachers from the states of Connecticut, Michigan, Minnesota, New York, Texas, Vermont, and Wisconsin, as well as teachers from the Coalition of Essential Schools located in Connecticut, Maryland, New Hampshire, and New York and several dozen scientists, mathematicians, psychologists, and psychometricians.

A second group of 90 teachers was added in July 1990. Two new groups were included. Project Relearning states were represented by Arkansas, Delaware, New Mexico, Pennsylvania, and Rhode Island. The American Federation of Teachers Urban Districts Leadership Consortium was represented by 16 large cities: Albuquerque, Cincinnati, Cleveland, Dade County (FL), Detroit, Hammond (IN), Kansas City (MO), Los Angeles, New York, New Orleans, Philadelphia, Pittsburgh, Rochester (NY), Saint Paul, San Francisco, and Washington, DC. The project work has been coordinated and crafted by a small group of Common Core of Learning Assessment staff members: in science, the task development effort was coordinated by Jeffrey Greig, with the assistance of Sigmund Abeles, Michal Lomask, and Daniel McGrail; in mathematics, the task development effort was coordinated by Bonnie Laird Hole, Judith Collison, and Susan Dixon with the assistance of Steven Leinwand and Leslie Paoletti. Douglas A. Rindone provided invaluable overall direction for the project, with very able assistance from Amy Shively, Hannah Kruglanski, Bruce Davey, Claire Harrison, and Steven Martin. Arlene Morrissey and Kathy Nugent provided the necessary clerical support.

2. Several teachers participated in developing these four tasks for the Connecticut Common Core of Learning Assessment. They are David Artt, Vermont; Susan Battersby, Connecticut; Gene Bourguin, Connecticut; George Hooker, Vermont; John Mangini, Connecticut; Daniel McGrail, a member of the Iowa Physics Task Force and a co-author of the PRISMS Teacher's Guide; Leona Truchan, Wisconsin; Gordon Turnbull, Connecticut; and Dale Wolfgram, Michigan.

3. National Assessment of Educational Progress, *Learning by Doing: A Manual for Teaching and Assessing Higher-Order Thinking in Science and Mathematics* (Princeton, NJ: Educational Testing Service, 1987).

4. National Association of State Boards of Education, *Restructuring Curriculum: A Call for Fundamental Reform,* A Report of the Curriculum Study Group (Washington, DC: National Association of State Boards of Education, 1988).

5. Lipman, M., "Some Thoughts on the Formation of Reflective Education," in J. B. Baron and R. J. Sternberg, eds., *Teaching Thinking Skills: Theory and Practice* (New York: W. H. Freeman, 1987), 151–161.

6. Burstall, C., "Update on the New National Assessment in Great

Britain," paper presented at the Large Scale Assessment Conference sponsored by the Education Commission of the States and the Colorado State Department of Education, Boulder, Colorado, June 1989 and June 1990.

7. Lovitt, G., and Clarke, D., *The Mathematics Curriculum and Teaching Program Activity Bank, Vols. 1 and 2* (Canberra, Australia: Curriculum Development Centre of the Australian Federal Department of Employment, Education and Training, 1988).

8. Clarke, D., *Assessment Alternatives in Mathematics* (Canberra, Australia: Curriculum Development Center, 1988).

9. Lange, Jzn. Jan de., *Mathematics Insight and Meaning* (The Netherlands: OW and OC, 1987).

10. Baron, J. B., "How Science Is Tested and Taught in Elementary School Science Classrooms: A Study of Classroom Observations and Interviews," paper presented at the annual meeting of the American Educational Research Association, Boston, April 1990.

11. Pecheone, R. L., Baron, J. B., Forgione, P. D., Jr., and Abeles, S., "A Comprehensive Approach to Teacher Assessment: Examples from Math and Science," in A. B. Champagne, ed., *This Year in School Science 1988: Science Teaching — Making the System Work* (Washington, DC: American Association for the Advancement of Science, 1988), 191–214.

12. Aronson, E., Blaney, M., Stephan, C., Sikes, J., and Snapp, M., *The Jigsaw Classroom* (Beverly Hills, CA: Sage Publications, 1978).

13. American Association for the Advancement of Science, *Science for All Americans: A Project 2061 Report on Literacy Goals in Science, Mathematics, and Technology* (Washington, DC: American Association for the Advancement of Science, 1989); reprinted as, Rutherford, F. J., and Ahlgren, A., *Science for All Americans* (NY: Oxford University Press, 1990).

14. National Council of Teachers of Mathematics, *Curriculum and Evaluation Standards for School Mathematics* (Reston, VA: National Council of Teachers of Mathematics, 1988).

15. Resnick, L. B., "Teaching Mathematics as an Ill-structured Discipline," in R. I. Charles and E. A. Silver, eds., *The Teaching and Assessing of Mathematical Problem Solving, Vol. 3* (Reston, VA: Lawrence Erlbaum Associates and the National Council of Teachers of Mathematics, 1989), 32–60.

16. Greeno, J., "A Study of Problem Solving," in R. Glaser, ed., *Advances in Instructional Psychology, Vol. 1* (Hillsdale, NJ: Lawrence Erlbaum Associates, 1978), 13–75.

17. Schoenfeld, A., ed., *Cognitive Science and Mathematics Education* (Hillsdale, NJ: Lawrence Erlbaum Associates, 1976).

18. Wiggins, G., "A True Test: Toward More Authentic and Equitable

Assessment," *Phi Delta Kappan* **70** (1989), 9.

19. Heath, S. B., "What No Bedtime Story Means: Narrative Skills at Home and School," *Language in Society* **11** (1982), 49–76.

20. Bransford, J. D., and Stein, B. S., *The Ideal Problem Solver: A Guide for Improving Thinking, Learning, and Creativity* (New York: W. H. Freeman, 1984).

21. Vygotsky, L. S., *Mind in Society: The Development of Higher Psychological Processes,* M. Cole, V. John-Steiner, S. Scribner, and E. Souberman, trans. and eds. (Cambridge, MA: Harvard University Press, 1978; original work published 1935).

22. Flavell, J. H., "Metacognitive Aspects of Problem Solving," in L. B. Resnick, ed., *The Nature of Intelligence (Hillsdale, NJ: Lawrence Erlbaum Associates, 1976), 231–235.*

23. Sternberg, R. J., " Teaching Intelligence: The Application of Cognitive Psychology to the Improvement of Intellectual Skills," in J. B. Baron and R. J. Sternberg, eds., *Teaching Thinking Skills: Theory and Practice* (New York: W. H. Freeman, 1987), 182–218.

24. Ames, R., and Ames, C., eds., *Research on Motivation in Education: Vol 1. Student Motivation* (San Diego, CA: Academic Press, 1984).

25. Ames, C., and Ames, R., eds., *Research on Motivation in Education: Vol 2. The Classroom Milieu* (Orlando, FL: Academic Press, 1985).

26. Ames, C., and Ames, R., eds., *Research on Motivation in Education: Vol 3. Goals and Cognitions* (San Diego, CA: Academic Press, 1989).

27. Covington, M. V., "The Motive for Self-worth," in R. Ames and C. Ames, eds., *Research on Motivation in Education: Vol. 1*, 78–108. See reference 24.

28. McCombs, B. I., and Marzano, R. J., "Putting the Self in Self-regulated Learning: The Self as Agent in Integrating Will and Skill," *Educational Psychologist* **25,** 1 (1990), 51–69.

29. Vygotsky, L. S. (1978). *Mind in Society,* 79–91. See reference 21.

30. Ibid., 86.

31. Ibid., 90.

32. Brown, A. L., and Palincsar, A. S., "Reciprocal Teaching of Comprehension Strategies: A Natural History of One Program for Enhancing Learning," in J. Borkowski and J. D. Day, eds., *Intelligence and Cognition in Special Children: Comparative Studies of Giftedness, Mental Retardation, and Learning Disabilities* (New York: Ablex, 1989), 81–132.

33. Campione, J. C., and Brown, A. L., "Linking Dynamic Assessment with School Achievement," in C. S. Lidz, ed., *Dynamic Assessment: An Interactional Approach to Evaluating Learning Potential* (New York: The Guilford Press, 1987), 82–115.

34. Feuerstein, R., *The Dynamic Assessment of Retarded Performers: The Learning Potential Assessment Device, Theory, Instruments, and Tech-*

niques (Baltimore: University Park Press, 1979).
35. Feucrstcin, R., Rand, Y., Jensen, M. R., Kaniel, S., and Tzuriel, D., "Prerequisites for Assessment of Learning Potential: The LPAD Model," in C. S. Lidz, ed., *Dynamic Assessment: An Interactional Approach to Evaluating Learning Potential* (New York: The Guilford Press, 1987), 35–51.
36. Deutch, M., "A Theory of Cooperation and Competition,"*Human Relations* (1949), 129–152.
37. Deutch, M., "An Experimental Study of the Effects of Cooperation and Competition," *Human Relations*, 2 (1949), 199–232.
38. Deutch, M., "Cooperation and Trust: Some Theoretical Notes," in M. R. Jones, ed., *Nebraska Symposium on Motivation* (Lincoln, NE: University of Nebraska Press, 1962), 275–319.
39. Hibbard, K. M., and Baron, J. B., "Assessing Students Working in Groups: Lessons from Cooperative and Collaborative Learning," paper presented at the annual meeting of the American Educational Research Association, Boston, April 1990.
40. Johnson, D. W., and Johnson, R. T., "Motivational Processes in Cooperative, Competitive, and Individualistic Learning Situations," in C. Ames and R. Ames, eds., *Research on Motivation in Education: Vol. 2. The Classroom Milieu* (Orlando, FL: Academic Press, 1985), 249–286.
41. Johnson, D. W., and Johnson, R. T., "Cooperative Learning in Achievement," in S. Sharan, ed., *Cooperative Learning: Theory and Research* (New York: Praeger, 1990), 249–286.
42. Sharan, S., and Sharan, Y., *Small Group Teaching* (Englewood Cliffs, NJ: Educational Technology Publications, 1976).
43. Slavin, R., *Cooperative Learning* (New York: Longman, 1983).
44. Johnson, D. W., Maruyama, G., Johnson, R., Nelson, D., and Skon, L., "Effects of Cooperative, Competitive, and Individualistic Goal Structures on Achievement: A Meta-analysis," *Psychological Bulletin 89* (1981), 47–62.
45. Johnson and Johnson, 233. See reference 41.
46. Slavin, R., *Cooperative Learning.* See reference 43.
47. Sharan, S., and Shaulov, A., "Cooperative Learning, Motivation to Learn, and Academic Achievement," in S. Sharan, ed., *Cooperative Learning: Theory and Research* (New York: Praeger, 1990), 173–202.
48. Ames, C., "Effective Motivation: The Contribution of the Learning Environment," in R. Feldman, ed., *The Social Psychology of Education* (Cambridge, England: Cambridge University Press, 1986), 253–256.
49. DeCharms, R., *Enhancing Motivation: Change in the Classroom* (New York: Irvington [Halsted-Wiley], 1976).
50. Ryan, M., Connell, J., and Deci, E., "A Motivational Analysis of

Self-determination and Self-regulation in Education," in C. Ames and R. Ames, eds., *Research on Motivation in Education: Vol. 2. The Classroom Milieu* (Orlando, FL: Academic Press, 1985), 13–47.

51. Weisz, J., and Cameron, A., "Individual Differences in the Students' Sense of Control," in C. Ames and R. Ames, eds., *Research on Motivation in Education: Vol. 2. Classroom Milieu* (Orlando, FL: Academic Press, 1985), 93–140.

52. Blum, W., and Niss, M., "Mathematical Problem Solving, Modeling, Applications, and Links to Other Subjects — State, Trends and Issues in Mathematics Instruction," paper presented at the Sixth International Congress on Mathematical Education, Budapest, 1988, and summarized in W. Blum, M. Niss, and I. Huntley, eds., *Modelling, Applications, and Applied Problem Solving — Teaching Mathematics in a Real Context* (Chichester, England: Horwood, 1988.)

53. Ibid., 19–20.

54. Ibid., 70.

55. Ibid., 33.

15

Assessing Accelerated Science for African-American and Hispanic Students in Elementary and Junior High School

Michael Johnson

In discussions about increasing the number of minority students in science and mathematics programs, it is often assumed that these students are below grade level or in some way inadequate academically. The term "disadvantaged" frequently conjures up a picture of a minority child not as a victim of a depressed economic situation or a poor school, but rather as a student with inferior intellectual ability. The latter image is far from true; however, many urban schools are not equipped to prepare most children to face the scientific challenges of the future. Teachers' low expectations of students, many inexperienced teachers, and a lack of materials, positive role models, and cultural support from other students and adults combine to create a situation in which a student's success depends more on luck than pluck. In other words, a child's success may depend largely on having a good teacher in a class that is held to at least modest expectations and includes supplementary parental education, which in many cases is also essential.

The public education system can be said to serve no student as well as could be desired. But perhaps the most ignored groups of students in the United States are the elementary and junior high school minority students who are doing well in school and who also are interested in science and mathematics. Many resources are spent on the minority students who do trickle through the science and mathematics pipeline. Few resources are provided to identify at an early stage students with an aptitude for science and mathematics and to finance a support system to ensure expansion of the number of young students choosing careers in science and mathematics. Often, their white counterparts, with whom they are equal in ability, will be able to participate in science enrichment activities. In minority communities extracurricular activities in science, where available, frequently focus on science literacy for all students, regardless of interest.

Many minority students do survive the education obstacle course, yet few pursue careers in science and mathematics. Academic ability is certainly not the issue, and remediation is not the solution. As a group, these students both like and are able to do science at a high level. In fact, the conclusion of the staff at the Science Skills Center in Brooklyn is that *most* minority students like science even if they have refused to take science courses in school.

The Science Skills Center was founded in 1979 by a group of African-American science professionals and science and mathematics teachers. Its objective was simple: to develop an institution that would encourage young minority students to pursue careers in science and mathematics. From its beginning as a Saturday program in the basement of the home of one of the founders, it has grown to occupy a wing of a public school and become an after school, Saturday, and full-day summer program.

Using Assessment to Promote and Document Achievement

The center's search for an evaluation tool started in 1988 when staff members pondered how to prove that what the center was doing with minority students was unique and special. For example, Denzil Thomas, an electrical engineer and instructor in the robotics program at the center, was teaching his students (third and fourth graders) advanced concepts such as voltage, current, circuitry, electromagnetism, and skills such as blueprint reading and drawing, radio construction, construction of robots that could be programmed through the Commodore 64, and use of the volt/ohmmeter and oscilloscope. At the same time, George E. Leonard, the center's life sciences teacher, was teaching his students (third through sixth graders) biology concepts directly from the high school biology syllabus. In both cases, the students (even to their teachers' surprise) were grasping the information and were able to demonstrate their knowledge on state Regents examination questions and in laboratory sessions. The center students were clearly extremely advanced in science and mathematics, but an objective measurement was needed.

Center staff asked themselves, What would be the results of the students undertaking a year-long course and then taking the examination offered by the state for that particular subject? It was obvious that it would be a waste of time for the students to take the sixth-grade science examination because their knowledge was far in advance of what that examination required. Further, the students had extensive laboratory skills (in measurement, microscopy, and dissection) that cannot be evaluated using an elementary school science examination. Of the two areas, robotics and biology, a statewide examination is offered only for biology in the tenth grade, so the New York State Regents biology examination was considered as a possible tool for measuring the achievement of the center and its students.

The purpose and meaning of these examinations stirred much internal debate at the center. Among the questions raised were these:

Would Regents tests measure all of the skills that the center teaches? If not, how could the tests be a suitable evaluation tool for the center's program?

- Given that the Regents examinations impose the direction and pace of the classroom work, in what ways would they interfere with the center's educational objectives?
- Would the examinations allow comparison for the first time between the center's students and their peers in the public school population?
- What statement would the center be making (and to whom) by permitting its students to take these accelerated courses and examinations?
- What would be the short- and long-term consequences — legal, psychological, educational — of the center's students taking the accelerated courses and examinations?

The decision to allow the center's students to study for the Regents biology examination was motivated in part by the desire to compel the public education system to raise the level of performance expected from minority students. If a fourth grader at the center could pass the tenth-grade New York State Regents biology examination, for example, then average tenth graders should be able to pass it also because they have to pass a ninth-grade science competency examination in order to enter the biology course. Finally, there was the prospect that success in the examinations might become a tool to empower students educationally, which is one of the center's most important objectives.

The initial request to the New York City Board of Education to permit the center's students to take the examination was denied. The board's argument was simple: If fourth or fifth graders take a tenth-grade examination, what does the system do with them afterwards? After intervention by the news media and a statement from the New York State Department of Education that there was no legal ruling that could bar the students from taking an accelerated course and examination, the Board of Education granted permission to the students to take the examination in January of 1989 and receive full Regents credit.

The next step was to prepare the students psychologically for the examination. Students who have been chronically underchallenged must be convinced that they can rise to a difficult occasion and succeed. They can be stimulated to do so by discussions and exercises designed to:

- Give the students a personal mission that tells them that, no matter what else is happening in their lives, they are in control of this particular challenge and thus the victory will be theirs. Adults may assist and friends may encourage, but ultimately, it is the students' decision to accept the challenge and to succeed or fail.

- Develop the students' sense of history and purpose so that they realize that they are not the first to take on a challenge, and that they represent the dreams and aspirations of all those with whom they have a cultural link and who, for whatever reason, have been unable to fulfill their own goals and objectives.
- Instill a strong sense of team effort so that, although each must pull his or her own weight in order to "move the mountain," all students understand that they must pull together. Therefore, winning does not mean coming in first; it means that all must finish the race.
- Create a strong sense of ability by inculcating in the students a positive attitude toward victory so that the more people doubt their ability to achieve their goal, the stronger and more determined they will become.

Another major challenge was that, in elementary school, children are accustomed to studying a "block" of information for a short period of time (often, for a week), taking a test on this information, and then forgetting it. This, of course, would not work in a year-long course in which what is learned in October is tested in June. Moreover, with perhaps the exception of mathematics, elementary school students are not compelled to organize seemingly diverse topics into a single unified understanding of a course. Taxonomy, reproduction, digestion, anatomy, and so on are all related and, during the year, the students must coalesce these apparently dissimilar ideas into a body of working knowledge called biology. In addition, they must integrate information from over 25 laboratory sessions done over a period of nine and one-half months. Teaching the students to retain and integrate information was a tremendous hurdle for the teachers, since this method of learning differed from the procedure to which the students were accustomed.

Preparing for the Examination

In October 1988, the students began studying biology for two hours after school on week days and for four hours on Saturday. They used the same textbook and review book that many New York City high school students use, *Concepts in Modern Biology*,[1] and they took the required laboratory sessions. Some difficulties arose in their public school classes, because the center's students read the biology textbook during class and some embarrassed their teachers. For example, one fourth-grader was asked, "What do plants need to live?" The student proceeded to give a detailed answer that included an explanation of

photosynthesis. The teacher was not amused. After several such inci-
dents, center staff talked with students, who then agreed not to read in
class, not to talk about the center at their "9-to-3" schools, and to give
what they termed "baby answers" to science questions in their regular
classrooms.

Results of the Examination

In May 1989, 16 students in the fourth through seventh grades took the
tenth-grade New York State Regents biology examination and passed, thus
becoming the youngest students[2] ever to take and pass the examination (see
Table 1). For the students, the most important result of taking these courses
and passing the Regents examination would be the "empowerment of
spirit." This positive attitude was enhanced by the fact that the subject was
science. Furthermore, having Regents biology passing grades on their
permanent record will enable them to enter the fast track when they reach
high school.

Preparing students to take the Regents examination forced the center's
staff to think a great deal about the art of taking tests — a separate issue

**Table 1. Scores and Grades of Science Skills Center Students
Who Took the 10th-Grade New York State Regents Biology
Examination in May 1989.**

Score	Grade
85	6th
83	6th
82	7th
78	7th
77	6th
73	5th
72	5th
72	7th
71	7th
70	7th
69	4th
68	7th
68	6th
68	5th
66	7th
65	5th

from learning biology. Perhaps many of the low scores earned by minority students in the public school system might be more a reflection of not understanding key elements in taking tests than of their ability to assimilate the subject matter. To prepare for the Regents examination, students practiced test-taking exercises, sometimes in a separate class and sometimes as part of a biology lesson. One of the most important points was learning to answer the questions posed by the examiner. This may sound simple, but even so-called "high achievers" took great liberties in restructuring questions. For example, this question was asked: "A school is planning a trip to the zoo. Each bus seats 40 students. There are 100 students. How many buses will be needed for the trip?" The possible answers were: 2, 2½, 3, and 4. Many students circled the answer 2. When asked why, they replied, "In order to save money, the extra 10 kids could be divided between the two buses and the smaller ones could sit three to a seat." This, of course, was a good answer and relevant to the economic environment of the students. However, it was explained to them that the examiner wanted the answer 3. Those students who circled 2½ were told that, although the calculation was correct, it was not the answer the examiner wanted because there is no such thing as half a bus. In this and other lessons, the students were advised that they must take a "bicultural" approach in order to understand the language of the test-maker. Although the students felt that this was basically unfair to African-American and Hispanic students, it was explained that this is the world in which they live and, therefore, they must accommodate to it for the present.

In other sessions, the students were taught how to dissect multiple-choice questions, filter out unnecessary data, analyze the remaining data (the givens), time themselves, relax before an examination, and, finally, check their answers after completing the test. It was made clear to the students that the techniques they learned would help them take the examination, but that there was no substitute for studying and that knowledge of the subject matter was the best preparation for a test.

The fact that center students passed these examinations suggests that two types of minority students in the public schools need to be reevaluated. The first group is minority students who are able to take accelerated courses and examinations. Unfortunately, many schools do not know what to do with minority students who do not need remediation. In most cases, such children are merely given more of the same work instead of different, more challenging work. As a result, their interest in science, mathematics, and other intellectual activities declines over the course of their school years. As long as these students produce passing grades and do not have behavioral problems, they continue to receive little attention. Smart students are acutely aware of this contradiction and will say that the only way to receive attention is to act in an unruly manner or to receive failing grades.[3] In schools with predominantly minority students, the few overworked guidance counselors spend most, if not all, of their time dealing with issues of student behavior and rarely are able to provide career and educational enrichment information.

Nor is there support in urban schools for minority elementary and junior high school students interested in science. The attendance, excitement, and energy seen in the Science Skills Center program attest to the responsiveness of these students and the need to overhaul public school teaching and curricula. To cite just a single example, one of the center's students was the valedictorian of her sixth-grade class and scored 85 on the tenth-grade New York State Regents biology examination in the same year. So the question is less What does the test reveal about the student? but, rather, What does it say about an educational system that clearly has underestimated the abilities of large numbers of its students?

The second group, underachieving minority students, are also underestimated. Testing goals for these students are set at the minimum level that will allow them to move on to the next grade. For these children, science and mathematics are not exciting ways of understanding how the world works, but instead are humiliating obstacles that exclude them. Science teachers must present their material to these students as if they were "educational salespersons" who want the consumer (student) to purchase (learn to love) their product. Using science and mathematics to segregate students in an educational caste system serves neither students nor science well. Furthermore, as at the Science Skills Center, there should be a place for "teams" of those students who have a tremendous inclination to do science, just as there are teams in basketball, the chorus, and the band.

Science Skills Center students are often catagorized as gifted or special. Center teachers use these terms because they believe that their students are gifted and special and tell them so in many ways. However, a positive mental framework is as important as innate ability in motivating students to take and pass accelerated science and mathematics programs and examinations. Although many of the center's students were good test-takers (as evidenced by the fact that they were able to survive the public education system), it was difficult to convince fourth graders that they could take a tenth-grade examination. The importance of teachers who believe in their students can be the best is often overlooked when the center is evaluated. This positive mental attitude can be developed in any school setting and can raise the level of achievement immediately, but doing so calls for a critical evaluation not so much of the syllabus — although there is much to criticize about the existing one — but of the preparation and attitude of the teachers.

The Science Skills Center Program

Assessing Potential Science Success

The students, teachers, and parents are the main components of the program. These three components working together are necessary for a child's success. First, the student must exhibit an interest or curiosity. Then the parents must notice that curiosity, enroll the student in the center, and support their child emotionally and psychologically during his or her time there. But most of all, dedicated and qualified teachers and cooperation among all of the parties are essential to ensure the program's success. (The center's programs and schedules are given in the appendix.)

The Science Skills Center's program is intended to provide a positive nurturing environment for learning science outside the public education system that encourages minority and female students to pursue careers in science and mathematics, and to develop students who, in the process of learning science, will build a positive attitude toward themselves, their families, fellow students, and the community and will use their knowledge of science as a tool for making the world a better place for all people. The center strives to be an institution of excellence where high academic achievement and hard work are expected from all involved. It attempts to ensure that children enrolled in the program will develop a positive self-image by understanding the historical and continuing contributions made to science by people with whom they share a cultural heritage. The program teaches skills, attitudes, concepts, problem solving, and positive work habits in science that are susceptible to evaluation by standard measures, and it provides a science education laboratory where innovative and alternative approaches to teaching science and mathematics can be developed. Other center activities include giving personal and career counseling to students who attend the center and functioning as a science awareness resource for the community through lectures, workshops, conferences, and other outreach activities covering such issues as health care, parenting, careers, and science topics. The center staff serves as a role model for teachers and administrators in the public education system so that they will reassess their approach to teaching. The staff also involve the student and the entire family in the educational process and, in particular, work to transform the parents' ideas about their role in developing a positive attitude in their children.

Recruitment activities are targeted at grades 2 through 6 for several reasons. For the most part, students from the ages of 7 to 13 are unaware that science is a subject they should avoid because of their ability, race, ethnicity, or gender. Also, more and more students seem to lose interest in science and mathematics in the early grades, so early recruitment is important. Experience has shown that younger students accommodate more easily to the center's pace and requirements than older students. Finally, the "bond" of negative peer pressure can be broken much more easily and positive peer pressure can be substituted as long as students feel that they have initiated it and can control it.

The recruitment procedure is simple. First, flyers and press releases are circulated throughout the city, especially in school districts that have a large minority population. Other methods include distributing leaflets at shopping centers and laundromats and speaking to church groups and civic organizations. Interested families are interviewed, with the interviews being arranged so that all families have equal access. These meetings take place on Saturdays and on weekday evenings. There is no cost for the interview and no entrance examination.

The Interview

Because of public awareness of the center's program through the print and electronic media, there is a degree of self-selection; that is, students understand that the program is serious and intense and that all of the teachers are very demanding. Students also are aware that they must give up a great deal of their free time to participate in Science Skills Center activities. In a sense, the idea is projected that the center is a "science team," which is hard to join and which requires continual hard work to stay in. Some direct recruiting is also done; for instance, a science coordinator from a Brooklyn school district brought all his elementary school science fair winners to a one-day workshop, at which time they were invited to join the center.

Often, parents bring reluctant youngsters to be interviewed. This is referred to as the "Little League syndrome" — parents want the program for their child, but their child does not want the program. These children are screened out at the initial meeting. Other parents, for whatever reason, will not bring their youngsters to an interview, although it may be known that these children would be excellent candidates for the program. A staff member may go to the child's home to explain the benefits of the program to the parents and seek the parents' written permission. Minority students who have participated in school and district science fairs are of especial interest, and they are actively recruited by school coordinators, principals, and teachers. A new outreach strategy targeted toward boys, whose numbers have dropped drastically in recent admissions, is being developed.

The interview questions are designed to reveal potential rather than current level of achievement. Some of the center's best students were very low achievers by public school standards. The interview frequently includes a "hands-on" activity, not to test the student's ability, but to disclose his or her interests. The setting is as informal as possible and parents may be present. There is also a separate meeting with the parents so that they can make additional comments. Essentially, the interview is intended to determine if the student has the following characteristics: (i) an interest, albeit informal, in science; (ii) a curiosity in knowing about things; (iii) the ability to work with different personalities; (iv) the endurance to withstand the Science Skills Center's rigorous schedule; and (v) an earnest desire to participate in the program. As noted earlier, there is no entrance examination. However, without effort by the center, more and more high-achieving students are applying for admission. This is taken into account when interviewing these students because they, along with the staff members, set the atmosphere in the center.

Developing and Monitoring a Positive Environment

There are several factors that characterize the center environment. Students learn that being smart is "okay" and that they will be praised for academic achievement. They study historical and contemporary African-American and Hispanic role models. They are taught that men and women are equal, that creative and different ideas are encouraged, that they should be the best they can possibly be, and that the center and its students are on a mission for themselves and their communities.

Inside the center, being smart is definitely acceptable. The students engage in friendly competition for the best test scores, the best laboratory results, the best-constructed robots, the best model bridges, and so on. The students also compete for the best school report card, which they must turn in each quarter. Frequently, the center's staff members are the only adults willing to listen and be sympathetic to a student who received 98 points instead of 100 on a mathematics examination. Their friends may say, "Well, it's only two points." Of course, a student outside the center who may be struggling to stay afloat in a mathematics class would resent a fellow student being upset over two points, but many of the center's students have been and are pursuing top honors in their primary schools. To such students two points may be critical. Attention from center staff members can make these particular students comfortable.

Praise for Academic Achievement

During orientation, the students are told very explicitly what is expected of them because most of them come to the center underchallenged. In fact, the cry heard most often in the first few weeks is, "But my teacher accepts this type of work!" However, it is made clear that whatever happens or is expected in their regular 9-to-3 school program has no bearing on what happens in the center; in time, students accept this. Center teachers give a good deal of positive reinforcement to those students who do well and make progress, no matter how little. "Students of the Month" awards are given to students who not only achieve academically, but those who project positive traits such as helping other students, showing leadership, and demonstrating a new idea. During the year, competitions are held in bridge-building, biology, mathematics, chess, and so on. The community is invited and awards are presented in the different categories. Student report cards are displayed on the "Report Card Wall." The concern is not whether a student passes a test, but, rather, the highest scores earned in each category. Science, mathematics, and language arts are considered more important than physical education and music. Most importantly, the students realize that they must do their best, whatever their best is.

African-American and Hispanic Role Models

All entering students are required to take a short course on the contributions to science, mathematics, and technology of the African-American and Hispanic peoples of the world. In so doing, the students study the rich cultures of these peoples. The present-day contributions of African-Americans and Hispanics to science and technology are included in the course, a highlight of which is monthly visits by minority scientists. The first class opens with this statement by the scientist Benjamin Banneker:[4] "The color of the skin is in no way connected with strength of the mind or intellectual powers." This emphasis is important both because the statement was made by an African-American scientist and because it raises an issue that must be confronted by the students early in their life at the center. They must believe that they can achieve and that their ethnicity and culture give them a strong foundation for that achievement.

Equal Rights for Men and Women

The center is not interested in merely teaching abstract scientific concepts detached from the cares of the world. Students are told that science must be used to improve the condition of all the inhabitants on the earth. To this end, qualities are modeled that it is hoped the students will have developed by the time they become participants in the scientific life of the future. In the center's first year of operation, all instructors and most of the students were male. This atmosphere, a sort of "male science club," was not conducive to encouraging girls to join, let alone stay. This situation was changed by bringing in female teachers and by frank and open discussions in classes. When a boy made a negative remark about girls, it was challenged. The center's teaching methods were critiqued to eliminate inherently different approaches to teaching males and females. (Videotapes were very helpful here.) The large number of girls now in the program has all but eliminated whatever gender problems there were in the past. The current goal is to encourage more girls to join the robotics and mathematics programs; they are already heavily represented in the biology classes.

Creative and Different Ideas

During a recent interview for the summer program, a young man was describing how he mixed vinegar and cleanser to see what would happen and how he took apart an old clock to see how it worked. His mother began to frown, thinking he was saying things that would not help him in his interview, but she could not have been more mistaken. Confessions of experimentation and interest are exactly what the center wants. Students are told that the center itself is an experiment or a laboratory and, therefore, things can change. Efforts are made to do things in new and different ways and, in fact, the center students frequently initiate changes. This means that a large part of the syllabus must be devoted to problem-solving activities. Students are encouraged to explore ideas with materials and through the format of science fiction storywriting and other art and science collaborations. All lessons are based on the scientific method, which describes science as a dynamic, unfolding discipline, not a study of abstract rules created by people in the past. This combination of scientific theory and a scientific approach to life and problem solving allows the students to begin placing their important ideas in the realm of present possibilities.

Doing One's Best

The process of self-realization at the center is an important aspect of the intellectual empowerment of the students. Often, students achieve a level of performance at the center of which they themselves are unaware. This careful, opening-up process is a major topic of discussion at staff meetings; a typical question is, "How do we get this student to function at a higher level?" There seems to be a "self-belief" gap that all underchallenged students face when they are stimulated by educators who will not give up on them and who will not expect less of them than their best. Once this chasm is crossed, the student becomes self-motivated and expects nothing less than the best from himself or herself. One of the most important discoveries made by staff members is that the high level of performance anticipated is still less than the achievements of some of the most gifted students.

Assessing the Program

The results of accelerated science courses and examinations are good ways of evaluating minority programs for high academic achievers. The center program demonstrated that participants in these and other programs should spend considerable time studying test-taking skills. Unfortunately, many test-taking skills projects do not discuss the possible gap between the test-taker's understanding of the question and the type of answer wanted by the examiner. The results of this accelerated science and mathematics program also demonstrate that not only is the public education system failing the students it claims have failed, but also it is not stimulating interest in science and mathematics in those minority students it claims are successful members of the system.

It is time to take a deeper look at how to motivate teachers and school administrators to create an atmosphere in schools and classrooms that conveys that learning is important, that all children can learn, and that all children are expected to learn. In many schools, the sports coach will fill student athletes with enthusiasm and excitement for a game. Academic teachers must also learn how to fill students with enthusiasm and excitement for academic achievement. Special science and mathematics programs — "science teams" — should be developed to identify interested elementary and junior high school students, and a "corridor of support" should be created for these students until they reach their career goals. These science teams should not apologize for existing. The student athlete band or chorus member has an entire year to feel special about his or her skills. The student scientist or mathematician may have only one opportunity a year to shine and that is at the school science fair. The school and the community should give these science students recognition, praise, and material benefits (special trips, luncheons, and so on), so that a healthy environment of academic achievement can be developed in the school and the community.

Finally, a creative study by urban education officials is needed to determine what to do with minority students who do not require remediation, particularly science and mathematics students. It is critical for the minority community and for the nation that present and future numbers of scientists and mathematicians be increased. This is not an equity issue, nor a question of raising the academic achievements of certain groups. Rather, it is a matter of redressing the crime of disregarding some of the nation's most talented citizens whose contributions, without positive intervention, will be lost forever.

References and Notes

1. Kraus, D., *Concepts in Modern Biology*, 5th ed. (New York: Globe, 1984).
2. Nineteen students in the same age group (9 through 13 years) took the tenth-grade New York State Regents biology examination in June 1990. That same month a group of fifth through seventh graders took

the New York State Sequential I mathematics (algebra, geometry, logic, problem solving) examination. A third group of students in the fourth through eighth grades took both the New York State Regents Biology examination and the New York State Regents Sequential I mathematics examination. All of these students, who are African-American and Hispanic, passed the examinations. This was the first year that students from the Science Skills Center were tested in Regents mathematics. The highest mathematics score (96) went to a fifth grader.

3. Fordham, S., and Ogbu, J. U., "Black Students: School Success Coping with the Burden of Acting White," *Urban Review* **18**(3) (1986), 176–206.
4. Benjamin Banneker (1731–1806) was born in Maryland of a free mother and a slave father. He built the first clock in the North American colonies in 1753. Banneker became famous for publishing an annual almanac, beginning in 1772. It included tide tables, the times of eclipses, and weather forecasts. Banneker was also a surveyor who was a member of the team that laid out the plans of Washington, D.C. When the chief surveyor resigned, Banneker reproduced the plans from memory.

Appendix

The Science Skills Center

Brooklyn, New York

Programs and Schedules

Afterschool and Saturday Program

First week in October until last week in June.

Monday–Friday, 3:30 p.m–7:00 p.m. (full dinner included)

Saturday, 12:00 p.m.–4:00 p.m.

A typical weekday schedule is:

3:30 p.m.–5:00 p.m. Homework and dinner

5:00 p.m.–7:00 p.m. Lecture or laboratory session

Saturday, 12:00 p.m.–4:00 p.m. Group study and review session

For new students, one day a week is set aside for the study of African-American and Hispanic contributions to science, mathematics, and technology.

Classes	Ages
General science I	7–8 years
General science II	8–9 years
Engineering and design I: Robotics	9–12 years
Engineering and design II: Robotics	10–13 years
Biology (preparation for New York State Regents examination)	9–12 years
Advanced placement biology	11–14 years
Mathematics, algebra, geometry, logic, problem solving	10–13 years
Chemistry (preparation for New York State Regents examination)	10–14 years

Summer Institute for Entering Students

Six weeks, July and August, 9:00 a.m.–3:00 p.m., Monday–Friday. New students spend 2 days a week (Mondays and Wednesdays) on science and mathematics activities in:

- Chemistry
- Biology (microscopy)
- Biology (botany)
- Computers
- Electronics
- Chess and mathematics

On these days, the mornings are devoted to teachers' demonstrations and discussions and the afternoons to students' "hands-on" projects and experiments. Students rotate through the six topics, spending 1 week on each area. Tuesdays and Thursdays are set aside for all-day trips to museums, engineering firms, colleges, etc. Fridays are reserved for day-long picnics.

Summer Science Institute for Biology

(For students who have been in the program one year and have passed Biology)

Six weeks, July and August, 9:00 a.m.–3:00 p.m., Monday–Friday. Classes are taught at the State University of New York Health Science Center in Brooklyn by the university faculty and Science Skills Center teachers. The 1990 schedule included:

Week 1 **(Jul 9-12)**	Monday	Orientation
	Tuesday	Review of safety procedures in the laboratory—the scientific method, laboratory procedures, and measurement techniques
	Wednesday	Review of anatomy
	Thursday	Review of physiology
Week 2 **(Jul 16-19)**	Monday	Welcome and tour of the campus (a.m.)
	Tuesday	Medical and scientific ethics
	Wednesday	Health and research administration, including a case study
	Thursday	Research skills and research methods
Week 3 **(Jul 23-26)**	Monday	Human anatomy
	Tuesday	Physical examination of a patient
	Wednesday	Hypertension and the circulatory system (a.m.) Planning a successful career (p.m.)
	Thursday	Neuropathology (a.m.) Neuromicroscopy practicum (p.m.)
Week 4 **(Jul 30-** **Aug 2)**	Monday	Microscopic organisms (a.m.) Electron microscopy (p.m.)
	Tuesday	Human immunodeficiency virus (HIV) and acquired immunodeficiency syndrome (AIDS)
	Wednesday	Nephrology
	Thursday	Molecular biology
Week 5 **(Aug 6-9)**	Monday	Genetics (a.m.) DNA sequencing laboratory (p.m.)
	Tuesday	Psychiatry (a.m.) Group discussion (p.m.)
	Wednesday	Obstetrics and gynecology, including female anatomy (a.m.) Study session (p.m.)
	Thursday	Alcoholism and heredity
Week 6 **(Aug 13-16)**	Monday	Cardiology, cardiovascular surgery
	Tuesday	Presentation of students' research papers
	Wednesday	Completion of journals Evaluation of program and interviews
	Thursday	Awards luncheon

Note: All Fridays are reserved for such recreational activities as a picnic and sports for relaxation and physical exercise.

The summer program provides an excellent opportunity to evaluate the students and for the students to decide if the Center is suitable for them. If both sides are in agreement, students usually continue in the fall program.

16

Group Assessment as an Aid to Science Instruction

David W. Johnson and Roger T. Johnson

The role of assessment in instruction is to improve the quality of both teaching and learning. By assessing what students have and have not learned, it is possible to modify teacher and student practices in order to achieve instructional objectives more satisfactorily. Assessment may be conducted in a wide variety of ways, including in groups. In group assessment several students complete a lesson, project, or test together while a teacher measures their level of performance.

Group assessment cannot take place when students are competing with each other or working alone.[1,2] When competitors are placed in small groups, interaction and communication become restricted. In order to "win," a student must outperform other group members and try to prevent them from performing better than he or she does. Thus, students hide their answers and reasoning from each other. When students are placed in groups and required to work as individuals, they typically ignore each other and work independently. Interaction and communication are minimal. Discussing the assignment with other group members takes valuable time away from one's own learning. Valid and reliable group assessment takes place only when students are assigned to carefully structured, cooperative groups. In order to understand how to use group assessment, it is necessary to define cooperative learning, the goals of assessment, the various assessment formats, and the usefulness of group assessment.

Cooperative Learning

Teachers may structure cooperation within a group by having students work together to accomplish shared goals.[3] In a cooperative situation, students' goal achievements are correlated positively; students perceive that they can reach their learning goals if, and only if, the other students in the group also reach their goals. Thus, students seek outcomes that are beneficial to all those with whom they are linked cooperatively. Students are assigned to small groups and given two responsibilities: to complete a task and to ensure that all the other group members also complete it. Students discuss the material with each other, help one another to understand it, and encourage each other to work hard. Individual performance is checked regularly to ensure that all students are contributing and learning. A criteria-referenced evaluation system is used. For this purpose, a well-structured, cooperative lesson should contain five essential components:

- Positive interdependence — the perception that one cannot succeed unless the other members of the group also succeed (and vice versa) and, therefore, that their work benefits one's own and one's own work benefits them. Positive interdependence may be structured by setting common goals or rewards, by creating dependencies on each other's resources, by assigning specific roles to each member, or by a division of labor.
- Face-to-face promotive interaction — students explain to each other how to solve problems, discuss the nature of the concepts being learned, teach their knowledge to classmates, and explore the connections between present and past learning. This face-to-face interaction is promotive in the sense that students help, assist, encourage, support, and reinforce each other's efforts to learn.
- Individual accountability — the performance of each student is assessed and the results are given back to the group and the individual.
- Small group and interpersonal skills — leadership, decision making, trust building, communication, and conflict management. These are the skills required for students to work together productively[4,5] and they have to be taught just as purposefully and precisely as academic skills.
- Group processing — members discuss how well they are achieving their goals and maintaining effective working relationships with each other. Participants need to describe what actions are helpful and detrimental and make decisions about what behaviors to continue or change.

Assessment Goals

Hundreds of studies conducted over the past 90 years demonstrate that, compared with competitive and individual learning situations, cooperation produces positive results in several domains:[6]

- Cooperative learning results in higher achievement and is superior to competitive and individual learning when tasks are conceptually oriented, require problem solving and creativity, involve higher-level reasoning and critical thinking, and require the application of information to the real world.
- In cooperative learning, more positive relationships develop among students and between students and faculty. Individuals tend to like others with whom they have worked cooperatively. This is true even

when students are from different ethnic and cultural backgrounds, social classes, and language groups. It is also true for students who have and do not have disabilities.

- Positive psychological well-being results from working with classmates cooperatively. Students experience greater self-esteem, self-sufficiency, social competencies, coping skills, and general psychological health.
- More positive attitudes toward science result when students work cooperatively, and they are more interested in taking advanced science courses. Also, after working cooperatively in science classes, students tend to perceive science as gender-neutral.

Besides being the instructional procedure of choice for science teachers (because of the superior achievement it promotes), cooperative learning provides a basis for more sophisticated assessment of student learning than competitive and individual learning situations permit.

Assessment Formats

Group assessment enables a wide variety of outcomes to be appraised in many flexible ways. With group assessments, teachers can compare a number of outcomes that they typically have difficulty evaluating with paper-and-pencil measures, such as the processes of reasoning and problem solving and metacognitive thinking.

Observational Procedures

Observational procedures are designed to describe and record behavior as it happens, so teachers can gain the information needed to make judgments about the current competence of the students and the success of the instructional program. For teachers to observe students engaging in scientific reasoning, scientific problem solving, and metacognitive thinking, students must be doing "thinking out loud" activities; in essence, cooperative learning groups are "windows into students' minds." Teachers assign students to small cooperative groups and give them a series of problems to solve. As the students work, the teacher moves from group to group, listening to the participants interact with each other. From listening to students explain how to solve the problems they are working on, the teacher can assess what they do and do not understand. By hearing how students question each other, the teacher can assess whether scientific problem-solving processes are occurring. Ideally, students will ask each other for predictions, for clarification of meaning, for justification of how a particular answer was derived, and for interpretations, explanations, and observations. In addition, the teacher can assess how frequently metacognitive strategies are being used. (Several observation sheets are given by Johnson,

Johnson, and Holubec.[7])

Students often know the "correct" answer to a question, but correct answers on tests and homework assignments tell teachers very little about students' reasoning processes or their understanding of science. The only way teachers can be sure that students really understand the science being taught is by listening to them explain what they know to each other step by step. This requires cooperative, not competitive, learning groups.

Interview Procedures

In addition to observing students work, teachers should interview them systematically to determine their levels of scientific reasoning, scientific problem solving, and metacognitive thinking. Students should be assigned to cooperative learning groups that are heterogeneous in terms of mathematics and reading ability.

A suggested format for a week's work is to give a set of questions to the groups on Monday, instruct the students to prepare all group members to respond to the questions, and allow time during each class period for the groups to practice their responses to the questions. On Thursday and Friday, the teacher meets with a group and chooses one member at random to give an explanation to a randomly selected question. When that member answers the question, other group members can add to the response. The teacher judges the answer to be "adequate" or "inadequate." Then the teacher asks another member a different question. This procedure is repeated until all the questions have been answered or until the teacher determines that the group is inadequately prepared. In this case, the group has to return to the assignment and practice until all members are better prepared. The teacher should provide guidance by identifying particular weaknesses and strengths in the members' answers. All members of the group are given equal credit for passing the test.

Individual Test Followed by Group Test

In addition to assessing students' scientific reasoning by listening to work sessions and by interviews, teachers can give individual tests to the students. Again, they should be assigned to cooperative learning groups that are heterogeneous in terms of reading and mathematics ability. They work on their assignments together during the week and take the exam on Friday. The students take the test individually, making two copies of their answers. They hand in one answer sheet to the teacher (who then scores it); they keep the other. After all members have finished the test, the group meets to take the test again. Their task is to answer each question correctly.

The cooperative goal is for all group members to understand the material covered by the test. If they disagree about any answer or are unsure of it, they must find the page and paragraph in the textbook that contain the answer. The teacher observes the groups randomly to ensure that they are following the procedure. If all members of the group score above a preset criterion (such as 90 percent correct) on the individual tests, then each member receives a designated number of bonus points, 5, for example. The bonus points are added to their individual scores to determine their individual grade for the test.

Weekly Group Tests and Individual Final Examinations

To maximize students' higher-level scientific reasoning and long-term retention of scientific knowledge, the following procedure may be used. Students should be assigned to cooperative learning groups of four members, again heterogeneous in terms of mathematics and reading ability. During the week, the groups should work on their assignments together. On Friday, an examination should be given in which each group is divided into two pairs and each pair takes the test, conferring on the answer to each question. The task is to answer each question correctly; the cooperative goal is to have one answer for each question that both agree upon and that both can explain. The pairs cannot proceed until they agree on the answer.

Once the two pairs have finished, the group of four meets and retakes the test. Their task is to answer each question correctly. The cooperative goal is for all members of the group to understand the material covered by the test. Group members confer on each question. If the two pairs have different answers to the same question or members are unsure of an answer, they find the page and paragraph in the textbook where the answer is explained. Each group is responsible for ensuring that all of its members understand the material they missed on the test. If necessary, group members assign review homework to each other. The teacher observes each group randomly to ensure that they are answering the questions correctly. Each group then hands in one answer sheet with a list of its members. Each member signs the answer sheet to verify not only that he or she understands the content of the test, but also that all the other members understand it, too. All members of the group are given equal credit for passing the test.

At the end of the grading period, each student takes an individual final examination. If any student scores below a preset criterion (such as 80 percent), then the cooperative group meets and reviews the content with the student until he or she can pass the test. This rarely happens because the group members have verified each week that they all are learning the content assigned.

Why Is Group Assessment Useful?

To summarize, assessment in science classes is a continuous process aimed at measuring what is important. What is measured is noticed and, in turn, influences what is taught. In overcoming the simplistic view of science classes as textbook-driven lecture courses, assessment procedures must focus on reasoning, problem solving, and metacognitive thinking. Assessment in cooperative learning groups is useful in this process for several reasons:

- This assessment method increases the learning it is designed to measure. The oral explanations required for group assessment result directly in higher-level reasoning, a deeper level of understanding, and long-term retention. In other words, tests become educational experiences as well as evaluations.
- The intellectual challenge, disagreement, and controversy that take place in cooperative learning groups stimulate higher-level reasoning, divergent thinking, creativity, and long-term retention.[8]
- Group assessment helps to build positive attitudes toward science while the assessment is taking place. Learning in cooperative groups results in more positive attitudes toward science than learning competitively or individually does.[9]
- The discussion inherent in group assessment emphasizes the understanding of science by deemphasizing reading ability. When group assessment is conducted, the material is reviewed orally. Students who have difficulty reading and writing can have the questions read to them and can explain to groupmates step by step how the answer may be derived.
- Creative thinking is enhanced by the reactions and input of groupmates. Oral tests allow the processes of thinking and problem solving to be public so that they can be improved. Feedback is immediate.
- Group assessments promote intrinsic, rather than extrinsic, rewards. Students do their best and most creative work when they find science interesting and fun, not when they are being graded. The intrinsic rewards of working with classmates outweigh concerns about extrinsic grades.
- Group assessment improves social skills and cognitive reasoning skills simultaneously. Through the group discussions, students develop and refine their interpersonal and small group skills while completing the academic assignments that allow the assessment to take place.

References

1. Johnson, D. W., and Johnson, R., *Learning Together and Alone:*

Cooperative, Competitive, and Individualistic Learning, 3rd ed. (Englewood Cliffs, NJ: Prentice-Hall, 1975).

2. Johnson, D. W., Johnson, R., and Holubec, E., *Cooperation in the Classroom* (Edina, MN: Interaction Book Company, 1990).

3. Ibid.

4. Johnson, D. W., *Reaching Out: Interpersonal Effectiveness and Self-actualization,* 4th ed. (Englewood Cliffs, NJ: Prentice-Hall, 1990).

5. Johnson, D. W., and Johnson, F., *Joining Together: Group Theory and Group Skills*, 4th ed. (Englewood Cliffs, NJ: Prentice-Hall, 1975.)

6. Johnson, D. W., and Johnson, R., *Cooperation and Competition: Theory and Research* (Edina, MN: Interaction Book Company, 1989).

7. Johnson, D. W., Johnson, R., and Holubec, E., *Cooperation in the Classroom*. See reference 2.

8. Johnson, D. W., and Johnson, R., *Creative Conflict* (Edina, MN: Interaction Book Company, 1987).

9. Johnson, D. W., and Johnson, R., *Cooperation and Competition*. See reference 6.

17

Portfolios for Assessing Student Learning in Science: A New Name for a Familiar Idea?

Angelo Collins

In 1989, Mr. Jackson began his ninth grade biology course with the following speech:

> This year in science, you will put your work in a three-ring binder. This notebook, which we will call a portfolio, will belong to you. It is to be kept in Room 25 at all times. In the first section of the portfolio, you will keep your personal record sheet and a calendar of what we do each day in class. Here are blank record sheets and a blank calendar for September, to begin your portfolio development process. You will also put all the handouts from this class in this first section of your portfolio. The second section will be a journal. Each day at the end of class, you will have a few minutes to write about what you did and learned and felt in your science class. I will look at your journal to make sure that you have written something every day, but I will not grade what you write in your journal. In the third section of your portfolio, you will keep copies of all the activities we do in science class — worksheets, laboratory notes, and group work, for example. The next section will be for homework. The fifth and last section will be for review sheets and quizzes. You will have to buy dividers for your portfolio.

In 1970, Mrs. Rock began her ninth grade biology class as follows:

> For this course, you will be required to keep a lab notebook. This notebook is a working notebook, not for show. You will use it a lot and, by May, it will be soiled and marked up. Since we will do our first lab tomorrow, be sure to purchase a sturdy spiral notebook tonight and bring it to class tomorrow. Your lab notebook will be collected each quarter and will be examined for evidence that you did each lab — preparation notes, data, questions, and comments — and

your work will be graded as acceptable or unacceptable. In addition to the lab notebook, once each quarter you will submit a formal lab report. When scientists do lab work, they want and need to share it with others. They talk to one another while they are preparing lab reports. Over the years, a very formal structure has been developed for presenting laboratory research. Writing a good formal lab report requires careful thinking. It also takes time. Writing one formal lab report each quarter will be enough to keep you busy. You will want to look at your lab notebook and choose a lab that you are interested in for your formal lab report. The directions for doing a formal lab report are on the green handout. There is also a copy of the directions hanging on the bulletin board. Please read the handout and bring it to class on Friday.

Both of the preceding scenarios introduce ninth grade biology students to their course. The first was used to initiate a class in which all of the students were recent immigrants with limited ability in the English language. This veteran teacher knew that these students were often overwhelmed by the experiences of an American high school. He intended that developing the portfolio would enable the students to organize their learning and thus to see relationships between different activities in the class. He hoped that as students saw the quality of their own work improve, they would begin to take pride in their school work and in their knowledge. He envisioned the portfolio as a tool to give the students confidence as well as competence. He used the term "portfolio" because he had an image of a portfolio as a collection of all the work done by one person. His purpose in designing the portfolio as he did was so that it would be responsive to the context in which his students were working.

The second scenario is a close approximation of the introduction to an honors biology class about 20 years ago. The teacher knew that these academically successful students would require challenges. She had decided to do laboratory work once a week and had selected a textbook with open-ended laboratory experiences. Her goal was to push these students to make choices, to express themselves clearly, and to be critical of their own work. (She also knew that reading and commenting on formal laboratory reports were time-consuming tasks, and that she would do a better job if she had fewer to read. Doing a few formal laboratory reports well fit the 1960s slogan that "less is more.") She did not use the term "portfolio," but she did have two images of a portfolio in mind: that the portfolio may be a collection of all work or that it may be a selection of the best work. She also had a purpose for the design of the laboratory notebook and the folder of four formal laboratory reports. The portfolio met the needs of a particular group of students.

Both of these scenarios highlight three important features of a portfolio: the purpose, the context, and the design. Using portfolios for assessment is an idea that is becoming more and more popular and, although the term is new, the idea may not necessarily be innovative to science teachers. Science teachers long have recognized that multiple-choice tests do not capture fully the richness of learning science and have looked to additional modes of assessment, frequently using notebooks, folders, and laboratory reports for student evaluation.[1]

The current interest in "authentic," that is, meaningful, assessment provides teachers with an opportunity to reexamine how portfolios might play a role in evaluation. Arts PROPEL, a program of Harvard Project Zero, has been exploring the use of portfolios for instruction and evaluation in the fine arts for over four years. Portfolios are cited frequently in directions for writing-across-the-curriculum projects. However, if science teachers are using portfolios for assessment to a considerable extent, reports of their use are not appearing in the literature or, alternatively (and this may be more likely), they are using processes that collect, organize, and display evidence, but do not call the product a portfolio. Therefore, in preparation for this chapter, seven high school biology teachers who had developed portfolios for teacher assessment, nine elementary school teachers, and five others concerned with improving school science instruction were invited to discuss the uses of portfolios for assessing students' science education. This chapter begins with a brief discussion of portfolios and authentic assessment. Then, the features of purpose, context, and design introduced in the scenarios provide a framework for exploring some issues of using portfolios for assessing science instruction.

Portfolios as Authentic Assessment Tools

What exactly is a portfolio, and why is it a form of authentic assessment? A portfolio is a container of evidence of someone's knowledge, skills, and dispositions. The evidence in the portfolio is used to make judgments about the quality of the performance of the person who developed it. Although portfolios are rather new forms of assessment to students and teachers, most people already have an image of a portfolio. One frequent image is the artist's, model's, or architect's portfolio. This notion is of a large leather folder containing selected samples of representative best work. Somehow, an artist knows just how many pieces of evidence to put in a portfolio — not too few, so that it is impossible to make a valid judgment, and not too many, so that there is no evidence of the ability to discern quality. An artist has confidence that these samples represent his or her best work because the pieces have been examined by peers. The contents of the portfolio vary depending on the purpose. A photographer competing for a position with a sports team will not include only still life photos in the portfolio. In selecting pieces for their portfolios, artists develop the skill of self-criticism.

Although the artist's portfolio is perhaps the most frequent image of a portfolio, there are others; for example, the list of securities maintained by an investment broker. This portfolio is not a record of best work, but of all work, of work in progress. This portfolio is maintained either in a notebook, a folder, or as a computer printout in the form of a log, rather than in a leather envelope. A salesperson's catalogue is also a portfolio. It is a promise to deliver the products of a third party. The form of a sales portfolio can be compared to a glossy magazine. Yet another image of a portfolio is the merit badges worn by boy or girl scouts. These badges are evidence of work completed, often in a group, with the help of a mentor. The symbol on the badge has special meaning to all scouts. The badge is awarded with ceremony and worn publicly, with the pride of accomplishment. Still other images of portfolios are the dissertation of a graduate student and the tenure certification of a professor. These images are more fully developed by Bird in his chapter on the teacher's portfolio.[2] This variety of images of a portfolio contributes to its potential as a mode or instrument of reliable assessment in science education.

Portfolios belong to the class of assessment termed authentic assessment. In *Portfolio,* the newsletter of Arts PROPEL, Ruth Mitchell says:

Authentic assessment means evaluating by asking for the behavior you want to produce.... Authentic assessment isn't a single method. It includes...portfolios, collections of student work. The list is limited only by the criterion of authenticity: Is this what we want students to know and be able to do?[3]

Grant Wiggins considers four sets of criteria that separate authentic assessment from inauthentic assessment.[4] Under the heading of "Structure and Logistics," he considers such points as the form used. He states that the criteria for success include public knowledge and the absence of artificial constraints, such as time. In authentic assessments, collaboration is acceptable; it is not cheating. The assumption that the artist's portfolio contains evidence that the contents have been judged by peers and that the merit badges of a scout imply the assistance of a mentor are examples of collaboration. Under the heading "Intellectual Design Features," Wiggins states that the assessments should be contextualized and should represent realistic but fair practices in the discipline. The understanding that portfolios' contents change depending on their intended use is another indicator that portfolios generally are regarded as authentic assessment indicators. In the section on "Standards of Grading and Scoring," he proposes that scoring should be as complex and multifaceted as the assessments themselves and appropriately contain some form of self-assessment. Again, this is a characteristic of an artist's portfolio, where the rating is based on professional judgment, and of a scout's badge, where the requirements increase in rigor as different skills are mastered. In the final section,

"Fairness and Equity," Wiggins asserts that authentic assessments should ferret out and identify strengths and enable students to show off what they do well. An art fair where the contents of an artist's portfolio are displayed and a campfire where scouts show off their badges are testimony that portfolios have this characteristic of authentic assessment. Accepting that portfolios do have the characteristics of authentic assessment, let us look at some issues of purpose, context, and design for using portfolios in student science assessment.

Important Features of a Portfolio

Purpose

If the objectives of learning science are to enable students to master some of the content of natural science, to understand and practice the processes by which this knowledge is constructed, and to develop the attitudes of a scientist, then, to be authentic, a portfolio would need to contain evidence that these are what a student knows and is able to do. The processes of determining the purpose, the clear goal, and the focus of the portfolio for student assessment provide a teacher or a group of teachers — or a group of teachers and students — with the opportunity to step back from the daily demands of a classroom and ask themselves what is the content, what are the processes, and what are the attitudes about which students will present evidence? In determining the purpose for the portfolio, students and teachers will ask such questions as, What is the content — the facts, laws, and theories — to be learned in this area? What should students be able to do — set up a pendulum, record an observation, pose a problem and solve it, structure an argument, or communicate with other students and with the teacher? What are the attitudes that the students should display as they develop their portfolios — comfort with ambiguity, accepting the tentative nature of science? (In setting the purpose of the portfolio, the design issue of what will count as evidence must be considered simultaneously.) Determining the purpose of a portfolio will, in turn, influence both curriculum and instruction. If comfort with ambiguity is an attitude that is to be learned in a science class and displayed in the portfolio (and if a laboratory report that has two alternative explanations for the data is acceptable evidence of this attitude), then instruction must include sufficient laboratory experiences of the type that allow for a variety of interpretations so that students can produce the necessary evidence to develop a portfolio.

In discussing the purposes of a portfolio, one high school teacher said,

> Since a portfolio consists of materials depicting a student's mastery
> of the course, it would be necessary to include a variety of items
> pertinent to biology such as skills in the following areas: writing, oral

presentation, drawing, planning, dissecting, testing, interaction with peers, interaction with the teacher, notetaking, discussion, and responding to questions.

Another said,

> Would portfolios help develop in students a better idea of the whole picture and the connections that a teacher is trying to get them to understand? I have some answers to this question, but I think the question is more important than the answers. As we are gradually changing what we expect to teach in the next century, we are obviously going to have to reconsider our methods of evaluation.

One of the elementary school teachers added, "You'd really have to define what you want the student to learn [in order] to do a portfolio!"

In addition to the desired results of the science instruction, other goals for the classroom will influence the portfolio. If a portfolio is going to be used for students to review their own growth and progress, then keeping all science work would be a reasonable decision. If the portfolio is to be used in lieu of science fairs or to demonstrate to parents and the community what the students in a school know and are able to do, then a "best work" portfolio model might be more appropriate. The uses of a portfolio are numerous. Three of the elementary school teachers responded almost simultaneously that they had been keeping portfolios for years — over 15 years — and had not known it. They described a process of keeping folders of students' work. One teacher has students review what is in the portfolio and select some pieces of work to show parents at parent-student-teacher conferences. This teacher encourages students to pick examples that illustrate their progress as well as their recent or best accomplishments. Another teacher asks students to select several items from the folder and then write a paragraph summarizing why these samples are evidence of what he or she has learned in science. The teacher then reads the paragraphs for evaluation purposes.

The teachers proposed other uses for portfolios. For example, the portfolio could provide the starting point for self-evaluation, for peer evaluation, and for cooperative group work. Although the elementary school teachers proposed using portfolios to show accumulated science experiences from year to year, so that "this student would not do volcanoes three years in a row," they eventually rejected the idea as too unwieldy for the teacher. In casting aside the idea of a cumulative portfolio as a source of information for teachers, they declared themselves in favor of a cumulative portfolio for students. Reviewing and revising their portfolios from year to year would provide students with the opportunity to recall the experiences they have had in science classes and to recognize how their work has changed and improved.

Context

A second feature of portfolios that makes them authentic for the assessment process is that they are sensitive to context. One example of context is the age of the students. As the elementary school teachers described above, with young children, the teacher maintains the portfolio while the students review it purposefully from time to time. The responsibility for maintaining the portfolio would begin to shift to the students themselves as they move into middle school and would become clearly the students' responsibility in high school. For example, one high school teacher stated, "I believe that many of us are actually having our students create a portfolio when we require semester notebooks or lab notebooks or journals of many types. Many projects require that students keep extensive records and present [the project] to the teacher...in essence, a mini portfolio."

Portfolios would no doubt differ from discipline to discipline. One teacher explained how he was trying to make his students sensitive to the concept that science changes depending on what the scientist knows — an element of context. He has students keep laboratory reports in a folder and rewrite the discussions later in the course to include new knowledge and skills. Another explained that he asked a teacher in another school to read a set of laboratory reports (and he reciprocated) in order to help students understand the need for clarity in such reports.

Student needs and interests are other elements of context that are important when designing and developing portfolios for assessment. The introductory scenarios in this chapter presented students with different needs and abilities. However, even within the same class, the criteria of acceptable evidence should be broad enough to meet the needs of a student who is a budding artist (evidence in the form of a sketch or diagram), a future playwright (evidence in the form of a script for role-playing), a mechanic (a soldered model of an abstract concept), or a banker (a cost-benefit analysis of the implementation of a technology). One of the high school teachers wrote, "One advantage for students might be that they can have a few bad days or even a [bad] week and probably the effect of the portfolio would be to wipe those days out of existence as far as the grade goes." Another teacher suggested that the portfolio would be a big help to students with limited ability in the English language and different learning styles.

Design

The third feature of the portfolio is the design, which includes the purpose and the context. The various images of the portfolio help in its design, but one of the elementary school teachers claimed that in collecting all her students' work as a log and then selecting some pieces of evidence for display, she had created a new image that was a combination of several types of portfolios.

Probably the most important design decision is what will count as evidence in the portfolio. One high school teacher wrote,

[From looking at my course goals] the possible portfolio entries may include: a notebook containing examples of homework, monthly calendars, a grade record sheet, class notes, lab reports, a videotape containing segments of students doing an oral presentation, examples of dissecting technique, cooperative learning situations, and a research paper.

Other important decisions are how much evidence to put in the portfolio, how the evidence will be organized, who will decide what evidence to include — which pieces of evidence will be prescribed by the teacher and which will be selected by the students. The determination of standards for the portfolio is another critical decision. Knowing the standards before the portfolio is developed is requisite for making the portfolio authentic for assessment purposes.

Several high school teachers reported that they had been designing portfolios unknowingly. One teacher described how he had been moving in that direction for several years:

As we go through a unit of study, I collect and grade papers, labs, assignments, and so on. These are returned to the students immediately. At the end of the unit, I describe a format for a table of contents. The students copy this down and arrange their papers in that format or arrangement…. It is very impressive to the students to see all the work they have done…. Their parents can view easily the work they are doing…they [the students] can review it easily for final exams.

Unlike the elementary school teachers who wanted to collect everything and then let students select samples, the high school teachers feared that unless a procedure for selecting evidence was developed early, the portfolio would take too much time to evaluate.

One high school teacher shared with great enthusiasm her conscious attempt to use portfolios in her biology class during a unit on animals. She modeled it after the portfolio development process that she had employed for teacher assessment. She explained how she stated learning goals that were problem-based, listed minimal requirements, presented optional activities, and required some synthesis. As she described her design process, it sounded as if she were calling on recent research in cooperative small groups[5] and on some of the old ideas in contract learning.[6] She designed a holistic scoring system based on standards of creativity, thoroughness, understanding, and overall quality. Her enthusiasm for portfolios came from how much the students enjoyed preparing them, how she had seen student strengths that had not been evident before, and how much fun she had had evaluating them. However, she warned that: (i) for the first try, she had demanded too much from her students and the portfolios took a lot of their time; (ii) she wished she had started portfolios earlier in the year so that they were a part of the students' routine; (iii) she realized that the directions for the portfolio, as for any good student assignment, needed to be explicit while allowing for creativity — as she said, "a real balance of structure and creativity;" and (iv) she would have to change the kinds of evidence in the portfolio for each unit in order to accommodate the topic, the goals, and the students' developing skills and interests. Her comments countered the fear of many of the high school teachers that portfolios would take too much time to grade. She said that they took no longer than other tests and papers and that they were fun. Other teachers echoed her comments, saying, "[It would] reduce time grading useless tests."

Conclusion

There is no guarantee that portfolios always will be a mode of authentic assessment; they can become folders in which teachers plunk the same tired stuff they have been doing because they have been instructed by the administration to use portfolios for assessment. On the other hand, under other names, creative science teachers have been using portfolios to capture the complexity of science instruction for many years. Portfolios, clearly defined as such, have the potential to influence how science is taught and assessed. The processes of conceptualizing and designing a portfolio present opportunities to teachers and students alike to identify clearly the instructional goals for the science class, to articulate the criteria for success, to negotiate publicly what will count as evidence, to participate in both the design and development processes, to express individual strengths and become self-reflective, and to become co-learners. This mode of assessment conjures up a different perception of a classroom — a portfolio cultural center where knowledge is constructed communally, individual strengths are developed, and learning becomes an adventure where success is displayed. As one teacher said about portfolios, "Is there any other way to assess science?"

Acknowledgments

Gratitude is expressed to the following persons for discussions and comments on the use of portfolios for assessing student learning in science: Emily Allen, Linda Austin, Ilan Chabay, Linda Clerkson, Gary Dillon, Susan Dutcher, June Fuji, Ricky Gease, Jenifer Helms, Janet Herman, Catherine Howard, Bruce King, Molly Lusignan, Cori McKenzie, Janice McKinley, Karen Miller, Joy Palmer, Ann Perrot, Nancy Rankin, Rosalie Shepherd, Tami Warr, and Doug Wong.

References

1. King, B., *A Report on the Assessment of Students' Portfolio Entry,* Technical Report No. B4 (Stanford University: Teacher Assessment Project, 1990).
2. Bird, T., "The Schoolteacher's Portfolio: An Essay on Possibilities," in J. Millman and L. Darling-Hammond, eds., *Handbook of Teacher Evaluation: Elementary and Secondary Personnel,* 2nd ed. (Newbury Park, CA: Sage, 1990), 241–256.
3. Mitchell, R., "What is "authentic assessment?" *Portfolio: The Newsletter of Arts PROPEL* (December 13, 1989, issue available from Harvard Project Zero, Harvard University, Cambridge, Massachusetts), 5.
4. Wiggins, G., "A True Test: Toward More Authentic and Equitable Assessment," *Phi Delta Kappan* (1989), 703–713.
5. Cohen, E., *Designing Groupwork: Strategies for the Heterogenous Classroom* (New York: Teachers College Press, 1986).
6. Romey, W., *Inquiry Techniques for Teaching Science* (Englewood Cliffs, NJ: Prentice-Hall, 1968).

18

Dynamic Assessment: Potential for Change as a Metric of Individual Readiness

Joseph C. Campione

Dynamic assessment (DA) procedures, while in their infancy in content area applications, already provide clear alternatives to traditional testing practices. These procedures were developed to overcome a number of shortcomings of standard practices such as: (i) classification and description of individual students; (ii) perception of students' abilities; (iii) identification of academic domains; and (iv) instructional practices.

This chapter will illustrate how DA approaches have dealt with these kinds of issues. Attention to different aspects of the problem of assessment has resulted in a family of approaches, the defining feature of which is reliance on evaluating process rather than product, information, and potential for change rather than merely current competence.

Interest in assessment procedures has been motivated historically by two different sets of questions. In one case, centered about special education programs, the major themes involve the evaluation and remediation of poorly performing students; the concern here is primarily with the interpretation and use of intelligence and ability tests. It is this topic that stimulated the original work on DA. In the second case, the concern is with general accountability for and success of the overall educational system, where the main focus is on content area achievement tests. Both kinds of tests have their detractors, and although the specific criticisms differ, the underlying causes are the same — a reliance on static, product-based tests; inappropriate levels of diagnosis; and the decontextualized nature of the evaluations. These features have clear effects on and reinforce the design of both regular and remedial instruction. In the next sections, the main problems with traditional assessment and instruction are outlined, and ways in which DA procedures can help correct them are indicated.

Analysis of Traditional Practices

Student shortcomings are seen in two general areas, knowledge and critical thinking. With regard to the first area, one worry is that students appear unlikely to have acquired deep and well-understood bases of knowledge in the major academic areas. Their acquaintance with history and geography is remarkably slender; their competence in scientific areas such as biology, chemistry, and physics is noteworthy primarily for its lack of deep understanding and the harboring of a variety of misconceptions about the disciplines. While students may acquire information about portions of a discipline, they lack strong relations among those portions.

In the second area, critical thinking, the problem of fragmented knowledge is exacerbated by the fact that students also seem to lack the skills needed to acquire and evaluate new information. Because of their limited knowledge, students are relatively unprepared to work effectively in a technological society when they leave school. Further, failure to master critical thinking skills makes it difficult for those students to adapt readily to new settings.

Instructional Shortcomings

The argument is that these shortcomings — impoverished knowledge bases and weak critical thinking skills — are related and are not a surprising outcome of traditional primary and secondary level curricula. In the major enabling literacies — reading, writing, and mathematics — the emphasis initially is on the acquisition of the major component subskills: decoding in reading, neatness and syntax in writing, and computation in mathematics, which are assumed to underlie successful performance. Even when "understanding" is purportedly the target of instruction, there continues to be a focus on teachers teaching and students practicing decomposed and decontextualized skills. Hence, students are not offered real opportunities to learn the critical thinking skills that permeate the cognitive repertoires of accomplished readers and writers. Thus, they do not amass the foundational skills needed to reinforce their later learning.

This situation is compounded by the nature of instruction in the higher grades, where the emphasis is on breadth of coverage. For example, science courses allow students — in fact, encourage them — to proceed without building strong relationships among the various concepts to which they are exposed. Students are not required to explore a subject in depth, and as a consequence, it is not easy for them to learn to evaluate new information critically. If instruction remains at a surface level and deep understanding is not required, processes of evaluation and reflection likely will not come into play and, therefore, probably will not be learned. Overall, the effect of the emphasis on decomposed and decontextualized skills practice in the early years, coupled with the focus on breadth rather than depth in the later years, means that students are placed continuously in settings where they are neither taught nor encouraged to evaluate critically what they are asked to learn.

Assessment Shortcomings

Static, product-based evaluation is the norm. Standard tests are geared to establishing students' current levels of performance; however, these tests yield no direct evidence about the thought processes that underlie that competence. In this sense, they provide at best a partial, and at worst a misleading, picture of student capabilities, a point made nicely by Vygotsky. He noted that static ability tests do not provide information about

> those functions that have not yet matured but are in the process of maturation, functions that will mature tomorrow but are in the embryonic stage. These functions could be called the "buds" or "flowers," rather than the fruits of development. The actual developmental level characterizes mental development retrospectively, while the zone of proximal development characterizes mental development prospectively.[1]

Vygotsky's notion is of a testing environment that incorporates some kind of social support, in which students will be able to demonstrate the embryonic skills not tapped by static test procedures. In his view, it is the observation of these nascent skills that provides a better estimate of an individual's potential for proceeding beyond current competence. Lacking such information, the likelihood of *underestimating* the capabilities of students and, hence, misclassifying them is increased. Particularly liable to be misclassified are students from culturally different backgrounds who have not had the opportunity to acquire the skills and knowledge assessed on standard tests.[2-4] In addition, without any way of articulating the processes that may have operated, or failed to operate, to produce a given level of performance, it is not possible to determine how to devise an intervention to improve that performance.

Similarly, a major criticism of content area tests is that, by resting on a purely product-based approach to assessment, they are silent on the processes involved in the acquisition of those products. There are several paths to getting a correct answer on a test, and unless they can be distinguished, the assessment has the potential for providing misleading information. In contrast to the situation with static, standard-ability tests, the greater concern is that content area tests can provide a distorted view of progress by *overestimating* the capabilities of many students.

Level of description and degree of contextualization are inappropriate and lead to incorrect assumptions about students' skills. Although standard ability tests are product-based, they are frequently interpreted in terms of sets of inferred psychological processes.[5] The problem is that the processes that emerge are regarded as quite general ones operating in many, if not all, academic domains. The belief is in the centrality of general, decontextualized reasoning skills. This invites the conclusion that instruction aimed directly at those processes will have widespread effects throughout the curriculum.[6] Unfortunately, there is no evidence that such instruction leads to improved reading, mathematics skills, and so on.

In contrast, if ability tests emphasize general global processes, content area tests go to the other extreme. They focus on the many subskills or discrete pieces of knowledge assumed to be involved in effective performance within the domain. In this way, they reinforce the standard instructional practices outlined above. For example, mathematics evaluations tap students' ability to run off algorithms, to solve problems displayed in a recognizable format, and the like, rather than assessing students' understanding of the procedures they are asked to carry out. In addition, it is assumed that what a child does *now* on a test is a reasonable reflection of his or her knowledge, and that that knowledge predicts, or is equivalent to, readiness to learn. Those are assumptions with which proponents of dynamic assessment quarrel. Students can be taught to run off algorithms in a purely rote, mechanical fashion, or they can be led to understand the rationale underlying algorithms and, therefore, to know something about the mathematical principles exemplified in them. The tests that are used to evaluate students clearly do not distinguish between these different "paths" to a correct answer.

Evaluation is static and assumes that individuals tested will not change. This final concern, particularly apparent in ability tests, stems from the conclusions that frequently are drawn. The results of assessment often are taken as providing a relatively permanent characterization of the individual in question. The classifications that result, already presumed to reflect "general" intellectual ability, are further regarded as fixed and unlikely to change. This provides a pessimistic view of the student and rationalizes institutional failures.

Approaches to Dynamic Assessment

To counter these problems, DA procedures focus on students' potential for change. The general approach is to observe and evaluate students in contexts where they are helped to refine, learn, and use new skills and information. Specific programs vary along a number of dimensions, including: (i) the method of evaluating potential — either by observing the actual improvement resulting from intervention or by specifying and assessing directly the processes assumed to underlie improvement; (ii) the nature of the interaction between the examiner and the student — it can be conducted in either a standardized or clinical fashion; (iii) the targets of the assessment — either general or domain-specific skills; and (iv) whether the assessment is conducted in a formal, one-on-one testing situation or in an informal teaching setting.[7] Choices along these dimensions are determined by the particular shortcomings of standard testing being addressed.

Classification and Prediction

For those who prefer a standardized approach, a major goal is to generate psychometrically defensible quantitative data that can be used to increase the predictive validity of the assessment process.[8] An example is to identify students whose performance is artificially reduced on a static test. In one method, a test-train-test format is used. Students take a particular test, are given practice and/or instruction on typical test items, and then take a posttest. The middle training portion is conducted according to a prearranged format. Resulting data sources include: (i) the pretest score; (ii) data from the instructional sessions; (iii) a change score, reflecting the amount of improvement from the pretest to the posttest; and (iv) the posttest score.

Depending upon their particular theoretical stance, different developers focus on different aspects of the data. Budoff, for example, is concerned with classification and prediction in the case of mentally retarded students. He argues that

> following suitable training, many low IQ children will function at a level similar to the child from more privileged circumstances. This posttraining score, regardless of pretraining level, represents the child's optimal level of performance following an optimizing procedure. It permits a comparison between his *presently* low level of functioning, as indicated by his IQ, and his *potential* level of functioning — the third score indicates the child's responsiveness to training, and, by extension, to the classroom regardless of his pretraining level. It is hypothesized to indicate the student's amenability to training given suitable curricula and school experience.[9]

Budoff's students took a standard ability test, were then given instruction on typical items, and took a posttest. Following such training on a variety of IQ-like test items, Budoff found three classes of children in his "educable mentally handicapped" population; these were: (i) "nongainers," subjects who demonstrated little or no gain following instruction; (ii) "gainers," subjects who showed quite marked gain; and (iii) "high scorers," subjects who performed quite adequately on the pretest.

Although there are some problems associated with employing gain scores as a metric, Budoff found interesting correlates of the gainer/nongainer status. For example, middle-class children in special education classes tend in the main to be "nongainers," whereas lower class children have a higher incidence of "gainers." Learning-potential status also predicts performance on a variety of laboratory concept-learning tasks and on a specially constructed mathematics curriculum. It also predicts successful adaptations to mainstreaming, the ability to find and hold jobs during adolescence, the mother's perception of the child, and several positive personality characteristics.

In Campione and Brown[10] and Ferrara, Brown, and Campione,[11] another approach is taken. The focus is on data from instructional sessions, concentrating on *learning* and *transfer efficiency* as major metrics of readiness and understanding. Students are pretested on their knowledge of the domain in question, provided instruction on the items, and then given a posttest. We are interested in determining which variables identify students likely to show the largest pretest to posttest gains; that is, which variables evaluate readiness. During the instructional/assessment sessions, students are asked to solve problems just beyond their reach; they work collaboratively with a tester/teacher who provides a series of graduated hints enabling them to solve the problems. This phase continues until the problems can be solved independently. The amount of help each student needs is taken as the estimate of his or her learning efficiency within that domain and at that particular time.

After achieving independent learning, students are given a series of "transfer problems" in order to evaluate the extent to which they can use what they have learned. The amount of help needed to deal with the "transfer problems" provides an estimate of the student's "transfer propensity,"[12] which is taken as an index of the extent to which the students understand the procedures they have been taught; that is, having learned the procedures, can they use and modify them in flexible ways?

These general procedures have been applied to the assessment of general intelligence test items[13-15] and of specific abilities such as early mathematics,[16] with consistent and positive results. Although static measures of general ability and task-specific competence do predict the amount of gain individuals achieve, the dynamic measures — learning and transfer scores — are better individual predictors of gain and account for significant additional variance in gain scores beyond ability and knowledge. The "transfer," or "understanding," scores are significantly more diagnostic than the "learning" scores. If one wishes to predict the learning trajectory of different students, the best indicator is not their IQ or how much they knew originally, or even how readily they acquire new procedures, but how well they understand and make flexible use of those procedures in solving novel problems.

Linking Assessment with Instruction

Those concerned primarily with increasing the link between assessment and instruction have opted for a different approach. They focus more directly on sets of cognitive processes assumed responsible for growth or performance in a particular domain, and they conduct the assessment in a clinical opportunistic fashion that combines assessment and instruction throughout. Within this camp, there are both formal and informal approaches.

Among formal approaches, the main program is that of Feuerstein,[17,18] which concentrates on the evaluation and teaching of general cognitive skills. He was motivated by skepticism about the suitability of traditional psychometric measures as indices of an individual's capacity to acquire skills, strategies, and knowledge. He developed the "Learning Potential Assessment Device" (LPAD) to measure low-achieving individuals' ability to benefit from instruction. The philosophy driving the LPAD is that structural change is possible regardless of etiology, stage of development, and degree of severity of condition.[19] Further, Feuerstein et al. argue that "*new* cognitive structures" can be produced through the intervention component of the LPAD, which serves a dual purpose. One purpose of the intervention is to bring about changes in the child's performance in order to be able to assess the child's degree of "modifiability" (ability to change). The second purpose of the intervention is to remedy difficulties in the child's problem solving and thus also to serve as the basis for the intervention program following assessment.

The goals of the assessment are to evaluate the modifiability of the processes and to remedy deficiencies in the student's problem solving. Toward these ends, Feuerstein employs a variety of specially devised tasks and materials designed to tap the processes and track change in their execution. He also changes the testing situation fundamentally. Rather than being standardized and static, the testing is flexible, individualized, and highly interactive. Feuerstein views the examinee as a learner/performer and the examiner as a teacher/observer; these are roles that require active involvement of the examiner as well as the student. The goal is not to obtain a quantitative index of performance that can be used to compare individuals or predict future trajectories, but rather to generate a rich picture of an individual's strengths and weaknesses, which can serve as a basis for directed interventions.

Although a detailed critique is impossible here, suffice it to say that Feuerstein's work has been extremely influential and his programs have been implemented widely in the United States. Feuerstein and his colleagues have used the LPAD with below-average adolescents and report great success with the methodology. Supporting evidence for Feuerstein's approach with this population has been found by other researchers working in the area of cognitive intervention.[20] The major questions about the approach concern the transfer effects. Do the skills taught during the assessment/intervention sessions transfer to other academic areas?[21]

To illustrate the informal approach, the reciprocal teaching of reading and listening comprehension skills are considered.[22] These researchers differ from Feuerstein in choosing to situate their program within the more constrained domain of reading comprehension. Working within a particular domain makes it considerably easier to contextualize the instruction in a reasonable way; the processes being honed are always practiced in the actual context of the academic task in question. In this procedure, the teacher and student take turns leading a discussion about a specific segment of text, with the goal of achieving joint understanding. When the teacher leads the discussion, he or she models a set of cognitive strategies (questioning, summarizing, predicting, and clarifying) that fosters comprehension. When students lead, they attempt to employ these same strategies in the search for meaning. As a consequence, the teacher is able to assess the extent to which individual students select appropriate occasions for using the strategies and execute them well. When their performance is less than optimal, the teacher is able to provide tailored feedback to individual students that helps them to improve their skills. In addition to the adult teacher, other students also provide encouragement and feedback to the current discussion leader.

As opposed to more formal assessment, the evaluation the teacher makes of individual student responses need not be precise. The virtue of this procedure is that it is a regular component of classroom activity, taking place on a daily basis. Because it is part of a familiar routine, the teacher has many opportunities to monitor each student, and his or her judgments can reflect aggregations of a number of different inputs. The fact that there are many opportunities for evaluation means that no single one is of undue significance.

This informal approach embodies some of the features of Feuerstein's program. Sets of processes involved in the task are specified; an environment is constructed in which students are observed as they engage in those activities; and the teacher acts as both evaluator and clinician, discovering strengths and weaknesses and responding to student input by providing feedback, practice, and support as needed. The informal approach is also different in fundamental ways. The processes that are targeted are chosen in reference to the academic domain in question, and the activities are always modeled and practiced in context. These two features work to minimize, or finesse, the transfer problem. As the activities are practiced in the context of reading for meaning, whether they will "transfer" to that task is not a concern. Also, if the program is successful, improvements are obtained directly on important school tasks rather than on processing skills that are assumed to be related to performance on those tasks.

Perception of Students' Abilities

As a final point, DA procedures focus on students' ability to learn and on their "modifiability."[23] In this way, they afford a picture different from that emerging from standard ability tests. For example, teachers often regard test scores as providing a relatively permanent characterization of individual students. These perceptions can be modified in interesting ways. Vye, Burns, Delclos, and Bransford[24] had teachers watch videotapes of children's performances before and after intervention. Teachers were asked to view both a segment of a standard static assessment and one of a dynamic assessment of the same child. After viewing the excerpts, the teachers' opinions of the child were elicited. Teachers who viewed only a static assessment consistently gave lower estimates of the child's abilities than those who saw both static and DA episodes. After dynamic events, teachers viewed that child as more competent and more likely to use strategic approaches to learning than they had thought. This positive influence has important implications for potential practical uses of dynamic assessment information by classroom teachers.

Summary

Dynamic assessment procedures have been devised as supplements to, or replacements for, standard testing practices. The main feature of dynamic assessment techniques is that they seek to evaluate students' potential for change either by observing their response to instruction or by determining the processes responsible for change. These procedures have been shown to be effective in providing richer and more accurate pictures of individual students, highlighting those who perform poorly on standard tests but who are nonetheless likely to improve quickly with opportunities to learn. They also have been useful in diagnosing students' understanding of emerging skills, again yielding helpful information about students' future trajectories. By focusing on the processes involved in change and growth, these procedures call attention to the fact that expertise in an academic domain involves more than the accumulation of factual or procedural knowledge. Finally, by emphasizing modifiability, they provide a more optimistic picture of student capabilities, and can influence teacher expectancies, which, in turn, can lead to increased educational attainments.

References

1. Vygotsky, L. S., *Mind in Society: The Development of Higher Psychological Processes*, in M. Cole, V. John-Steiner, S. Scribner, and E. Souberman, trans. and eds. (Cambridge, MA: Harvard University Press, 1978; original work published 1935), 86–87.

2. Campione, J. C., Brown, A. L., and Ferrara, R. A., "Mental Retardation and Intelligence," in R. J. Sternberg, ed., *Handbook of Human Intelligence* (Cambridge: Cambridge University Press, 1982), 392–490.

3. Feuerstein, R., *The Dynamic Assessment of Retarded Performers: The Learning Potential Assessment Device, Theory, Instruments, and Techniques* (Baltimore: University Park Press, 1979).
4. Vygotsky, *Mind in Society.* See reference 1.
5. Brown, A. L., and Campione, J. C., "Psychological Theory and the Study of Learning Disabilities," *American Psychologist* **41** (10) (1986), 1059–1068.
6. Ibid.
7. Campione, J. C., "Assisted Assessment: A Taxonomy of Approaches and an Outline of Strengths and Weaknesses," *Journal of Learning Disabilities* **22** (1989), 151–165.
8. See, for example, Budoff, M., "The Validity of Learning Potential Assessment," in C. S. Lidz, ed., *Dynamic Assessment: An Interactional Approach to Evaluating Learning Potential* (New York: The Guilford Press, 1987), 52–81; Campione, J. C., and Brown, A. L., "Linking Dynamic Assessment With School Achievement," in C. S. Lidz, ed., *Dynamic Assessment: An Interactional Approach to Evaluating Learning Potential* (New York: The Guilford Press, 1987), 82–115; Campione, J. C., and Brown, A. L., "Guided Learning and Transfer: Implications for Approaches to Assessment," in N. Frederiksen, R. Glaser, A. Lesgold, and M. Shafto, eds., *Diagnostic Monitoring of Skill and Knowledge Acquisition* (Hillsdale, NJ: Erlbaum, 1990), 141–172; Carlson, J. S., and Wiedl, K. H., "The Use of Testing-the-Limits Procedures in the Assessment of Intellectual Capabilities in Children with Learning Difficulties, *American Journal of Mental Deficiency* **82** *(1978), 559–564; Carlson, J. S., and Wiedl, K. H., 'Toward a Differential Testing Approach: Testing-the-Limits Employing the Raven Matrices," Intelligence* **3** (1979), 323–344; Embretson, S. E., "Improving the Measurement of Spatial Aptitude by Dynamic Testing," *Intelligence* **11** (1987), 333–358.
9. Budoff, M., *Learning Potential and Educability Among the Educable Mentally Retarded*, Final Report Project No. 312312 (Cambridge, MA: Research Institute for Educational Problems, Cambridge Mental Health Association, 1974), 33.
10. Campione, J. C., and Brown, A. L., "Learning Ability and Transfer Propensity as Sources of Individual Differences in Intelligence," in P. H. Brooks, R. Sperber, and C. McCauley, eds., *Learning and Cognition in the Mentally Retarded* (Baltimore: University Park Press, 1984), 265–294; Campione, and Brown, "Linking Dynamic Assessment"; and Campione and Brown, "Guided Learning and Transfer." See reference 8.
11. Ferrara, R. A., Brown, A. L., and Campione, J. C., "Children's Learning and Transfer of Inductive Reasoning Rules: Studies in Proximal Development," *Child Development* **57** (1986), 1087–1099.

12. Campione and Brown, "Learning Ability." See reference 10.
13. Bryant, N. R., Brown, A. L., and Campione, J. C., "Preschool Children's Learning and Transfer of Matrices Problems: Potential for Improvement," paper presented at the meeting of the Society for Research in Child Development, Detroit, Michigan, April, 1983.
14. Campione, J. C., Brown, A. L., Ferrara, R. A., Jones, R. S., and Steinberg, E., "Breakdowns in Flexible Use of Information: Differences Between Retarded and Nonretarded Children in Transfer Following Equivalent Learning Performance," *Intelligence* **9** (1985), 297–315.
15. Ferrara, Brown, and Campione, "Children's Inductive Reasoning Rules." See reference 11.
16. Ferrara, R. A., "Learning Mathematics in the Zone of Proximal Development: The Importance of Flexible Use of Knowledge," unpublished doctoral dissertation, University of Illinois, Urbana-Champaign, IL, 1987.
17. Feuerstein, *Assessment of Retarded Performers.* See reference 3.
18. Feuerstein, R., *Instrumental Enrichment: An Intervention Program for Cognitive Modifiability* (Baltimore: University Park Press, 1980).
19. Feuerstein, R., Rand, Y., Jensen, M. R., Kaniel, S., and Tzuriel, D., "Prerequisites for Assessment of Learning Potential: The LPAD Model," in C. S. Lidz, ed., *Dynamic Assessment: An Interactional Approach to Evaluating Learning Potential* (New York: The Guilford Press, 1987), 35–51.
20. Ballester, L., "Feuerstein's Model of Cognitive Functioning Applied to Preschool Children: A Study of the Relationship Between the Specific Cognitive Strategies and Learning," unpublished doctoral dissertation, Temple University, Philadelphia, 1984; Haywood, H. C., Filler, J. W., Jr., Shifman, M. A., and Chatelanat, G., "Behavioral Assessment in Mental Retardation," in P. McReynolds, ed., *Advances in Psychological Assessment, 3* (Palo Alto, CA: Science and Behavior Books, 1975), 96–136; Hobbs, N., "Feuerstein's Instrumental Enrichment: Teaching Intelligence to Adolescents," *Educational Leadership* **37** (1980), 566–568; Meltzer, L., "Cognitive Assessment in the Diagnosis of Learning Problems," in M. D. Levine and P. Satz, eds., *Middle Childhood: Development Dysfunction* (Baltimore: University Park Press, 1984), 131–152.
21. Bransford, J. D., Arbitman-Smith, R., Stein, B. S., and Vye, N. J., "Improving Thinking and Learning Skills: An Analysis of Three Approaches," in J. W. Segal, S. F. Chipman, and R. Glaser, eds., *Thinking and Learning Skills, 1* (Hillsdale, NJ: Erlbaum, 1985), 133–206.
22. See, for example, Brown, A. L., and Palincsar, A. S., "Guided Cooperative Learning and Individual Knowledge Acquisition," in L. B.

Resnick, ed., *Knowing, Learning, and Instruction: Essays in Honor of Robert Glaser* (Hillsdale, NJ: Erlbaum, 1989), 393–451; Palincsar, A. S., and Brown, A. L., "Reciprocal Teaching of Comprehension-Fostering and Comprehension-Monitoring Activities," *Cognition and Instruction* **1** (2) (1984), 117–175.

23. Feuerstein, *Assessment of Retarded Performers*. See reference 3.
24. Vye, N. J., Burns, M. S., Delclos, V. R., and Bransford, J. D., "Dynamic Assessment of Intellectually Handicapped Children," in C. S. Lidz, ed., *Dynamic Assessment: An Interactional Approach to Evaluating Learning Potential* (New York: The Guilford Press, 1987), 327–359.

19

Equity and Excellence Through Authentic Science Assessment

Shirley M. Malcom

Perhaps the greatest challenge we face in education is to provide quality education to all children regardless of race, ethnicity, gender, socioeconomic status, ability, disability, or geography. Many proposals have been advanced for how equity in education might be achieved. The discrepancy in performance on tests of basic skills between haves and have nots led to the passage of landmark education legislation during the 1960s in the form of Title I of the Elementary and Secondary Education Act (compensatory education for disadvantaged students). The 1992 proposed budget for Chapter I programs, the successor to Title I, is over $6 billion.

The earlier policy focus on enhancing basic skills, which characterized the educational reforms of the "War on Poverty," is being replaced by emphasis on higher-order skills. From the mid 1970s into the 1980s, students from the most impoverished regions of the country (such as the Southeast) and from the poorest subpopulations (including Blacks and Hispanics) showed improvement on the science and mathematics assessments of the National Assessment of Educational Progress (NAEP). While the trend in NAEP scores shows that performance gaps in science and mathematics have been narrowed at the lowest skills levels, little has changed at advanced levels. The kinds of capabilities that are valued in a knowledge-based economy are not equally available to all through the present mechanisms by which we educate our children.

As evidence mounts that it makes a big difference in life chances whether students attain the skills and knowledge transmitted in mathematics, science, and technology education, success for all students in these areas takes on more meaning and achieving such success takes on more urgency. But enabling all students to achieve at the highest levels possible will require a rethinking of current goals for education and a reinvention of schooling in America. Raising standards for all children and helping all children achieve much higher standards will take major reform in the way schooling is organized. Achieving excellence and equity in science and mathematics education will require a revamping of curriculum goals and standards, enhanced instruction for all students, and very different ways of measuring our trajectory toward these higher standards.

In 1990 the President and governors articulated six national education goals for the country including the goal that "by the year 2000, U.S. students will be first in the world in science and mathematics achievement." The very phrasing of that goal places assessment in a prominent (though not dominant) role. There are three objectives listed under that goal: strengthening math and science education, especially in the early grades; increasing the number of teachers with a substantive background in science and mathematics; and significantly increasing the number of women and minorities who study and complete degrees in mathematics, science, and engineering.

Reform is the big idea in which assessment is embedded. Concerns for standards and equity are controlling the discussion of reform. The challenge we face as Americans is to achieve both without compromising either.

Assessment in the Service of Excellence and Equity

While concern for equity has been a component of the expressed rationale for reform and a major civil rights and psychometric concern within the testing community, less attention has been given to the notion of how one might design assessment to drive equity in the reform of science and mathematics education. In the past, advocates for equity in education have used assessment results to argue for the need for reforms, such as expanding the mission of Chapter I funding to include emphasis on science as well as mathematics and to focus on higher as well as basic skills. Michael Johnson (chapter 15) reported on the use of high-stakes testing to build student esteem and confidence, and to raise teacher, community, and parent expectations for the student participants in programs of the Science Skills Center in Brooklyn, N.Y. While serving these purposes, assessment also documented the quality of the center's programs in the eyes of the world. But Johnson's work also called attention to the fact that the standards set in examinations that can be barriers to high school students may be attainable by students at an earlier level; that is, the current standards may actually be too low because expectations for all children are likely too low, but especially expectations are too low for the African-American and Hispanic children whom the center serves.

How can these students be so successful in intervention programs outside of school when such discrepant results are obtained within the typical school setting? A study conducted by American Association for the Advancement of Science in 1983 looked at programs that work outside of school to improve the precollege education of women and minorities in mathematics, science, and technology and synthesized the findings obtained from questionnaires, site visits, and interviews. The characteristics of exemplary programs that have implications for school reform are these:

- Strong academic emphasis exists in mathematics, science, and language arts.
- Teachers are highly competent in their subject-matter areas.
- Teachers believe that students can learn the material.
- The focus is on enrichment rather than remediation.
- Emphasis is on applications of science and mathematics and on careers.
- Instruction is integrative (across subjects) and focused on real-world problems and projects.
- Opportunities are provided for hands-on instruction and work with computers.
- There is strong organizational leadership, with low staff turnover.
- Resources are adequate.
- School and area resources are linked.
- Opportunities are offered for in-school and out-of-school learning experiences.
- Parents are involved; the community is supportive.
- Specific attention is paid to removing race- and gender-based inequities in instruction.
- Female and/or minority professionals and staff are involved.
- Peer-supported systems (communities of learners) are developed.

Evaluation was critical to determining the effectiveness of these programs. The primary questions focused on success of students in the program, their attitudes toward science and mathematics, and their later success in school as evidenced by grades and tests of student performance, college admissions, degree attainment, and awards.

Cole and Griffin draw on the AAAS study in their report "Contextual Factors in Education: Improving Science and Mathematics Education for Minorities and Women," and extend its implications to all schooling by building on the educational research base.[1] Other studies of successful intervention projects by Stage et al. and by Clewell support the findings by AAAS.[2,3] One could easily surmise that if the schools attended by minorities and females more closely approximated in standards, character and intent the most successful of these projects, American education would be well on the way to reinvention.

Many of the elements in the list above relate to general organizational concerns, such as the need for able leadership or low staff turnover; others are specific to science and mathematics because of the structure of practice or definitions of mastery, such as the need for hands-on activities or working with real-world problems and big ideas in contexts appropriate for the learners. If one is persuaded by Baron's paper (chapter 14), one could argue that the context for assessment should reflect the context of curriculum and instruction. And by extension, if the context described above is one that works for female and minority students, it should also be the context for assessment that serves equity. Baron describes (in Table 1) the characteristics of enriched performance assessment tasks that appear to be consistent with equity-enhancing instructional models. By contrast, in Johnson's report, the need to explain the intent of examiners and question writers to students indicates assessment that is not equity-enhancing since all the onus for translation of intent is transferred to examinees, and students have little opportunity or encouragement to provide answers in a context that makes sense to them. Without a translator the students cannot show what they know because they have difficulty determining what the questioner is getting at.

Assessment for Accountability

Most current discussions about examinations center on those tests intended to measure progress toward educational reform or those aimed at ensuring accountability. The debate has focused primarily on the extent to which the multiple purposes of assessment can be served by various proposed assessment designs, and on the purposes of the different kinds of assessment. It is here that the battle line has been drawn: is the issue reform *versus* equity, or is it reform *to achieve* equity?

If assessment is viewed through the lenses of reform and equity, two very different perspectives emerge, as the following comparisons show:

Reform	**Equity**
Change will occur within the educational system only when people (teachers, students, principals, etc.) are held accountable. Assessment holds them accountable, and a new system of assessment can serve as the driver in the system.	Introducing a new system of assessment may make a bad situation worse since the testing will be in place long before any changes in the schools that affect performance levels can occur. It could have a chilling effect on students who once again bear the bulk of the burden and blame for any performance deficiencies.
The United States is the only major industrialized country that doesn't have a regular set of assessments in place, especially a school-leaving examination.	It is difficult to take international comparisons seriously since assessment for school leaving has had little effect on equity or on gender, racial/ethnic, and/or class-based disparities in school performance, educational access, and life chances.

Students have little reason to work hard since there is no penalty for not working hard. A test with real consequences to the students' life chances will serve as a motivating force.

Many of the disparities in motivation and deficiencies in effort are related to disparities in opportunities for challenging, meaningful coursework. It is difficult to motivate students with material and pedagogical techniques that are contextually poor, that reek of low expectations, and that are not engaging.

Employers need a way to know whether students have mastered the goals for K-12 education. We need indicators that tell us what students know and are able to do.

Students also need feedback on their performances early enough to correct deficiencies. But it is difficult to learn what you are not taught. There is ample evidence to verify that the schooling experience of haves and have nots is sufficiently different to maintain the gap. Employers need to support meaningful reform before a system is put in place to further restrict students' employment chances.

Assessment can affect what happens in the classroom, especially what and how teachers teach. Teachers teach those things that are tested and those things for which they are held accountable.

The level of staff development that teachers need in order to accommodate the new assessment is staggering. Unless we get the goals right and the assessment right, we may end up where we don't want to be.

Those who speak against the current assessment movement are self-interested, do not want standards, and/or do not want to be held accountable.

Those who push for assessment first are naive, and are looking for a simple solution to complex problems, including those of resource allocation, low expectations, and unequal opportunity to learn.

Clearly, there is considerable disagreement about means, ends, and who is to be held accountable in the current discussion of assessment in the service of reform.

Equity in Standard Setting

The gravest inequity in standard setting would be to set different *levels* of standards for different subpopulations of students. Ideally, standards would evolve from a common core developed through a wide-ranging, consensus-building process that leaves room for locally expressed requirements. While standards would satisfy the test of scholarship (best knowledge, skills of the disciplines, and habits of mind), there should be recognition of the legitimacy of needs within standards (the interest of early adolescents in and their need to be informed about the workings of their own bodies, for example).

Equity in Implementing Standards and Designing Assessment

It is not sufficient to put in place standards that address national, local, and personal needs; resources must be provided to implement these within schools. No assessment can be considered equitable for students if there has been differential opportunity to access the material upon which the assessment is based. If students have been provided opportunity to learn, assessing how well they have used that opportunity is best done at a level closest to the student; that is, in the classroom setting.

To determine whether schools have provided students with opportunity to learn, district/system level testing against the system's stated standards is needed to show that the material has been covered to a sufficient level of mastery. District standards can be compared with others in the state, region, and country. Student assessment based on school-generated techniques can be calibrated to external testing of samples of students, providing benchmarks for curriculum, instruction, and assessment for each school and guiding quality control for the system. Thus, the student would be held accountable for performance specified by the school, and the school would be held accountable for providing access to curriculum and instruction necessary to perform well in external testing. The district would be held accountable for devising standards, and providing resources, incentives, and assistance to schools, while the state would be accountable for providing an adequate resource base for each of its systems depending on the job that system must do to meet the standards. "Pass through accountability" would replace "pass down blaming of the victim."

Many elements must be considered in the design of external examinations or systems of examinations, including those in the following list:

- What is being tested; specifically, what aspects within each subject matter domain are being assessed? How well do these track with the way the content domains were taught, the way they appear in actual practice, and how the knowledge and skills within those fields are valued?

- What specific goals for education are being tested (for example, the ability to recall, or the ability to use information once provided)? How are these assessed goals related to the real-world use of the knowledge and skills being assessed?

- What exactly will be measured (for example, processes, products, skills)? To what extent will measurement be direct ("Student will demonstrate ability to fly a plane by flying a plane") or indirect ("Student will demonstrate ability to fly a plane by answering questions about flying in a written or oral exam"), sequential (answering questions and then flying) or simulated. How different, similar, or interdependent is the information we obtain through each type of measurement?

- Who will develop and score the assessment? What will be the relationship among the teacher, assessment developer, and assessment scorer, and how much do these groups interact with and reflect the race/ethnic, gender, regional, and other characteristics of the learners?

- To what uses will the assessments be put? What will be the balance between student, instructional, monitoring, policy, and accountability purposes? Who needs to know the outcomes and for what purposes?

- What will be the elements of the assessments and of the specific examination units within the assessments (for example, short answer, performance, some combination)? How will these elements themselves be evaluated over time for their ability to predict the desired characteristics or to indicate mastery for different segments of the student population?

- What will be the specific mechanisms of assessment? To what extent will testing occur in the context in which the behavior will ultimately be judged? In familiar or unfamiliar settings? With or without technology? With familiar or unfamiliar testers? Individually or in groups?

- How will we interpret the results? What will be judged an acceptable, unacceptable, or exceptional response; how will we respond to "outliers" or to "nonstandard" answers? How will standards and ranges be set (relationship to learners, to judgments of content "experts," and, especially, where context-dependent)?

- How will subject assessment be packaged? To what extent will assessments be made "one subject at a time" or across subject-matter domains (that is, as the student is likely to encounter the material in the real world or as it was encountered in the learning situation)?

- How will we balance nationally agreed upon and locally expressed curriculum standards with national and local needs for reporting on student performance? On how an individual's performance maps against these, or how the performance of different groups of students map against these standards?

- How much difference from current modes of assessment and mechanisms of reporting will the public tolerate? How close to world-class standards do we dare assess? How much bad news can we stand, especially about the standards achieved by students whom we consider advantaged?

Are we better off with the flawed system now in place or with an unknown examination system that could bring even greater problems? What differences in opportunity to learn and achieve will flow from assessment? Will it help students, teachers, or parents do something different to promote learning? Will it lead to reallocation of resources to enhance learning; for example, by moving the best teachers to the neediest students or providing summer instruction for students not at grade level at the end of the school year? And does better assessment increase our responsibility for intervention, as better technology in medicine has increased the demand and the ethical dilemmas we face in determining use of that technology in treatment? If we are prepared to *do* more, once we know more, perhaps the dangers of inequity possible in new assessment are worth the risk. But absent the resolve to intervene, one could argue that assessment becomes little more than voyeurism.

What is the proper role of assessment in reform? If we accept the premise that the primary purpose of assessment should be to improve learning, then the primary audience for the data must be those who can directly affect learning (teachers, students, and parents). Many other stakeholders can argue that they need information – the investors (the public and their representatives, policymakers, elected and appointed officials), and the end users (employers, colleges, and universities). The operative question is, How much information and about what? To move from improving learning for *a* student in *a* classroom, to changing school systems to improve learning for many students, to improving learning for all students nationally, shifts the purpose and structure of assessment as well as its costs in time and money. Who shall be assessed (every student or a sample of students?), how often (every day or two to three times in 12 years?), what (progress in mastering small competencies or ability to understand large ideas and demonstrate articulated skills?), and why (to improve learning or to hold educators or policymakers accountable?) are decisions that ultimately hinge on our concept of and belief in the need for equity for all our children.

Reform reports such as *A Nation Prepared: Teachers for the 21st Century*[4] and *America's Choice: High Skills or Low Wages*[5] insist that we must have educational reform because we must provide for the many the level of education in this country now provided to the few so that the floor of learning can be raised to support a knowledge-based economy. Concerns for equity are the cornerstone of this wave of educational reform. Equity concerns must then be at the heart of assessment design if we expect assessment to play a central role in moving the reform agenda.

What Do We Mean by Equity?

Equity is a meaningful term only when we talk about difference. "Difference" has meaning in many areas that bear on a discussion of learning and assessment. We can talk about differences of biology, culture, policy, and education and what effects these might have on learning. In biology, difference refers to the natural variation that occurs in the genetic makeup, and outward expression of that genetic makeup, as it develops in a particular environment. Difference is a matter of perspective: compared to living organisms in other phyla, humans seem very like each other. Compared to members of one's own family, the human family seems very diverse indeed. Humans have a common genetic core that defines them as human and different genetic aspects; for example, those that define them as female or male. Humans develop over time as genes and environment interact; growth and development affect body and brain, allowing students to construct and handle more abstract ideas as they "mature." While we are far more alike than different, the variation available to nature renders each of us unique.

Looking at difference from a cultural perspective, female and male children may have certain roles that have been ascribed to them by virtue of custom and independent of capability; that is, where expectations, attitudes, and opportunities are based first or primarily on gender. In a political context difference may affect the opportunities that have not been (and are or are not being) made available to particular groups. As a nation we have sometimes highlighted and celebrated differences; at other times we have tried to downplay or obscure them. While the metaphor for our population has shifted from melting pot to salad bowl, the clear intent of education in America remains that of providing a common core of shared experiences, values, and knowledge to hold the country together. How one addresses the differences around that common core (or even how one defines the core specifically to include pluralism) remains the great debate of educators and scholars alike, from among both "lumpers" and "splitters."

Political considerations aside, the diversity within the classroom is real and requires a pedagogical response. Different sets of skills, aptitudes, and interests, (and, in all likelihood, preferred ways of learning) tend to cluster around particular kinds of intelligences, ranges, and styles that are observable. These skills and interests may also cluster around different sets of shared experiences, many of which are guided by traditions, customs, or shared institutions which are products of where we live, our racial/ethnic group, the structure of the language we speak, or even whether or not we have a computer at home. Good teachers know how to pull all these differences into their pedagogical toolboxes, to serve as launching pads and cognitive anchors upon which to build the shared core of learning.

If our societal goal is to allow everyone to achieve her or his full potential — if we are to nurture the talent of all our people — then we must do two things concurrently: provide the common core that we have as a community decided we want all students to have, and provide individuals with what they need to attain mastery as they develop individual talents. This argues for a broad education that includes the humanities, arts, sciences, sports, mathematics, and technology as students explore the possibilities. The common core includes those elements that define the values of our country; our human needs; our local, regional, and national workforce and citizenship needs; and our needs for personal empowerment and working cooperatively to achieve shared goals; and for building a common vision of global responsibility. Standards must recognize the consensual core of required knowledge and skills while at the same time taking a pluralistic approach to that which surrounds the core. In teaching, one necessarily moves in the opposite direction – through individual interests, talents, experiences, and culture — to build the core. Most importantly, each student must be given what she or he will need to continue to learn. Since the school is not the only place that education takes place, a crucial question is, What should school provide? The home, community, and family are also central places for learning. What should they be expected to provide? When we assess that which the school has provided, how do we do so without contamination from the effects, good or bad, of the larger environment? Yet, if the core is centrally important, school or some societally structured equivalent must ensure that it is provided somewhere for everyone. As society has decided that more and more topics are central — are necessary for the well-being of the individual and the nation — it has added them onto the mandate of the schools. Until recently we have added more and more content expectations without asking whether the core really ought to be expanded or whether it should be re-examined or even reduced.

Standard Setting Amid Diversity

The establishment of standards has to date largely been controlled by content specialists looking mostly at their own part of the curriculum. Their statements tend to become accepted or rejected depending on the extent to which those who achieve the standards also are able to achieve larger societal goals such as earning a living, raising a family, voting, or carrying out other such roles. How much of the curriculum is negotiable; how much of assessment is negotiable? Do the differences within our population require greater choice than a single examination for the entire country can accommodate?

Concern for equity requires us to ask how much the standards should depend on the needs of the learner as opposed to relying solely on the "absolutes" articulated by the disciplines.

If we accept the need for pluralism, then we would tend to favor a system of external examinations to accommodate the political reality of decentralization and local control as well as the cultural reality of differences in lifestyles and values in different regions of the country. If we accept such pluralism within curriculum and assessment, are we also willing to accept some level of flexibility in the medium that students use to demonstrate their knowledge and skills?

Challenging Current Practice

Equity concerns cause us to look at many current testing practices in a new light, to explore what we are measuring, what we think we are measuring, and what we want to measure.

A recent publication, *Assessment Alternatives in Mathematics,* notes, for example, some of the problems with timed tests:

> The use of time as a factor in testing, whether for standardized tests or for practicing basic addition facts, is an equity issue. Many students suffer from enough anxiety in testing situations to keep them from showing their true understanding. Others have physical handicaps, vision problems, or difficulty in writing. Even the slowest student should have as much time to finish work as he or she needs, especially when being assessed. Preanswered timed tests cannot allow for mathematical investigations and for working on problems for which the solution is not immediately apparent. Timing of performance tends to discourage persistence and to promote thoughtlessness and jumping to conclusions. Quickest does not equal most talented. The use of time in standardized tests is mainly for the purpose of creating a [normal distribution] by preventing some children from answering all the questions. Speed is less important to real-life success than reasoning, accurate analysis of situations, and the willingness to tackle tough problems.[6]

Other equity concerns noted in this document were reliance on standardized testing and especially the issues related to the norming of such tests (with what populations and with what deemed appropriate responses); and the language of testing (especially compared to the primary language of instruction and the language of those tested).

The most fundamental issue, as noted earlier, is whether all students have equitable access to the courses and teachers (the curriculum and instruction) needed to perform on the assessment. This is a serious problem for many subpopulations of students in the United States who live in inner city or rural areas, who are poor, who are female, who are members of traditionally undereducated minority groups, or who are some combination of these characteristics. Studies by Jeannie Oakes point out the devastating effects of tracking and the distribution of intellectual and material resources in science and mathematics on opportunity to learn.[7] As science and especially mathematics play critical gatekeeping functions for employment and education, specific equity concerns arise with regard to access before assessment issues can ever be broached.

Science and Mathematics: The Great Equalizers?

In 1983 in an article in *Daedelus,* David Hawkins argued that mathematics and science are the "great equalizers."[8] In contrast to language (reading, writing, and speaking), which is highly related to experience and thus to socioeconomic status, Hawkins noted that science and mathematics, depending, at least initially, on formal instruction in school, are potential levellers among young children who come from widely different social conditions. If Hawkins is correct in his assertions about the equalizing *power* of science and mathematics, why do we see such unequal outcomes of NAEP performance, even early within schooling, between students in inner city or rural schools and those in suburban schools? Why is there such discrepancy between the scores of Black and Hispanic students on the one hand and those of White and Asian students on the other? Or why, by the time female students are tested at 17 years of age, have science scores diverged from those of males students?

Most observers of these results would likely agree that these outcomes reflect different educational experiences for these different groups. So the equalizing potential of science and mathematics is lost in the reality of schooling. We have learned of these differences through testing, but what do they tell us about the students themselves, what they know and can do, or the educational experiences they have had? And what have we done with the information we have gained? As the assessment stakes grow, as reports call for testing that affects employment and college opportunities, as the movement toward a national examination (or examination system) escalates, fears of bias and concerns for equity should increase.

Assessment in the Service of Equity

Advocacy groups and testing companies alike can agree on the need to avoid bias in testing. What will be tougher than avoiding bias will be building a system from the ground up that supports equity. What considerations would have to be incorporated to construct an equity-supporting assessment? Assessment can be equitable without yielding the same results for everyone. What it must do is, give all groups a fair chance to demonstrate what they know. Assessment will be equitable if the following conditions are met:

- The rules about what is to be known (and therefore what is to be assessed) are clear to all.
- The resources needed to achieve are available to all.

- The ways of demonstrating knowledge are many and varied.
- Some of the things that are valued by different groups are reflected in the original statement of what is to be known (or what skills one should have), along with societal or expert community statements of what everyone should know and be able to do.

The issues related to assessment become more focused when we look at assessment of specific disciplines or activities. We assess writing by having students write. Write what? In what setting? With rewrite opportunities? Under the conditions we give them or under conditions like those in which they are most likely to have to write in authentic situations? The community of people who teach writing have agreed on certain ideas about what constitutes a fair assessment of writing. But since writing is a competency about which we all have views, opinions, and standards, no matter what teachers might say, parents can form judgments about the quality of their children's effort and output. Mathematics is also a competency, but the distance between school mathematics and real-world mathematics, and the uneasiness of much of the public in debating specific aspects of quality of effort or outcome, combine to place the onus for reform and assessment squarely within the mathematics and mathematics education communities. Parents may look for the basic indicators: can my children add, subtract, multiply, divide, and do simple fractions? But beyond that many parents do not know how to assess the quality of the mathematics instruction their children receive. Over a period of some years, standards have been derived for what mathematics should be taught and learned by students at different grade or developmental levels.

These standards should be disseminated to the public in understandable language so that they, too, may make judgments about the adequacy of the mathematics being provided in the schools they support.

Contemporary reformers point to the leveraging role of examinations and examination systems and suggest a role for such exams in the United States. The examination systems in the European Community can provide important historical, cultural, and political insights to the U.S. assessment debate (see chapter 12). A major feature in the origins of examination systems in Europe relates to their "selective function;" for gaining access to university, for example. But the use of examinations for selection for the civil service and professions predated their use to determine admission to the university. As Madaus and Kellaghan note, examinations were used for reasons of equity, "to replace the old system of patronage and nepotism for making appointments to the civil service which had secured the dominance of the aristocracy;" that is, to move away from a system where who you knew rather than what you knew got you a job. The authors note the prevalence of many of the same arguments (motivating force on teachers and students; connections among exams, achievement, and competitiveness) among 19th century reformers. At the time when examinations were flourishing in 19th century Britain, the authors note that "the state was recognized to have only limited obligations to secure the education of its citizens." There was no ethos operating to support the idea of a right to a publicly funded education for all. In addition, examinations were perceived as a "cheap and effective method" to cause classroom and local agencies alike to invest more resources in support of education.

Reinventing Assessment

The European examination systems were developed as instruments to sort, select, and direct the distribution of scarce resources (jobs, education, university positions). How do we use examination systems if we want to improve learning for all in a country where education is accepted as a right; indeed, as an imperative? In reality, examinations are also used in the United States to sort, select, and distribute resources. We would have to *reinvent* assessment for it to work as a cultivating, standard-raising tool that is effective in a highly decentralized educational system, with an extremely heterogeneous school population. These are challenges the European systems have never addressed. I believe the "reinvention" would look something like this:

- Assessment would be aligned with the nationally agreed upon and locally available curriculum.
- Curriculum would be based on standards that are rigorous but not rigid; diverse but without compromising the highest standards of excellence; reflective of student and societal needs; derived through interaction of public, discipline, teaching, and user communities; world class but distinctly developed to fit America's values and needs.
- Assessment would inform resource allocation rather than simply reflect it. Students would be provided with various modes of assessment, and if a particular mode is found to provide the best instructional information, its use for a particular student could increase.
- Assessment would focus on higher-order skills, use of subject matter, and demonstration of competencies, especially in settings that are contextually meaningful to students.
- Assessment would be continuous and usually indistinguishable from instruction; it would prompt reinstruction and reassessment until the standards are achieved.
- Assessment would change as the curriculum changes toward subject-matter instruction according to the way the discipline is conducted by its practitioners, and as it connects to other fields.
- Assessment would inform student initiative and learning outside of school.
- Teachers would be prominent in the assessment development and scoring process.

- Teachers, developers, and scorers would be reflective of the tested populations and would include males and females and persons from different regions and different racial/ethnic groups.
- Acceptable ''outlier'' responses to assessment questions would be captured and, where appropriate, would be included as training examples for teachers to illustrate the range of appropriate responses.
- Consistency, inconsistency, and redundancy among assessment results would be captured and used to improve the assessment process.
- Assessment of individuals would be primarily for improving instruction and to direct reteaching, intervention, and career counselling.
- Anonymous, aggregated assessment results would provide information for quality control and accountability of schools and school systems.
- Technology would become a more prominent tool of assessment, especially in providing contextually richer questions; assistance to poor readers, non-English speakers, and students with disabilities; and in diagnosing learning problems and misconceptions.
- The use of timed testing would decrease.
- Students would be able to see their own scored assessments (with comments) as well as examples of exceptional and acceptable responses.
- An appeal (or explanation) process would allow students to challenge scoring.
- Assessment results for quality control would be publicly reported, and resources and technical assistance would be provided to assist schools reach appropriate levels of performance.
- Standards for school performance would be set at world-class levels, and schools given some period of time, resources, and technical assistance to raise their performance to meet those standards.
- Regular subpopulation sampling and separate reporting would be a part of the assessment process, and appropriate intervention would be put in place where necessary to improve subpopulation scores.
- As in sports performance (in gymnastics and diving, for example), students would be encouraged to demonstrate superior performance in tasks they select, with degrees of difficulty attached. Creativity would be rewarded with additional recognition.
- Single-score reporting would be replaced with scoring that provides more information about competencies, skills, and knowledge attained.
- Scoring would be rich enough to inform teachers of children who move from grade to grade, school to school, or region to region, about what these students have been taught and what level of knowledge they have achieved.

Reinventing assessment to support equity can only be achieved through reinventing schooling; reinventing schooling can only be achieved by reinventing assessment to support equity. What must come first is the will to do both.

References

1. Cole, M., and Griffin, P., *Contextual Factors in Education: Improving Science and Mathematics Education for Minorities and Women* (Madison, WI: University of Wisconsin, 1987).
2. Clewell, T. T., and Anderson, B., *Intervention Programs in Math and Science and Computer Science for Minority and Female Students* (Princeton, NJ: Educational Testing Service, 1987).
3. Stage, E.K., Kreinberg, N., Eccles, J., Becker, J.R., "Increasing the Participation and Achievement of Girls and Women in Mathematics, Science, and Engineering," in S.S. Klein, ed., *Handbook for Achieving Sex Equity Through Education* (Baltimore, MD: Johns Hopkins University Press, 1985).
4. Carnegie Forum on Education and the Economy's Task Force on Teaching as a Profession, *A Nation Prepared: Teachers for the 21st Century* (New York: Carnegie Corporation of New York, 1986).
5. National Center on Education and the Economy, Commission on the Skills of the American Workforce, 1990.
6. EQUALS staff and the Assessment Committee of the California Mathematics Council, *Assessment Alternatives in Mathematics: An Overview of Assessment Techniques That Promote Learning*, 1989.
7. Oakes, J., with Ormseth, T., Bell, R., and Camp, P., *Multiplying Inequalities: The Effects of Race, Social Class, and Tracking on Opportunities to Learn Mathematics and Science* (Santa Monica, CA: The Rand Corporation, 1990).
8. Hawkins, D., "Nature Closely Observed," in *Daedalus* **112** (2) (Spring, 1983), 65–89.

Appendix:

Examples From the Field

The appendix of this volume is a collection of assessment ideas and examples from schools and universities. Even though there are a great number of suggestions for reforming science and mathematics assessment, including those in this book, there is not yet a published record of examples of school-based assessment practice. Many of the available examples of performance or authentic assessment come from large-scale tests, or from examples of innovative ideas for items that might be used in the place of multiple-choice tests. Few of these examples provide information for how they might be integrated with the curriculum or how a teacher might use them. Practitioners and teachers currently have few examples of specific approaches that work at the classroom level. The collection that is presented here is an initial step in providing concrete examples developed, used, and described by teachers.

The examples range from elementary grades through college and span several areas, including life, earth, and physical sciences as well as mathematics. The assessment approaches incorporate strategies that include journals, creative drama, interviews, construction of models, and laboratory performance. All of them represent attempts to develop multiple assessment processes that reflect both individual and group achievement. Many of them also look beyond simple, objective, content-oriented assessment, working hard to develop innovative strategies that include affective, motor, and social aspects of science learning. These strategies are what sets the examples apart from traditional assessment, reflecting a holistic and highly individualized view of student science learning.

As the reform of science and mathematics curriculum and teaching begins to find its way into the schools, and classroom practice, it is important that new assessment approaches accompany these changes. If that parallel activity is to take place, it is important that a literature of field-based practice, developed and implemented in schools, be accumulated. Policy-makers and researchers need to work together with teachers to develop this collective experience of what works in science and mathematics assessment. The following examples represent a first step toward this goal and toward developing ways to communicate and share ideas among the larger community that is interested in advancing science and mathematics education.

1

Students Showing What They Know: A Look at Alternative Assessments

Thomas M. Dana, Anthony W. Lorsbach,
Karl Hook, and Carol Briscoe

In this paper we present examples of assessment activities we have developed and used with our students at the Florida State University School, an urban K–12 school located in Tallahassee, Florida. The classes represented in this report of our assessment practices are middle- and high-school–level physical science, biology, and chemistry classes.

Assessment in school science classes is viewed as an opportunity for students to show what they know about science in terms not only of specific content knowledge, but of the processes of science as well. We see this knowledge as more than a set of isolated ideas and facts. For students to truly know science, they must know it with understanding, in a way that allows them to connect new knowledge with what they already know, and to apply their knowledge to problematic situations inside and outside of the classroom. Indeed, assessment practices, especially in science classes, must integrate the best of what we know about student learning with the nature of science itself.

The assessment ideas included here should be looked at as a way to diversify and humanize the assessment process by having students show what they know. By reasoning, inferring, talking, and writing, students express their knowledge in a variety of ways. The ideas included in this report should be considered components of an ongoing assessment program. Taken together, the assessments are part of a portfolio that form a holistic profile of what our students have learned through their experiences and thinking in our science classes.

Theoretical Rationale

Two major theories drive the way we approach assessment. The first is constructivism, an epistemological theory about the nature and origin of knowledge. The constructivist view recognizes that individuals construct knowledge as they interpret new information and reconstruct what they already know. The other major theory, cognitive interests, provides a framework to consider the purposes behind curricular processes. Each of these theories is described below in relation to our assessment practices.

Constructivism[1]

Learning in science is a search for viable solutions to problematic situations. To learn science is to understand the process of constructing scientific knowledge, to identify problems and work out solutions with others in the classroom, and to explore current scientific knowledge and understand why it is accepted by scientists as viable. Constructivists believe that knowledge is personally constructed by students and that learning occurs as meaning is given to experiences in light of existing knowledge. Viewing scientific knowledge in this manner leads to the establishment of a classroom environment where students are actively involved in building scientific meanings. At the Florida State University School, we want our students to understand and to accept responsibility for their own learning. Accordingly, our assessment techniques allow students to express their personal understanding of scientific concepts in a way that is uniquely theirs.

Theory of Cognitive Interests

Another way we conceptualize classroom assessment practices is through Habermas' theory of technical, practical, and emancipatory cognitive interests,[2] as described by Grundy.[3] Assessment based in technical interests helps control and manage student learning. Educators with technical interests are themselves keenly interested in specific facts and algorithms, and they tend to focus on precise behavioral objectives, such as those found in many state and school district curriculum frameworks. Such teachers feel the need to "teach to the test" to ensure that students will learn what the teacher, or the framework objectives, designate as being important. When assessment is based solely on this interest, it is difficult to assess the extent to which students understand scientific concepts.

Practical interests focus on the understanding and communication of ideas. Learning science is considered a matter of making sense of one's own scientific ideas in the social context of the classroom. Emancipatory interests are evident when students take control of the assessment process, rising above the traditional teacher-dominated classroom to ensure that the teachers know what they have learned. Our assessments follow practical and emancipatory interests as they give our students the responsibility to demonstrate to us the scientific concepts and relationships they have constructed.

Applications

Samples of the alternative assessment techniques we have employed are described in the following sections. Student examples of each technique are included where appropriate.

Concept Mapping[4]

Concept mapping helps students to organize and represent concepts in meaningful ways. After identifying concepts relevant for a particular topic, students organize these concepts in hierarchical relationships. Through concept mapping, students can connect concepts in a variety of ways and can represent the personal meanings they hold for concepts. We have adapted concept mapping as a means of alternative assessment. For teachers, concept maps provide periodic "windows into the minds" of learners as they construct their understandings of science concepts. As summative assessment tools, concept maps provide teachers with a richer view of students' knowledge than is possible using conventional tests. We often use pre- and postunit concept maps to document student growth. Because concept maps assess meaningful learning, both students and teachers can direct their energies away from rote learning and toward learning for understanding. The examples presented here (Figures 1a and 1b) are from

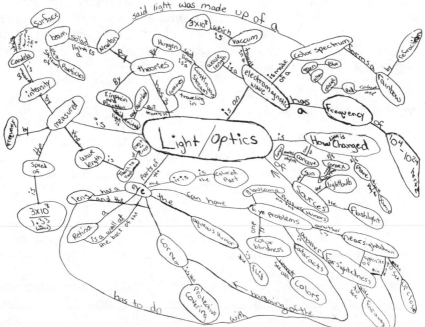

Figure 1a.

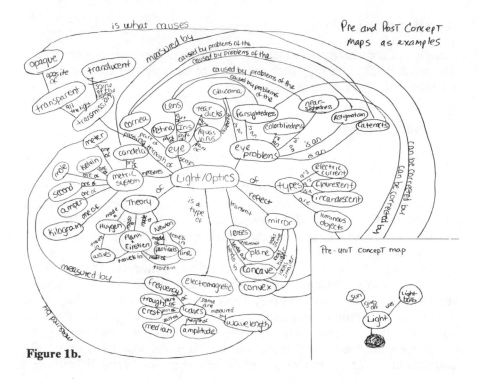

Figure 1b.

an 8th grade physical science class. The students were asked to generate a list of ideas associated with the topic of light and then to organize the ideas in a concept map. This was done both prior to and after the learning activities. These students have been using concept maps as a means for organizing ideas for two years.

Creative Assessments

These assessment techniques are used by students to represent what they have learned during a particular "unit" in a creative manner. Rather than just recalling unrelated facts, the students show what they have learned in a context that makes sense to them. They might use scrap books, comic books, or home videos, for example, to demonstrate particular concepts. This helps students make particular scientific concepts relevant to their own lives. This assessment technique also emphasizes the application of higher-level thinking skills in a context that many students find highly enjoyable.

In the scrap books, 8th grade students were asked to choose ten concepts they had learned about light and color. For each concept, the student prepared a page of the scrap book. Using the model of a child's "how-and-why type" science book, the students wrote questions and answers they thought best showed what they had learned. Each page was augmented with an illustration or magazine cut-out to represent the concept. The scrap books were evaluated by both the teacher and fellow students according to a scoring scheme negotiated by the teacher and the class. Comic books and home videos were produced in a high school chemistry class by students showing what they had learned about atomic theory and the history of its development. Teams of three students cooperated to reach consensus on the meaning of the concepts they wanted to represent and how they wanted to present them.

This Is What I Know...

Our classrooms become a place for students to tell others what they have learned when we conduct a simulation of a professional scientific meeting. The presentations help the students to understand that scientists are more than laboratory researchers, stereotypically represented with white laboratory coats, protective glasses, and Einstein-like hairstyles. The students learn that communication is an integral part of the work of scientists, and that scientific progress would be quickly curtailed should scientists not communicate their ideas to their peers.

In all of our classes, students work in small cooperative groups to prepare presentations to be made to fellow students. Groups of students become class experts on a specific concept and synthesize different perspectives and ideas in the presentations. All students do not have the same knowledge, even though they all "studied" the same topic, so the presentations become sessions where peers teach peers.

In the discussions after each presentation, both the teacher and the students provide questions and comments to the presenters. Presentations to the class are evaluated by the teacher, the presenters themselves, and fellow classmates according to a scheme worked out prior to the presentations. Content knowledge, individual understanding, and presentation effectiveness were considered to be important factors by one of our recent high school classes. As a conclusion to a "unit," presentations have been ideal ways for our students to synthesize and evaluate the scientific concepts they have constructed.

Journals

Journals are a way for students, teachers, and parents to keep track of the development of a student's knowledge. Before, during, and after class activities and often as a homework assignment, students record thoughts, feelings, and ideas related to their science learning.

Journal writing has been used in a variety of ways with our classes. We often provide a sentence stem or question to help students focus their thinking. Enablers such as "When I think of atoms, I wonder..." or "What do you think about making recycling mandatory?" have been successful ways to have the students reflect on their scientific beliefs and values. Students are sometimes asked to look back over their journal entries and write another entry that focuses on the growth of their knowledge. At other times, the teacher and the student maintain a personal dialogue by writing back and forth about the scientific, social, and technological relevance of ideas raised during class.

Journals, or other records of reflective thought, are a way to capture growth over a period of time. Students can become more thoughtful assessors of their personal learning histories. Further, by having our students keep written journals, we are better able to understand the whole of a student's knowledge, not just a sampling of what he or she has learned.

Oral Interviews

Not all students can adequately express what they know in writing. Others have difficulty reading and understanding written test questions. Oral interviews focusing on the communicative act between teacher and student are a way to overcome these problems.

In oral interviews teacher and student are jointly responsible for determining what a student knows. Through personalized communication, both teacher and student negotiate the meaning of what is being said, with the student empowered to make sure meanings are expressed thoroughly and understood.

Students learn that the teacher respects what they have to say. Because students know they can communicate with everyday language rather than formal scientific jargon, test anxiety decreases and self-esteem blossoms. Similarly, a teacher can ask students to elaborate on statements to determine the breadth and depth of students' understanding. The focus of the task is to determine conceptual understanding rather than vocabulary knowledge.

An interview turns the assessment process into a learning opportunity for both teacher and student. In our efforts to understand what students know about science, we have found oral interviews to be an effective way for both teachers and students to communicate what is known.

Implications of Alternative Assessments

Tests constructed separately from students and the learning process pervade much of educational assessment today. Science educators must ask themselves if they are satisfied with an assessment program that is almost solely concerned with whether decontextualized scientific facts can be recalled from students' short-term memories. Far more valid are assessments that allow students to show what they know in a variety of contexts. We view alternative assessment techniques to be a part of the classroom curriculum because they continually create opportunities for students to clarify the meanings they have constructed. We feel that utilizing only one type of assessment is insufficient for educators to ascertain the nature and extent of students' knowledge. Therefore, we advocate the use of a variety of classroom-based assessments in order to develop a more extensive profile of what students know.

References

1. vonGlasersfeld, E., "Cognition, Construction of Knowledge and Teaching," *Synthese* **80** (1) (1989), 121–140.
2. Habermas, J., *Knowledge and Human Interests,* 2nd ed., J. Shapiro, trans. (London: Heinemann, 1972).
3. Grundy, S., *Curriculum: Product or Praxis* (London: Falmer Press, 1972).
4. Novak, J. D., and Gowin, D. B. , *Learning How to Learn* (New York: Cambridge University Press, 1984).

2

Use of Creative Drama to Evaluate Elementary School Students' Understanding of Science Concepts

Michael Kamen

There is a great deal of concern regarding the quality of elementary science education. The typical science lesson in the elementary classroom requires students to read a chapter and choose the "right" answers to a set of questions. Science is usually not presented in ways that encourage students to construct models and maximize their conceptual understanding of the phenomena. Students generally have to memorize the correct answers. Even when a hands-on approach is used, students tend to hold on to their original conceptions. Students may use the correct words but fail to understand those words.

Scientific knowledge is viewed by many science educators and philosophers as being a human construction.[1] This view is very different from the positivist view of science. The positivist perspective treats science as a process of revealing the truth, with the scientist seen as being objective and apart from the "truth" that is being investigated. In this approach the scientific method is considered *the* process to reveal this truth. The constructivist perspective does not treat science as static information to be discovered; instead, science is viewed as a fluid process of inquiry. From this inquiry, models are developed that explain the world in useful ways; however, these models are influenced by a person's experience and information.[2]

If we accept the notion that knowledge is constructed by people, then we must find ways to focus on our students' current conceptions and help them construct more sophisticated models. It does not work to simply try to teach the concepts.[3]

Assessment is a vital ingredient for the teacher who is trying to help children construct meaningful models. It is not the final step of a lesson; rather, it is an integral part of the teaching process.

Creative Drama as Assessment

Creative drama can be used as an effective assessment instrument within the context of a lesson.[4] A definition of creative drama will help to clarify its use. Cottrell defines creative drama as "an art form for children in which they involve their whole selves in experiential learning that requires imaginative thinking and creative expression. Through movement and pantomime, improvisation, role-playing and characterization, and more, children explore what it means to be a human being."[5]

The use of creative drama to help students understand concepts is not new. Drama educators have been making claims that it is an effective strategy in many curriculum areas. In an ethnographic pilot study, I examined one fifth grade teacher using creative drama in her science lessons.[6] The teacher discovered that creative drama was a powerful assessment instrument. She made the comment that "it [creative drama] tells you what they don't know. It's a great evaluation." She asked the children to act out interactions and adjusted her lesson based on the way the children presented their ideas.

Examples

Creative drama can be used in many areas of science in this way. Some examples will illustrate how it works:

A class has been studying air pressure. All of the students can tell you that if a gas is heated, it expands. The teacher then asks the children to stand in a sectioned-off area of the classroom. They are told that they are each an individual molecule of air. They pantomime the action of the molecules as the teacher controls the temperature. Some children curl up into the fetal position and slowly move to a position with their arms and legs extended as the temperature rises. They are presenting their model which shows individual air molecules expanding as the temperature increases. The teacher now can address those children's conceptions (or misconceptions).

A fifth grade class is studying solar energy. They have watched balloons inflate on top of bottles placed in front of a light bulb. The students observed that the balloon on the black bottle inflated first, followed by the clear and silver bottles. The children discuss what light does when it strikes the different colors. They use the terms "heat" and "light" interchangeably. The teacher wants to help them understand that there is a difference between heat and light. Three children are asked to be molecules in the clear bottle. These children line up shoulder to shoulder. A fourth child is labeled a photon. The photon (running at 186,000 miles/per second) heads toward the molecules. The four children pantomime what happens when the photon hits the clear bottle. This is repeated for the other bottles. The teacher gains clear insights into how the children are thinking about these concepts and can then address the students' conceptions in their own terms.

A third grade class is studying the behavior of land snails. They test the snails with a variety of foods, classifying the snails' behavior as positive, neutral, or negative. Yet the results are not consistent. The teacher asks the students to pair up. One student is a piece of food and the other is a snail. The students act out positive, neutral, and negative interactions. Very quickly the teacher sees how the students have operationalized the three kinds of interactions. The teacher can then use the pantomimes as a way to help the students see the need to have consistent operational definitions.

A teacher wants to see what her fifth grade students have learned about wavelength, amplitude, and resonance. She sets up the class in one long row of chairs. The children are instructed to show a wave. She then asks them to change the wavelength and amplitude. They stand up higher when the amplitude is increased. More students stand up at the same time when the wavelength is increased. There is an interesting result. When the wavelength is too, long the wave dies out. A student eagerly raises his hand to announce that if they added more students to the line, there could be a longer wavelength. This gives the teacher insight into the children's construction about waves and why objects of different length make different sounds.

Creative Drama Compared to Other Forms of Assessment

The power of using creative drama is that it encourages the students to construct clear models that can be observed directly by the teacher and other students. Many assessment procedures require students to restate a definition or concept or to demonstrate that they have achieved a skill by performing a specific task. Creative drama is unique in that it requires a student to demonstrate his or her understanding of the concept or idea in a way that is meaningful to the student. By using creative drama the teacher is gaining insight into the students' understanding of the concepts while the lesson is in progress. This allows the teacher to respond to the students' needs during the lesson.

The students can also learn directly from the assessment. The process of designing and presenting a representation of their conceptions will help them to think about the concept in a way that is meaningful to them; they will "own" the idea. They will also have the opportunity to watch and discuss the ways in which other students present the same concept. In this way abstract ideas can be discussed in concrete terms. As a fifth grade student described the creative drama in his science class, "You may not learn as much from acting, but you remember it better. It's sort of like the secondary in football. If the first one doesn't get him, the secondary will come around and tackle him."

A final point is that creative drama is a dynamic activity. In addition to providing the teacher with information about what the students understand, it is a great deal of fun for the students and elicits an extremely high level of engagement for the entire class.

References

1. Fosnot, C. T., *Inquiring Teachers, Inquiring Learners: A Constructivist Approach for Teaching* (New York: Teachers College Press, 1989).
2. Posner, G., Strike, K. A., Hewson, P. W., and Gertzog, W. A., "Accommodation of a Scientific Conception: Toward a Theory of Conceptual Change," *Science Education* **66** (2) (1982), 211-227.
3. Smith, E. L., and Lott, G. W., "Ways of Going Wrong in Teaching for Conceptual Change: Report on the Conceptual Change Project," East Lansing Institute for Research on Teaching, Michigan State University, 1983 (ERIC Document Reproduction Service No. ED 242 769).
4. Heing, R. B., and Stillwell, L., *Creative Drama for the Classroom Teacher* (Englewood Cliffs, NJ: Prentice-Hall, 1981).
5. Cottrell, J., *Creative Drama in the Classroom* (Lincolnwood, IL: National Textbook Company, 1987).
6. Kamen, M., "Whatever We Study, She Finds Something to Make it Fun," unpublished pilot study, University of Texas at Austin, 1989.

3

Alternative Assessments in Elementary Science

Mary Dolan Thiel

Briggs Elementary School is located in Maquoketa, Iowa, a rural community of 8,000 in central-eastern Iowa about 20 miles west of the Mississippi River. During the past three years, the curriculum taught at Briggs has been rewritten according to Iowa's Standards for Approved Schools. This paper will outline particular changes in science and, more specifically, changes in how we now assess students in this area.

I am part of a team of three third grade teachers who teach all subject areas except physical education, music, and art. My students range from gifted to learning disabled, at-risk to out-of-school enriched. They are "Everyman's" children. Our team is finding that alternative assessment in science may be the most dynamic current resource currently available for developing a positive and relevant image of science for both teachers and students!

Elementary teachers often feel inadequately prepared to teach science. We frequently relied on text-driven science curricula, which generally have a low-level cognitive focus that ignores the affective and psychomotor domains. Standardized tests reinforced the practice of teaching students factual information. As a result, students seldom went beyond the "knowing and understanding" phase. With so much information to process, they had no time for "doing," although some students did get to watch their teachers do an experiment from time to time.

Now, however, Iowa's Standards for Approved Schools mandate *nontest* assessment in all curricular areas. Curriculum writing teams determine expected student performance outcomes at various grade levels; as a result, students may gauge their success in fulfilling clear, specifically stated performance requirements.

Assessment has been integrated with learning tasks, and I have come to view it as multidimensional and continuous. This has not been an easy change, nor has it been completed. At this time I am continuing to utilize standardized tests, but I also consider performance assessments, Iowa Tests of Basis Skills data analyses, library circulation records, parent input, pupil self-evaluations, modifications of book tests, comments from people visited on field trips, students' daily journals, and grade-book checklists in determining student achievement. Several techniques used in other content areas can be adapted for science assessment. Both students' cognitive and affective domains can be addressed by using "KWL" charts ("What I *K*now, What I *W*ant to Know, What I *L*earned; " see Table 1). Similarly, personal journals describe class activities, then allow personal comments about feelings and attitudes.

The sometimes neglected psychomotor domain can be addressed by teacher observation and visual records (video tapes and photos) of student manipulative activities. Student creativity and application of knowledge can be assessed in a similar manner by examining student products, such as a musical instrument made as a part of a sound unit. Student enthusiasm for the subject can be recorded by charting items of interest brought and shared with the class. I have used several methods for recording anecdotal comments, but I find clipboards with individual flipcards or post-it notes to be most effective. I hang the clipboards near my desk so I have a constant visual reminder about recording student behaviors, reactions, and comments relevant to the science unit.

The success I have had has primarily been in the area of "authentic assessment." Grant Wiggins, a leading expert on student assessment, sees this type of evaluation as real-life representation that concentrates on both teaching and learning to ensure that mastery is genuine. I found the most difficult task was to create clear, specific performance outcomes that could be understood by third graders. The first authentic activity I assigned was to prepare a booklet reporting a three-day animal observation. I wanted the students to observe the same animal over a period of time, make comments about specific behaviors, and show higher-order thinking skills by using qualifying numbers, inferences, or interpretation of current observation compared to knowledge of a similar, previous occurrence. I attempted to convey these desired outcomes to the students on a "rubric" (a sheet of guide rules; see Table 2). I indicated to the students what was expected for a "3," the best "grade"; what would be considered "2" quality; and finally, what would be graded as a "1."

This rubric was not specific enough for third-graders, except for the number of days that observations were required. Performance outcomes should have been stated in observable terms, such as, "the student will write a sentence describing at least one thing the pet is doing, tell if the animal has done this thing in the past, and make a comment telling whether this is new or old behavior." This would be considered a "2," and to get a "3," the student would have to write more than one sentence and describe more than one thing the pet was doing. Besides indicating whether the behavior was old or new, the student would have to use number qualifiers (how many times), relationships (cause and effect), or comments about how today's behavior reminded the student of something from the past.

Table 1. The KWL Chart

The KWL Chart is a group assessment technique. Before instruction, students list what they already Know (or think they know) and what they Want to Learn. Afterwards, students list what they have learned, checking back to see that what they listed under "K" is factual and whether they answered the questions in the "W" section. This technique can be used to assess pupils, instruction, or curricular materials. In this example, the subject was dinosaurs.

What We *Know*

Some were meat-eaters (carnivore).

Some were plant-eaters (herbivore).

Some were large — some small — some size of Mrs. Thiel (medium).

They are extinct.

Some were the size of dog or chicken.

They had four stages when they died.

Some could fly.

Some had sharp teeth.

Some lived in water.

Some had sharp horns.

Some had flat teeth (plant-eaters).

Most hatched out of eggs.

Some used their back-spikes for protection.

Some lived 70 million years ago.

Some lived by swamps.

Some had very big feet.

Most of them didn't like each other.

Some had long tails.

They died over 60 million years ago.

Some were just big lizards.

Some were egg-stealers.

Some have long necks.

T-Rex was tyrant king.

Some weighed more than 12 elephants.

Dinosaur means thunder lizard.   .

What We *Want* to Know.

Why did they die? (become extinct)

How did they stay alive for so long?

Really — how long ago did they live?

Why did they hate each other?

How much did they weigh?

How did they live their life — what did they do?

How did they swim?

Is there still one alive? (like Loch Ness Monster)

Why were some plant eaters, while others were meat eaters?

What are all their names?

Why are some so big?

Why did the meat eaters eat the plant eaters — or did they eat each other?

How many bones did they have?

Why did some have such sharp teeth?

How long were their tails?

Did Triceratops really kill T-Rex?

What We *Learned*

Dinosaur fossils were first discovered about 200 years ago.

Richard Owen called the fossils Dinosauria.

Scientists find dinosaur bones in steep canyons, by riversides, in quarries.

They carefully dig so they won't break the bones.

They shellac the bones to keep them preserved.

Scientists cover the bones with plaster casts so they won't break on the way to the museum.

Little bones are safe in tissue paper or match boxes.

Molds are made so other bones can be made from this — if they break or to make a model.

Scientists number the bones and take pictures to make sure the dinosaur goes together correctly in the museum.

Table 2. Three-day Animal Observation Rubric

Day 1	Day 2	Day 3
Student completes observation of pet/nature for one day.	Student completes observation of pet/nature for at least two days.	Student completes observation of pet/nature for all three days.
Student documents a number of pet/nature behaviors.	Student demonstrates a number of pet/nature behaviors and indicates some insight about the behavior.	Student documents a number of pet/nature behaviors thoroughly and shows great insight about the behavior. (Example: compares/contrasts, indicates numbers of behaviors; relates to past experiences that are similar; gives specific details, etc., for more than one thing the pet was doing. Besides indicating whether the behavior was old or new, the student would have to use number qualifiers (how many times), relationships (cause and effect), or comments about how today's behavior reminded the student of something from the past.

The rubric format was difficult for the third grade students to understand. I had to explain that the two main expectations applied to all levels of activity. (A matrix layout would be an improvement.) Further, I explained that students could get more points for thorough, explicit sentences about their observations. (The third graders had previously received S's and U's, so the idea of ranking their performances was vague.) In the original rubric I used terms like "insight" and "observable behaviors." Many third graders had difficulty with this vocabulary. A significant amount of explaining was needed to convey the meaning to some students. A more controlled vocabulary should be used.

If the rubric had been written in terms of observable outcomes, evaluation would have been easier. I was disappointed that I still was subjective in distinguishing the "1-2-3" performance levels. This is a problem that must be addressed if there is to be true inter-rater reliability among a group of teachers. This first attempt taught me much about the process and the need to be explicit about desired outcomes while leaving room for creativity.

The students did a good, perhaps intuitive, job of interpreting the requirements. Still, the majority of students received "1's" and "2's." In a one-sentence self-assessment, the students indicated they enjoyed this type of activity, especially after having taken the Cognitive Abilities Test and the Iowa Test of Basic Skills earlier that same month. Further, by giving the students the opportunity to present their observations orally, I also found that some students knew more than their written products indicated. I recorded an anecdotal comment for each of those students, which I placed in their science portfolios.

The rubric helped me focus on what was truly important; it also helped me focus on how I determined student understanding through evidence of a performance activity. It was useful to students because they were able to see exactly what was expected and what had to be done to achieve the objectives.

Alternative assessment has positively changed my attitude about evaluating student performance. I no longer view "tests" as definitive statements of what students know. I see a need for clarity in teacher/student communication. As a teacher I need to become more aware of the variety of student learning styles and the need to address them within the five domains of education — concept, process, creativity, attitude, and application of knowledge.

There is still a need for standardized tests as indicators and predictors of group behavior. However, alternative, authentic assessments are vital if we are to give individual students the opportunity to create, elaborate, demonstrate, and evaluate what they are capable of learning.

4

Math: The Tool We Use to Study Science; or, There's More to Assessment than Computation

Kirstin Lebert

The purpose of an assessment is twofold: it is a means by which children's grasp of knowledge is measured, and a means by which educators can determine the instructional needs of children. Above all, assessments must give a clear picture of what children actually understand. Recently, the District of Columbia Committee on Public Education (COPE) published "Our Children, Our Future: Revitalizing the District of Columbia School System." One of COPE's recommendations for elementary education is "to replace current tests for mathematics with state-of-the-art tests that are integral to instruction itself." For a recommendation of this magnitude to be implemented, a cadre of educators is required who have a deep understanding of the discipline, are committed to change, will accept the short-term risks associated with change (such as a possible drop in their classes' computational test scores), and who will assume leadership roles. Only with dedicated and committed teachers can we make such a drastic change in the way we teach and assess the mathematics and science curriculum.

After I had been in the classroom a number of years, it became apparent to me that we were not accurately assessing children's math and science concepts. Children were fooling us. The model that I developed over the years while teaching third, fourth, fifth, and sixth graders was not so much a collection of lesson plans and accompanying assessments as it was a procedure that allowed the children (with guidance) first to discover an algorithm and then to apply that mathematical procedure to solve science problems.

The Program

The philosophy on which this model is based is that assessments are to be used for determining children's understanding of the concepts taught, as shown by their active participation in problem-solving activities. The program was developed at Ben W. Murch Elementary School in Washington, D.C. The school serves children in prekindergarten through the sixth grade. The teaching staff believe that children learn best when they are active participants in their own instruction with developmentally appropriate learning experiences.[1] The success of this teaching strategy for grades 3 through 6 is in large part the result of many of the children's problem-solving experiences in the early grades.

This program, which does not group according to academic achievement, was developed to compliment the District of Columbia Competency-based Curriculum (CBC) in science and mathematics. The CBC was originally developed and implemented in 1980. The science curriculum was revised in 1984, and the mathematics curriculum is being revised now (1990–1991). The CBC is the instrument by which the math and science skills are closely organized among the various levels, kindergarten through grade 12.

Instructional Practices

The teaching techniques used with this model include cooperative work by the children, questioning, discussion, justification of answers, and reading about mathematics and science. Children are placed in working groups so that there is a range of academic achievement. After group work, there is discussion time when children are required to justify their answers or results. The classes are required to do outside reading about mathematics and science and to turn in book reports.

Assessment Techniques

Children's understanding of the concepts taught can be assessed through their oral responses, model construction, illustrations, and written statements. Correct answers to math problems alone do not fully assess children's understanding of the concepts taught. Asking children to justify their answers puts the responsibility of learning on the child. When youngsters are asked to write statements and use illustrations to describe science concepts, teachers can determine more easily how much the children really understand, as the following examples illustrate.

   Balance of Forces. This instructional sequence introduced and reinforced children's understanding of force and how weight and counter-weights can be arranged on a lever arm so that they are in balance. Children experimented by balancing pennies on a ruler to determine the rule that force times lever arm on one side of the pivot point must equal force times lever arm on the other side. Children were instructed to balance the ruler on the edge of a book at the six-inch mark. The students developed different possibilities for balancing the

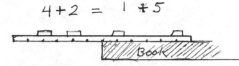

Figure 1.

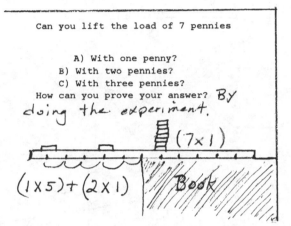

Figure 2.

weights (pennies) on the ruler. At the end of the session, there was a discussion about the possible configurations and how one could state a mathematical rule that would fit them.

A class of third graders initially stacked only one penny at each mark on the ruler. After seeing several patterns that balanced, one third grader determined the mathematical rule for this must be 4+2=1+5 (Figure 1). Later the class was asked to stack more than one penny at a marker. They learned that the initial rule was insufficient. When fifth and sixth graders experimented with this model, several identified the correct rule. On an assessment question a fifth grader clearly indicated his grasp of the concept by his illustration (Figure 2). Children were then required to apply this mathematical rule by designing mobiles using straws and paperfolded designs (Figure 3). The end-of-unit assessment included multiple-choice questions with illustrations. In some of the questions, the children were required to justify their answers.

Seasonal Change. To determine first-semester sixth graders' understanding of seasonal change, an assessment was developed in which the children wrote in narrative form their reasons for the change and

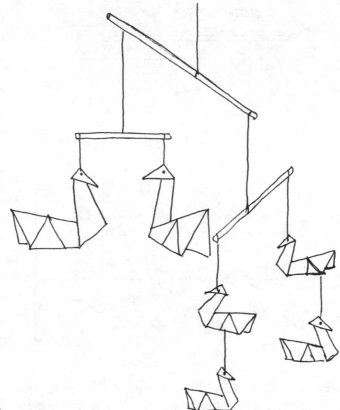

Figure 3.

drew an illustration for clarification. This assessment was given after approximately two weeks of unit study that included a math unit on degrees in a circle. The papers were rated 1 (understanding), 2 (possible understanding), or 3 (little or no understanding). The response shown in Figure 4 was rated "understanding." Although there is no reference to the earth's tilt, the paragraph clearly states the reason, and the illustration matches the description. Figure 5 shows a response rated "little or no understanding." The student has excellent writing skills and clearly described the zones but gives no reference to the earth's tilt. The illustration shows a possible confusion with night/day and seasonal change. The child with little understanding also illustrates the sun revolving around the earth.

Ratios. The District of Columbia curriculum includes the study of ratios in the sixth grade. To develop children's understanding beyond a superficial level, assessments require children to construct scale drawings and models of their homes or other buildings. They were

We have summers and winters where we live because in the summer the more direct rays are hitting where we live and in the the way the earth rotates the sun is shining on us for a longer period of the day. In the winter the direct rays of the sun are shining Somewhere else and the day don't last as long because as the earth rotates the sun hits us for a shorter amount of time.

Figure 4.

Describing why We Have Seasonal Changes

Why do we have temprature changes.
We have changes in temprature where we live because we live in the middle latitud which means we are between the Tropic of Cancer and Anartartica where most seasonal changes take place, and in seasonal changes temprature changes too. As you can see on the chart we live in a temprate zone where we get all seasonl / temprature changes. The Torrid zone only gets summer and the Fridged zone only gets winter.

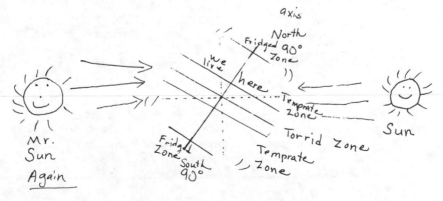

Figure 5.

asked to calculate floor space and room volumes in their respective projects. By doing this, the children had to review and use concepts of area and volume learned earlier. Interest was sustained since most students were visualizing and quantifying space that was very familiar or significant to them personally. They could also see what numbers do when they calculate perimeter, area, and volume of their models.

Erosion. In this unit children use math skills to calculate percent of slope in their study of erosion. Each working group had to identify erosion in the immediate area and what might have caused the problem. Each child was responsible for sketching the area, determining percent of slope, and writing a short narrative on how erosion might be prevented (Figure 6).

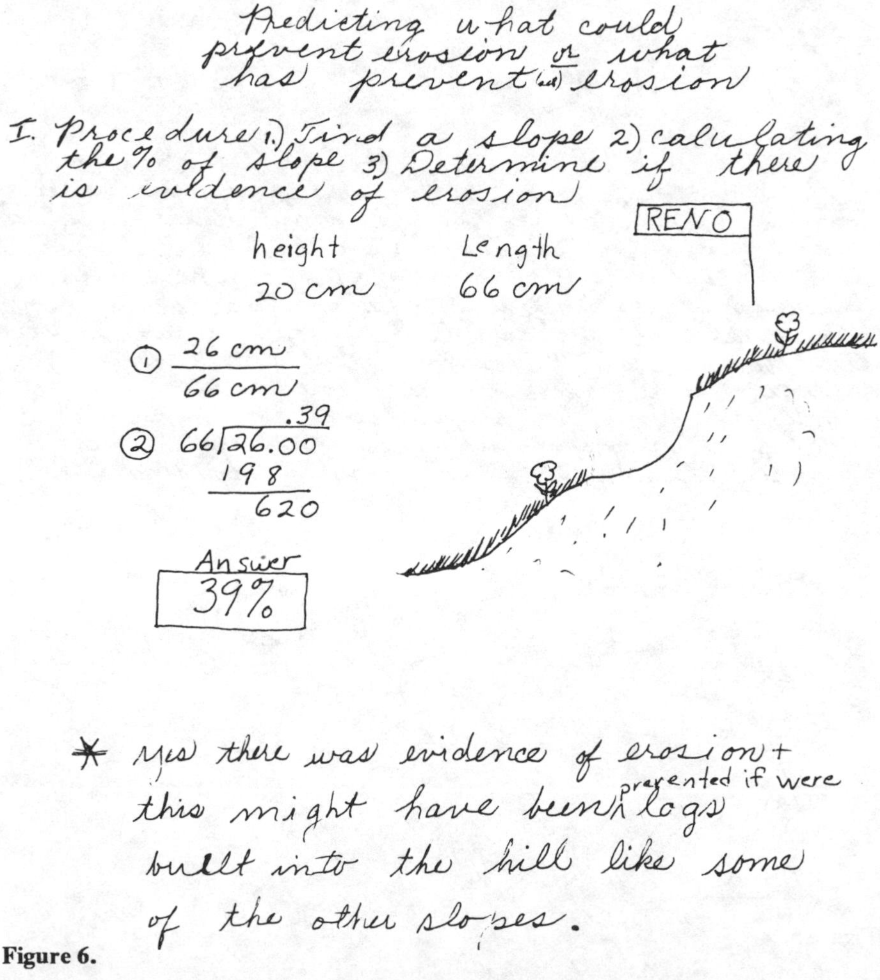

Figure 6.

A Final Word

In the preceding discussion I have illustrated how I believe children should be taught and how learning should be assessed. For instructional practices, I rely heavily on cooperative work, questions, engaging the children in discussions, and requiring them to justify their answers. I assess children's knowledge using written statements and oral responses. I also assess their knowledge by requiring them to apply the skills learned in constructing models or solving problems using the scientific method. Some of the lesson plans used I have developed; most have been shared by others.

This program is based on the concept that the child must be given the opportunity to formulate math rules using concrete objects and then must apply the skills learned. Science is used as the vehicle in which children apply math skills. By doing this, teachers are better able to assess children's understanding of both science and math concepts.

Reference

1. *Ben W. Murch Elementary School Middle States Self Study, 1990,* (District of Columbia Public Schools, Washington, D.C.)

5

Alternative Assessment in Measurement Studies

Donald C. Snyder, Jr.

This alternative science assessment project is an interdisciplinary approach that incorporates science and mathematics in the instruction of measurement. Students learn the dimensions that are measured (length, time, mass, weight, temperature, and volume), the units in which they are expressed (gram, meter, liter, second, degree, etc.), and the instruments utilized for their measurement (balance, meter stick, graduated cylinder, cubes, vernier calipers, measuring tape, thermometers, spring scales, beakers, and stop watches). The unit was developed for inner city (urban) students in grades 7 through 9. The science classes were heterogeneously grouped for instruction.

Measurement is an area that is confronted every day by students, and this project is based on student interest and curiosity. Questions such as, "What time is it?" "How far away is the subway station?" and "How cold is it?" are frequently discussed in hallway and classroom. Students often ask how distance, volume, and mass are measured. They also want to know what instruments are utilized to make measurements in these areas. This alternative assessment project is intended to strengthen students' belief in the relevancy of the curriculum in their daily lives.

This assessment project permits extensive integration of concepts in science and mathematics based on the hands-on approach. (Examples of measurement projects are shown in Figures 1, 2, and 3.) Upon completion of these laboratory assignments, students should have enhanced their process skills (ability to make observations, manipulate apparatus, interpret data, prepare graphs, record results, draw conclusions based on experimental data, and classify), their problem-solving techniques (mental computation, estimation, approximation, and geometric relations), and their critical thinking skills (ability to formulate key comprehension and mathematical questions such as, "What is the problem?" "What is being asked?" "What are the unknowns?" "What are the data?"). Assessment was based on the results of the completion of a series of measurement laboratories. Students utilized these skills throughout the school year and did well

Figure 1. Metric System Measurements.

I. Problem: The meter is the basic unit of length. We will apply this idea to the measurement of various objects.

II. Materials: meter stick, paper, pencil, and crayons.

III. Sketch:
    A. objects measured
    B. materials

IV. Procedures:
    A. Measure various objects in the classroom.
    B. Record the measurements in a table.
    C. Sketch the objects measured.

V. Results:

objects	width		length	
textbook	mms	cms	mms	cms
desk				
chair				
height				
arm				
hand				
finger				
class space door				
classroom				

VI. Conclusions:
    A. Define these terms:
        1. milli-
        2. centi-
        3. kilo-
        4. circumference
        5. meter
        6. distance
    B. What is the base unit of length?
    C. Develop a bar graph of your class height.
    D. What is a graph?

Figure 2. Measurement of Volume, Mass, and Temperature.

I. Problem: Illustrate the measurement of volume, mass, and temperature.

II. Materials: graduated cylinder, beaker, florence flask, Erlenmeyer flask, rock, string, balance, thermometer, ice cubes, water, and various objects (rock, rubber stopper, beaker, pencil, pen, and crayon).

III. Sketches:
    A. Sketch objects that can hold and measure liquids.
    B. Sketch a balance and thermometer.

IV. Procedures:
    A. Record the amount of water in your graduated cylinder.
    B. Measure the volume of water needed to fill a beaker.
    C. Diagram objects that hold and measure liquids.
    D. Measure the volume of a rock by water displacement.
    E. Find the mass of objects using a balance beam.
    F. Find the mass of a dry beaker and a beaker filled with water.
    G. Take the temperature of cold and warm water.

V. Results:
    A. Volume: 1. Amount of water in the graduated cylinder: _____
                         2. The volume of water in the beaker: _____

    B. Mass:

1. object	mass
a. rock	
b. rubber stopper	
c. beaker	
d. pencil	
e. pen	
f. crayon	

                 2. Mass of beaker and water _____ beaker _____
    C. Temperature: 1. freezing point_____ 2. boiling point _____
                       3. warm water: _____ 4. cold water:_____

VI. Conclusions:
    A. Define the following terms: 1. thermometer; 2. temperature;
         3. meniscus; 4. volume; 5. area; 6. mass; 7. Celsius scale;
         8. circumference.
    B. What are the units of mass and volume in the Metric System?
    C. What method did we use to measure the volume of the rock?
    D. Water displacement: 1. water and rock: _____
         2.water:_____ 3. rock:_____

Figure 3. Density Laboratory: What is Density?

I. Problem: To determine the density of various objects.

II. Materials: block of wood, rubber stopper, cork stopper, graduated cylinder, string, balance, water, and graph paper.

III. Sketch:
> A. Illustrate a block of wood and all items that had their densities determined.
> B. Illustrate the balance and graduated cylinder.

IV: Procedures:
> A. Measure the length, width, and height of a block of wood. What has this determined?
> B. Measure the mass of each object.
> C. Use the formula D (density) = M (mass) / V (volume) to determine the density of the objects.
> D. Calculate the densities of a rubber and cork stopper by water displacement.
> E. Use a graph to determine the density of water.

V: Results:
> A. Record the mass and density of the various objects in the following charts.

Object	Mass	Density
wood		
rubber stopper		
cork stopper		
graduated cylinder		

> B. Graph the density of water.

VI: Conclusions:
> A. Define these terms: 1. density; 2. volume; 3. mass; 4. measurement.
> B. What units are used to express density?
> C. Why is the density of a bowling ball greater than the density of a volleyball?
> D. What is a graph?

in measurement competition (State champions, Division B, grades 7 and 8, "Measurement," in 1988).

This type of assessment offered students an opportunity to apply the units, manipulate apparatus, and measure a variety of items, and they were eager to do so. They understood the relevancy of activities beyond just memorizing photos, reading textbooks, and answering questions. This project enhanced instruction because students became "active participants" in science.

6

Earth Science Final Exam:
A New Approach

David L. Stevens

The interdisciplinary, contemporary earth science class at Glenelg High School in Maryland is one which adheres very closely to the philosophy and content of the American Geological Institute (AGI) program, "Investigating the Earth." The major topics covered during the school year include energy, matter, and change; the water cycle; the rock cycle; earth's biography; and plate tectonics.

The program is intended for ninth grade students whose ability level may range anywhere from average to gifted and talented. With the exception of talent-pool students, most classes are homogeneously grouped. Class sizes usually range from 28 to 34 students. The classroom/laboratory setting is a self-contained facility generally suitable for any introductory-level science class.

A major goal of the earth science program is to involve students in the inquiry process of science. Throughout the school year this involvement is accomplished through numerous laboratory investigations, classroom demonstrations, and home activities known as "actions." The final exam is merely the last stage of a year-long continuum — students learning by doing, students assimilating by asking, and students retaining by applying. This final exam encourages cooperative learning as well as the use of higher-level thinking skills such as prediction, deduction, synthesis, and problem solving.

The Exam

For the exam (see Figure 1), the classroom becomes a geologic location which the students must investigate and describe. Students start the final exam by spending two to three days measuring, identifying, and recording data taken from meter-long plastic tubes that have been placed at specific locations around the classroom. Each tube represents a hypothetical core sample of material taken from that location. The tubes contain various rock and fossil specimens that will be used to draw conclusions about climate, geologic age, geologic events, and geologic history. Students may collect

Figure 1. Earth Science Honors Final Examination

Part A:_____
Part B:_____
Part C:_____
Total score on final exam:_____
  A B C D E
Teacher: Mr. Stevens
Date:
Name_____ Period _____

Directions: You may use any source of information you wish. You may freely converse with and/or work together during the initial two and one-half days of observation. YOU must formulate your own theory and be prepared to write it in your own words on the day of the final examination.
Your job is threefold:
A. Take all necessary measurements and observations about the area **during the two and one-half days of observations.** It is your responsibility to devise data charts that will enable you to systematically record all measurements.
B. Using all of the data collected in class, devise a model of the area. Your model is to be prepared at home and brought to class on the day of your final examination. Be sure to include all observed rock types and be sure to be accurate in showing the thicknesses of each rock layer. Keep in mind the relative distances of each observation point when constructing your model. HINT: You have used several kinds of models this year. Among those have been topographic maps (Lab 1-10), cross-sections on graph paper (Labs 9-1 and 11-10 A), and three dimensional models (Lab 3-1). You may find some and or all of these helpful when preparing your model. Be mindful of each sedimentary rock and its metamorphic counterpart. Pay particular attention to the effects of different cooling rates on magma and the specific igneous rock that is produced. You may find it helpful to refer to the class notes section of this quarter's notebook, specifically that section entitled "Rocks."
C. The day of your final examination, come to class prepared to write your theory explaining what the general area looks like today, during the past, and at its very beginning. Be sure to validate each part of your theory with the data you have collected. Include geologic time periods as to when specific events took place. Be sure to describe each event thoroughly. Leave nothing for granted. Assume I am totally unfamiliar with this area and therefore need a very detailed commentary.

Special notes:
  1. Horizontal Scale: 1 cm = 1 km
  2. Vertical Scale: 1 cm = 10 m
  3. Lab Table Top = 1 km
  4. All height readings are taken from the table top.
  5. Fossils belong to the layer of rock below them.
  6. The thickness of the fossil is to be added to the thickness of the layer of rock below the fossil.

BE THOROUGH AND GOOD LUCK!

their data individually, in pairs, or in small groups. The data collected and the method by which the student represents the data comprise one-third of the final exam grade.

In the second part of the final exam, each student takes his or her data home and constructs a "model" of the area represented in the classroom. (Previous examples of models are withheld from the students.) Finished models usually take the form of geologic maps, topographic maps, cross-sections, or three-dimensional replicas. The model comprises another third of the final exam grade.

Last, each student is required to write a geologic history of the area. The theoretical geologic history must be plausible, and the theory should identify as many of the area's unique geologic features as possible. There is no required length for the theory, which constitutes the remaining third of the final exam grade.

Summary

Alternative methods of assessment such as these provide the teacher with additional, nonconventional methods for evaluating a student's progress. Each year I encounter students who are consistently unsuccessful with traditional, paper-and-pencil, objective tests. Many of these students desire to succeed and are willing to try. They just don't get passing grades on conventional forms of assessment. Their apprehension is overwhelming when confronted with an end-of-the-year objective examination that is supposed to cover the entire course and take 80 minutes to complete. Taking and passing such an exam is virtually inconceivable to these students. It seems that failure is inevitable. Once my students start working on their final exam, however, most of their apprehension disappears and the opportunity for success becomes a reality.

Within the last three years, my school district has undergone a complete revision of all science curricula. Every topic or unit in the newly revised curriculum guides contains an "alternative assessment" section.

The final exam described here is an alternative method for demonstrating student competence. It is a method of assessment that offers the same chance of success for the average ability student as it does for the gifted and talented student. It gives each student a unique opportunity not just to recall facts but to solve problems and generate ideas. It allows everyone an opportunity to employ his or her individual talents or personal strengths to complete a task. Since the final grade on this exam is derived from three separate parts, the successful completion of any two parts can ensure the student a passing grade. The successful completion of all three parts can ensure the student an outstanding grade. In my classroom, the use of alternative assessments has improved both the level of student interest and the likelihood of student success.

7

Project Learning Assessment

Glenn Fay, Jr.

At Champlain Valley Union High School (CVU), a school of roughly 1,000 students in rural–suburban Burlington, Vermont, student projects are used to assess achievement in knowledge, behaviors, and skills. The components of project learning assessment are shown in Figure 1. In project learning assessment, students design, implement, present, and evaluate projects. (A sample project is shown in Figure 2.) Students involved with project learning range from high-risk to gifted.

Over a three-year period, two teachers collaborate and collect projects to be used as indicators of student progress. Students in freshman level 2 biology move on to earth science as sophomores and to environmental studies as juniors or seniors. Students in freshman level 1 biology go on to chemistry level 1 as sophomores. The same two teachers work with these students and compile projectfolios and other assessment data.

These courses offer a somewhat traditional scope and sequence, but with significant differences from conventional courses. First, learning is focused on the essential skills and content from CVU's mission. As a result, the content is more integrated with other programs such as the humanities, and demonstrating skills receives more attention. Second, students are empowered to a large extent. The focus of classes is growth and demonstrating achievement. The entire school and community are used as resources, including three nearby colleges, a university with a moderate-sized medical center, and local businesses and industry.

The philosophy behind project learning assessment is that, in a changing world, learners need to be assessed authentically as active participants, not as passive observers. The goal of project learning assessment is to evaluate students on the basis of their ability to demonstrate essential skills, behaviors, and knowledge. (The points covered in evaluation are shown in Figure 3.)

Results and Significance

Students develop a performance record based on project folios, performance assessments, quizzes, and self-evaluations as they work on projects.

PROJECT LEARNING ASSESSMENT

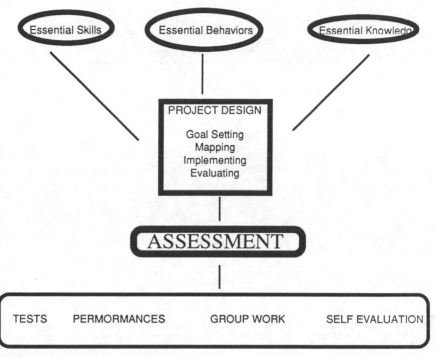

Figure 1.

For each student there is a record showing application of knowledge, skills, and behaviors. High-risk students often do not take or choose not to finish standardized tests. They find relevance in projects and take pride in their performances. They also remember what they have learned.

In order for this type of assessment to be accomplished, the teacher's role must change from lecturer to manager. This role becomes one of planning, gathering resources, helping develop skills, and working with individual students, parents, and other teachers rather than drilling students on content. Experience with project learning assessment has shown that there are a number of implications for curriculum and instruction, including the following:

- If the mission is to be achieved, it is necessary to design "down" from the mission to goals, objectives, learning resources, and activities to ensure mastery.
- Essential skills, behaviors, and knowledge that all students should acquire need to be defined.
- Disciplines should be integrated so students can integrate knowledge

PROJECT - FUNKY CAREERS IN CHEMISTRY

Objective:
Research a career in or related to the field of chemistry, interview a person in that field, and bring back information to class for some type of presentation. Presentation and one page abstract should include job title, description, salary range, potential risk factors, job demand, educational qualifications assets or rewards, and a small original graphic. Oral presentation could be with the help of visual aids or anything that will get the audience's attention.

Grading:

Career information	15
Name and title of Interviewee	5
Class presentation	20
Written Abstract	10
Total	50

Examples
Chemistry Teacher
Biochemist
Associate Scientist
Chemical Engineer
Technical Consultant
Environmental Chemist
Industrial Chemist
Radiologist
Medical Technologist
Nuclear Scientist
Medical Doctor
Psychiatrist
Registered Medical Nurse
Pharmacist

Figure 2.

EVALUATION FORM

Student name

Project name

Total class days preparation

Were your goals achieved?

Did your plan work fully? Explain.

Skills you learned

Grade for Project

What would you do differently for your next project?

Figure 3.

and find a context for learning similar to the real world, making learning more relevant.

- The schedule should be designed with longer time periods available to students to learn.
- Teachers should be provided with training that promotes alternative assessments.
- Teachers should be evaluated using alternative assessment as a criterion.
- There should be exit surveys and college and business surveys to evaluate student growth using alternative assessments as compared to traditional assessment.

Finally, with project learning assignments students, teachers, college admissions officers, parents, and others have a comprehensive history of students' demonstrated growth, strengths, and weaknesses in skill areas that have been identified by business people and educators as being essential for success. Assessment will demonstrate what students know and what they can do with what they know. Project assessment provides a method of evaluation; it can also be a source of feedback to improve instruction.

8

Performance Assessment in an Elementary Science Methods and Materials Course

Ronald K. Atwood

The elementary science methods and materials course at the University of Kentucky is one of six courses in a methods, materials, and practicum block taken by education students prior to student teaching. The entire elementary education component, including this course, reflects a constructivist perspective, since that approach has been shown to be more effective than more traditional approaches.[1,2] The central thrust of the course involves critically analyzing a variety of curricular goals, instructional materials, and instructional, management, and evaluation strategies.

The evaluation of preservice teachers (and elementary students) in science has both informal and formal components. The informal component includes utilizing good questioning techniques and observing students interacting with concrete materials and with each other. Evaluative data obtained through more formal means complement data obtained informally by providing a more nearly complete picture of instructional effectiveness, and such data are probably more defensible in arriving at grades.

Some of the formal evaluation tasks used to measure preservice teachers' understanding of instructional materials and science content would probably be viewed as performance assessment by the majority of professionals currently working with that concept. Like many terms used in education, however, performance assessment has not been sharply defined, although the criteria suggested by Wiggins[3] are quite helpful, especially the view that performance assessment should be authentic.

Even the authentic criterion is subject to interpretation; that is, a task could be viewed as authentic by one teacher but not by another teacher, student, or legislator. An analysis of examples of what others consider to be performance assessment tasks is also helpful.[4,5] Emerging leaders in performance assessment may soon be willing to take the risk of providing numerous examples of what should be excluded. The debate that could be expected to follow should help sharpen the concept.

The examples below are examples of performance assessment tasks used by the author in an elementary science methods and materials course to assess students' understanding of science content and investigative skills.

Example 1: Assessing Ability to Observe and Infer

Expected behavior: Students will be able to state three to six observations of some phenomena and one inference that follows logically from the observations.

Performance activity: Students are told that one object to be used in a demonstration by the instructor is a black steel cabinet and that they may use this information in observation statements if they wish. Students are also told to write from three to six observations based on the demonstration and one inference that follows logically from their observations. The instructor, without further comment, brings two small, shiny objects, which appear to be essentially identical from where the students are sitting, into obvious contact with a vertical surface of the steel cabinet. When hands are removed from the objects, one object falls and the other object remains where it was placed.

Scoring rubric:

Outstanding performance occurs if the student focuses on the most interesting event, that is, the different results for the two shiny objects. The student should make observations that provide a basis for inferring an explanation of the different results. The observations must indicate that one object remained on a vertical surface of the steel cabinet with no visible means of support.

Acceptable performance occurs if the student makes one minor error such as stating that one object remains against the side of the cabinet, but not explicitly stating that the object has no visible means of support.

Unacceptable performance occurs if the observations are unrelated to the inference or do not clearly support it, or if statements presented as observations contain inferences.

Example 2: Assessing Ability to Infer Properties of Sets

Expected behavior: Students will be able to infer the properties used to set up two intersecting sets of "A-blocks."

Performance activity: Students are shown two intersecting sets of

A-blocks placed in two intersecting string loops. One string loop is black and one is white. No A-block is either black or white. A-blocks that do not have the property of either set are located in a cluster outside the two loops. Students are asked to infer the one property used to place the objects in the black loop and the one different property used to place the objects in the white loop.

Scoring rubric:

Outstanding performance occurs if the correct property for each set is identified, such as "yellow" or "not triangle."

Acceptable performance occurs if the student describes the entire set of objects rather than the property used to form the set. An example would be "all yellow objects."

Unacceptable performance occurs if the student uses an incorrect property or more than one property to identify either set.

Example 3: Assessing Ability to Assemble a Circuit Tester and Ability to Test a Circuit Puzzle

Expected behavior: Students will be able to assemble a circuit tester from D-cells, a flashlight bulb, and wires and use it to infer four plausible circuit patterns for a circuit puzzle.

Performance activity: A student is presented with two flashlight batteries (D-cells), each in a battery holder with metal clips at either end of the battery; a flashlight bulb in a bulb holder; several wires with alligator clips on either end to facilitate repeated and rapid assembly and disassembly of a circuit tester; a circuit puzzle with more than four plausible circuit patterns. The student is asked to include both batteries in making a circuit tester and to use the tester in the task of inferring (and drawing) four different circuit patterns (each plausible for this particular circuit puzzle), showing as little wire as possible on the inferred patterns.

Scoring rubric for part 1, assembling the circuit tester:

Outstanding performance occurs if the student arranges the two batteries and light bulb appropriately in a series circuit with two wire test leads for completing the circuit.

Acceptable performance occurs if the student arranges two batteries and a light bulb in a series circuit which can function as a circuit tester but does not use two wire test leads for that purpose.

Unacceptable performance occurs if the student can not assemble a functional circuit tester in the allotted time.

Scoring rubric for part 2, testing the circuit puzzle:

Outstanding performance occurs if the student is able to use the circuit tester to obtain the appropriate data in an efficient manner and infer four plausible circuit patterns, each of which shows as little wire as possible.

Acceptable performance occurs if (i) the student has one pair of terminals improperly classified as either a "will light" or a "won't light" combination but completes the task, based on the data he/she has collected, with no more than one extra wire on one inferred pattern, or (ii) the student works with accurate data and infers four plausible patterns but one or two patterns include redundant wires or one pattern has a needed wire missing.

Unacceptable performance occurs if (i) the student improperly classifies more than one pair of terminals as either a "will light" or "won't light" combination, or (ii) the student works with accurate data but includes more than two redundant wires or leaves out more than one needed wire.

While each of these three tasks requires the preservice teacher to use paper and pencil in recording responses, concrete materials that also can be used with elementary students are utilized as the major data source for the performance. In examples 1 and 2 it is the instructor who is directly manipulating the materials to provide concrete data sources for the students. In example 3 the students are manipulating the materials, as required by the nature of the task. All three examples are considered by the author to represent authentic performance assessment tasks for the elementary science methods and materials course.

In addition to the demonstration and hands-on task formats used to administer assessments of some desired course outcomes, student understanding of other course content, such as issues and trends, is assessed with flexible essay questions, exclusively a paper-and-pencil format. For all assessment tasks, regardless of the administrative format, the focus is on assessing the student's ability to demonstrate understanding, usually through application and/or analysis. Consistent with this focus, students are allowed to use their own notes and copies of required readings on all assessments. They are also allowed to retest on comparable but not identical tasks when their initial performance is not satisfactory. An evaluation system with these characteristics has been associated with improved student performance.[6] Further, student ratings and comments on the college's evaluation of teaching forms indicate:

- Many students judge the scoring rubric of the performance tasks to be too rigorous.
- The retest option is highly valued by students.

It appears graduates of our program have accepted positions in schools typical of those described in recent surveys,[7] where science tends to be a low-priority curriculum area. Experience suggests that in schools where instruction in investigative science does occur, formal performance assessment strategies comparable to those modeled in the methods and materials course are seldom used.

In Kentucky, elementary science historically has not been included in the required state assessment program. However, the sweeping Kentucky Educational Reform Act (KERA) of 1990 includes provisions that will require science to be evaluated in determining the effectiveness of each elementary school. The state will utilize performance assessment tasks to determine each school's effectiveness, and rewards or sanctions will be assigned to personnel at the building level based on the results. Further, each school will be required to implement a performance evaluation system as a component of its instructional program. Finally, the six broad goals of KERA require elementary science instruction consistent with the kind of science advocated in the methods and materials course, and the science education profession.[8] Thus, the demand for performance assessment in teacher education programs in Kentucky can be expected to increase, and massive experimentation with performance assessment can be expected in the state's schools. Other states seem to be pursuing or contemplating similar moves. Looking beyond the multitude of problems to be worked out and pitfalls to be avoided, the end result should be more effective teacher preparation programs and young people who are much better prepared for life in the 21st century.

References

1. Shymansky, J.A., "What Research Says About ESS, SCIS and SAPA," *Science and Children*, **26** (7) (1989), 33–35.
2. Bredderman, T., "Effects of Activity-based Elementary Science on Student Outcomes: A Quantitative Synthesis," *Review of Educational Research* **53** (4) (1983), 499–518.
3. Wiggins, G., "A True Test: Toward More Authentic and Equitable Assessment," *Phi Delta Kappan* **70** (9) (1989), 703–713.
4. Mitchell, R., "A Sampler of Authentic Assessment: What It Is and What It Looks Like," paper presented at 1989 curriculum/assessment alignment conferences in Sacramento and Long Beach, California Department of Education, 1989.
5. National Association of Educational Progress, *Learning by Doing: A Manual for Teaching and Assessing Higher-Order Thinking in Science and Mathematics* (Princeton, NJ: Educational Testing Service, 1987).
6. Atwood, R.K., and Atwood, V.A., "Relationship Among Initial Test, Retest, and Delayed Retest Performance on Three Competency

Tasks," *School Science and Mathematics* **78,** (1978), 465–471.

7. Weiss, I., *Science and Mathematics Education Briefing Book* (Chapel Hill, NC: Horizon Research, 1989).

8 *Science for all Americans: A Project 2061 Report on Literacy Goals in Science, Mathematics, and Technology* (Washington, DC: American Association for the Advancement of Science, 1989); reprinted as, Rutherford, F. J., and Ahlgren, A., *Science for All Americans* (NY: Oxford University Press, 1990).

9

Preparing Preservice Kentucky Science Teachers for a Performance-based, Nongraded School System

John Guyton

Kentucky's 1990 Education Reform Law (HB 940) mandated performance-based evaluations and nongraded primary schools as a means of radically improving the quality of education in the state. At Murray State University, students preparing to teach elementary science are being trained and evaluated in a performance-based mastery system that will lead to success under the new system. Pre- and posttest measures include attitude, knowledge of science, and knowledge of methodology. The underlying philosophy of our approach includes the reduction of fear of failure in science activities and the use of performance assessments to prepare individuals with minimal science experience to teach science at the elementary school level.

Teachers have been found to model the techniques they are exposed to as students. Preservice elementary teachers in Kentucky have had little if any experience with hands-on science. Surveys of preservice elementary teachers in west Kentucky indicate they have had fewer than nine opportunities to participate in hands-on science prior to methods class. One reason reported by the author's students for hesitating to teach science using discovery activities is fear of failure in front of their students. On the pretest, over 75 percent of students reported fearing that their hands-on experiments and demonstrations would fail.

Preservice teachers as well as recently graduated teachers reported making more extensive use of hands-on activities and performance assessments after experience with them and with the Model for Failure/Success (described below) in their methods course. The positive experiences and enthusiasm for teaching science while learning it are passed from new teacher to elementary children. When preservice teachers were exposed to these experiences, they reported less stress when using hands-on methods.

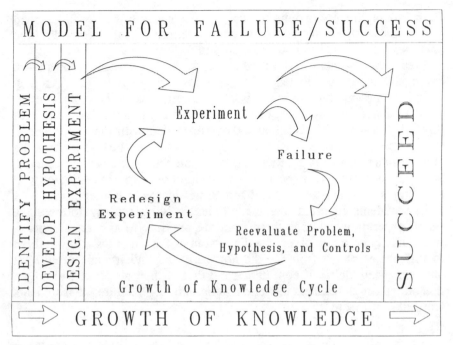

Figure 1.

Performance assessments have been incorporated as an integral component of diagnostic, formative, and summative evaluations; they have also effectively measured the cognitive, psychomotor, and affective domains. However, in this elementary science methods course, the diagnostic assessment has been used to identify strengths and weaknesses within the class, and the summative assessment has been used to establish an exit grade. Traditionally, these have been pencil-and-paper tests, not performance-based in the truest sense. The formative evaluations, including cognitive, psychomotor, and affective measures, have been performance assessed.

The Model for Failure/Success

The Model for Failure/Success (diagrammed in Figure 1) is based on the path researchers in science follow, including the failures. The system progresses from identification of a problem to developing a hypothesis and designing an experiment. The experiment may succeed or fail. If it fails, knowledge still grows as the problem, hypothesis, and controls are re-examined. The experiment is redesigned and repeated. This cycle is repeated as many times as necessary, hence the model's subtitle, "Growth of Knowledge Cycle." Once teachers understand the frequent failure of science experiments and the need to prepare students to analyze and profit from failure, anxiety is reduced. The teacher's fear of failure is lessened by the knowledge that some experiments *should* fail, thus providing the students with a model for success through failure.

Preservice teachers in the methods class were oriented to the failure cycle and then participated in experiments designed to fail. As they examined the variables and exercised more control over them, they developed an understanding of the concept that they could never have obtained through reading only. Eventually they succeeded with the experiments, thus becoming "expert." Using this approach releases the teacher from the tension inherent in having to appear as a science expert before students, and permits him or her to become a consultant or fellow investigator who sometimes leads the students and sometimes learns from them.

My students reported that the problem-solving activity following a failed experiment released them from the pressure to appear as experts. Immediately being involved in a discussion of the variables allowed them to shift the emphasis from "It did not work" to "What went wrong?" It also enhanced their self-confidence and that of their students. Moreover, it is only through teachers modeling and practicing "failure" that students learn to profit from it.

Table 1. Other Performance Assessed Activities in Teaching of Science

Activity	Performance Criteria
Peer teaching	Accuracy; involvement; proper use of discovery
Teaching in area schools	Accuracy; teacher evaluation; independence; participation; activity level
Computer competencies	Accuracy; organization; application to other task; involvement
Science kits (SAPA, ESS, SCIS, PALS)	Accuracy; thoroughness; use
Environmental education weekend	Involvement; use of new ideas
Current event or TV discovery	Accuracy; appearance; discovery; relevance
Egg drop or invention	Originality; performance; relevance

Performance Assessment of Class Activities

All evaluations of preservice teachers in the class (except for pre- and posttests) are performance-based. Performance assessment allows preservice teachers to refine their psychomotor skills and appraise their cognitive development; it also allows observation of preservice teachers' affective domain. Sample performance assessment activities are shown below and an overview of others is provided in Table 1. By the end of the semester, students' performances have been assessed in over 12 categories. Any deficiencies identified during performance assessments receive attention to prevent an unprepared teacher from entering the profession.

Sample Performance Assessment No. 1 "Rockets"

Goal: Basic skills and core concept.

Performance outcome: Skills in constructing models and ranging techniques developed; safe launch procedures practiced.

Assessment tasks: Students are observed constructing and launching rockets.

Performance criteria: 1. Proper construction (yes/no). 2. Level 2 or higher (yes/no). 3. Aesthetically pleasing (yes/no). 4. "With-it-ness" (yes/no). 5. Safe launch (yes/no). 6. True flight (aerodynamic stability) (yes/no). 7. Parachute deployment and recovery (yes/no). 8. Proper ranging and decision (yes/no). Totals: number of affirmatives; number of negatives.

Scoring rubric: *Excellent*: 6 affirmatives. *Good:* 5 affirmatives. *Below standard:* fewer than 5 affirmatives.

Sample Performance Assessment No. 2 "Circuits and Motors"

Goal: Core concept.

Performance outcome: Understanding basic characteristics associated with series and parallel circuits, motors, and diodes.

Assessment tasks: Students are guided through several discovery activities involving small electric motors, batteries, bulbs, and wires.

Tasks include: Creating series and parallel circuits; reversing direction motor shaft turns; connecting motors and batteries with motor shafts turning in opposite directions.

Performance criteria: Students are given a diode and allowed to investigate. They are then asked to describe the nature and characteristics of the diode.

Sample Performance Assessment No. 3 "Surface Tension"

Goal: Basic skills and core concept.

Performance outcome: Demonstrate understanding of control of variables after failed activity and knowledge of surface tension.

Assessment tasks: Students learn to examine variables after experiencing failure in an activity and learn some characteristics of surface tension in the process.

Performance criteria: Students float and refloat wire screen boats and investigate problems encountered. Students propel soap boats back and forth across a pan of water and investigate problems encountered.

Scoring rubric: *Excellent:* correct description of surface tension and understanding how to recover from failure. *Good:* understanding surface tension or benefiting from failure. *Additional experiences needed:* confusion, no conclusion.

Sample Performance Assessment No. 4 "Weather"

Goal: Basic skills and core concept.

Performance outcome: Knowledge of weather and skills appropriate to teaching weather concepts.

Assessment tasks: Design a 3-day unit on weather to teach children operating on a variety of cognitive levels. Utilize the materials provided (cotton balls, matches, ice, coffee, salt, test tube, plate, drinking glass, 2 thermometers). Include a task designed to evaluate the students' knowledge of condensation. If students do not demonstrate mastery, what strategy will be employed to remedy the situation?

Performance criteria: 1. Relevance of selected activities (yes/no). 2. Appropriateness of activities for cognitive levels (yes/no). 3. Adequacy of evaluation (yes/no). 4. Remediation plan (yes/no).

Scoring rubric: *Excellent:* 4 affirmatives. *Additional experiences needed:* 3 or fewer affirmatives.

Mastery in a Performance-based System

Preservice teachers learn how to teach in the new nongraded elementary level classes through participation in a performance-based mastery system. Their mastery of the science methods course is necessary since it is their only experience with teaching hands-on science. If their experience is poor, their teaching will be compromised, so they must master the skills of directing hands-on activities. Students participate in numerous activities that ensure maximum exposure to and saturation with the methods of teaching science, and they are evaluated in the cognitive, psychomotor, and affective domains. When all components of an objective have been successfully completed, students advance to the next objective.

Three essential elements — the mastery approach, performance assessment and the Model for Failure/Success — are combined in this system for preparing elementary science teachers. The system allows preservice teachers with limited backgrounds in science to overcome their fear of failing in science activities as they master techniques of teaching and using discovery. Comprehensive performance evaluations chart their progress through the objectives and document the skills, knowledge, and motivation of new science teachers.

Recommendations

The approach described here has two uses. First, teachers can increase their confidence and skills through practicing "Failure/Success" while actively engaging in hands-on science. Second, performance evaluations provide both a mechanism for students to master the subject matter and for teachers to assess the level of attainment of each student. These techniques are recommended to all science teachers. Too often students see the teacher as authority, not leader or guide. Practicing failure/success transfers the responsibility of doing science to the student and allows performance evaluations of student work. Such evaluations give teachers a clear view of the level of attainment and the skills the student is developing. This contrasts with traditional evaluations that yield only superficial ranking of students and ignore areas where the student's knowledge is deficient.

10

Assessment of Collaborative Learning in Chemistry

Judith A. Kelley

Applied Chemistry for the Nonscientist is a one-semester chemistry course at the University of Massachusetts at Lowell taken by liberal arts, management science, and fine arts students as a part of their core requirements. The students range from freshmen to seniors and from 17-year olds to an occasional 30- or 40-year old. Their majors are diverse and include art, music, management science, English, criminal justice, and psychology. The text used was *ChemCom: Chemistry in the Community*.[1] A collaborative learning approach is used in the course, and in 1989 and 1990, the assessments of learning turned out to be a major contributor to the students' learning activities.

Course Content and Structure

The course goals are for each student to learn how to approach problem solving scientifically, to be able to apply basic principles of chemistry to solving theoretical and practical problems, and to become adept at locating resources and applying learning to the understanding of contemporary issues involving chemistry. Considerable emphasis is placed on hands-on learning, searching out information, and distinguishing between fact and opinion.

In addition to the text, students use the Sargent-Welsh periodic table of the elements, a computer-conferencing network,[2] current news articles, and a calculator with logarithms (for calculations involving pH). In 1989 a class of 26 students met once a week for 3 hours (no laboratory facilities). The 1990 class of 25 met twice a week, for one hour in a regular classroom and for two hours in a classroom with tables (no sinks). The 1991 class returned to one 3-hour time period and had a laboratory available.

The collaborative learning aspects involved students working in groups to: (i) learn techniques for solving problems involving moles, heat gain or loss, pH, and so on; (ii) comprehend concepts such as solubility and isomerism; and (iii) acquire skill in distinguishing fact from opinion, recording observations, and graphing and interpreting data. Students worked in groups to go over their homework, do experiments, take weekly tests, and carry out projects. Students collectively decided the grade weight to be given to each component of the course and which units or topics were to be covered in the course. (Topics chosen in 1989 were water, nuclear chemistry, food, and health; in 1990 they were water, petroleum, air, and climate.) Students also participated in computer conferencing that provided for dialogue between weekly classes and sharing information on resources, library readings, and current events.

As homework, students read sections of the text and responded to its questions, sought information from the library, and prepared themselves for hands-on work. There were no prepared lectures in the usual sense, but each week students' questions were answered and requested information about theories and concepts was provided. The students took some tests individually and some in groups. Test content ranged from traditional questions and problems through carrying out experiments (always in groups); pooling, plotting, and analyzing data; reporting on information learned in individualized library assignments; and applying knowledge to comprehension of news articles.

Assessment

The chemistry concepts tested in the various units were the following:

Water unit: measurement and the metric system; physical versus chemical properties; symbols, formulas, and equations; nonpolar, polar, and ionic substances; solutions, solubility, concentrations, and pH.

Chemical resources unit: moles, grams, conservation of matter, periodicity of physical and chemical properties.

Petroleum unit: structure and properties of organic molecules, basis for methods of separation, and energy from reactions.

Nuclear chemistry (1989): types of radiation, nuclear atoms, isotopes; radioactive decay, half-life; fission, fusion, and nuclear power; ionizing radiation.

Air and climate (1989): breathing and photosynthesis, composition of the atmosphere, gas laws and kinetic molecular theory, temperature control of climate.

Health (1989): homeostasis; proteins, carbohydrates, and lipids; types and roles of proteins; how enzymes work; interparticle forces controlling protein shape; buffers; rate of alcohol metabolism.

The skills assessed were the following:

Conducting experiments: observing, collecting, recording, and analyzing data; graphing and interpreting graphs; drawing conclusions; constructing hypotheses; and designing experiments.

Studying and doing projects: distinguishing fact from opinion, using logic, locating and digesting information needed for analysis and problem solving, organizing and presenting information in oral and written form, assessing risks versus benefits, learning how to learn, learning how to solve problems.

In addition to the assessment of skills listed above, through discussions with students, their answers to open-ended questions, and observing their work, it was possible to assess students' personal growth in the areas of: (i) development of a sense of mastery of content, materials, and computers; (ii) enjoyment of learning and science; and (iii) development of a sense of responsibility for their own learning. At the same time, it was possible to understand something about how learning really occurs, the variety of ways in which students learn, and the conceptual and other blocks that prevent students from learning.

Grade Performance

In comparing student grade performance in 1989 and 1990 with that of previous years (Table 1), it was clear that the collaborative approach provided performance at least equivalent to that of preceding years. Grades were based on student performance on weekly in-class assessments, homework, and the final examination, as well as on student participation in class, group, and telecommunications activities. It appeared that the greatest amount of learning took place when students took tests and did experiments in groups of three or four. Because they were being graded on the output of the group, and could submit only one set of answers to questions and problems and one report for experiments, they took these group activities seriously. They taught each other how to solve problems. They argued ideas and issues out when they were in disagreement. Some learned to argue for what they believed and others learned how to listen. Because the composition of the groups varied each week, any particular strengths or weaknesses of individuals (in ability, skill, personality, commitment to the task, and so on) were shared among all the students.

Assessment of groups rather than individuals was a major teaching/learning tool. Students helped each other gain an understanding of chemistry theory and its applications. At the same time, most students

Table 1. Course grades earned by students in 1990 and 1989 using the *Chem-Com*/Collaborative learning approach, and in 1987 and 1986 using the precollaborative learning approach.

Grades		Number of Students			
		1990	1989	1987	1986
4.0	A	6	4	2	3
3.5	AB	4	5	0	3
3.0	B	3	4	1	5
2.5	BC	0	3	1	4
2.0	C	3	3	2	5
1.5	CD	4	0	1	2
1.0	D	3	2	0	3
	Pass		1		
0.0	F	1	2	4	1
	W	1	2	4	3

	1990	1989	1987	1986
No. of students enrolled	25	26	15	29
% earning credit	92%	85%	47%	86%

acquired attitudes toward science and themselves that are an essential part of our current goals in science education and education in general. Several students reported at the beginning of the course either that they had never had chemistry or that they were afraid of science. By the end of the semester, student responses to open-ended items on a questionnaire indicated that many students felt that they had acquired both general and specific knowledge, had learned about learning; had a greater understanding or awareness of science issues, were empowered by the collaborative approach, felt free to contribute, enjoyed the course, had acquired skills, found science interesting and were less afraid of it, and were more willing to question things. Some students reported that they still felt inadequate with science or felt that they had not made any permanent gains.

Student Creativity

Students were encouraged to truly experiment, rather than just follow cookbook descriptions of demonstrations of chemical phenomena. For the first time in my 25 years of teaching introductory chemistry courses, a group of beginning chemistry students came up with a unique idea for an experiment and were able to test it out. What follows is the copy of the announcement of this experiment sent to class members via telecommunications:

[appchem/announce #250, jkelley, 1221 chars, 26-Apr-90 23:07]

TITLE: ANNOUNCING THE INVENTION OF A NEW EXPERIMENT.

The Air and Climate Group (you all should ID yourselves on here for the benefit of posterity) has invented a wonderful new experiment:

As part of testing CO_2 in bromphenolblue, group members brought in 4 samples of exhaust in 250 ml Erlenmeyers:

1925 Model T

U Lowell diesel bus

'76? Kawasaki

'86 Caprice

10 drops of the indicator were added to 125 mL distilled water. 10 mL of this brew were added to each 250mL Erlenmeyer in which the exhaust sample had been collected. The flask was restoppered and shaken vigorously. The indicator solution was added in 10mL increments till flask were nearly full (total vol. 220 mL). By color comparison it could be seen that the list as given above goes from least to most efficient in burning.

If anyone else wants to try this experiment, the students will tell you more about it, but if you decide to try it, be sure to give credit to these students for their invention: Deb Maue-Sprague, Bill McWhinnie, Jeff Dorandi, Heather Boudoin, Joseph Abdulmassih, Bob Charette, Jennifer Alexander, Joe Fernald, Arthur Smith, Tim Sprissler, Kirk Ross (who missed the best part today), Kevin Bibeau.

Conclusions

The factors that worked well in this collaborative learning situation included the weekly, in-class assessments; group problem solving; the final examination; and group decision making. In fact, all group activities enhanced the learning process except the computer conferencing.

Computer conferencing has been known to make significant contributions to collaborative learning, but it did not do so in 1989 and 1990 although student contributions improved with time. It is likely that the conferencing would be improved if students were required to contribute to on-line discussions of topics they perceived as important, and if the students were convinced that they were talking to each other as well as to the instructor.

Overall, with collaborative learning, students learned basic chemistry content and were exposed to other approaches to chemistry teaching. In addition, most students acquired awareness and understanding of significant contemporary issues, and many learned about learning. Group tests and experiments turned out to be valuable teaching/learning tools. Finally, collaborative learning created an environment in which students felt free to ask questions and enjoy learning and felt empowered by their participation in significant decisions about course content and grading. It is also likely that experimenting with collaborative learning will permanently alter a teacher's conception of knowledge and how it is acquired and shared.

Reference and Note

1. American Chemical Society, ed., *ChemCom: Chemistry in the Community* (Dubuque, IA: Kendall/Hunt, 1988).
2. The computer network used is called "CoSy," developed by the University of Guelph, Ontario, available from Softwords in British Columbia.

11

Figural Response in Science and Technology Testing

Michael E. Martinez

The changing role of testing demands new assessment methods. One such method is the figural response item format. Figural response items differ from traditional test items in two fundamental ways. First, they are not multiple-choice items; rather, they call for constructed responses — answers generated by the student rather than chosen from a list of options. Second, the student demonstrates understanding by carrying out some operation on figural media such as graphs, illustrations, and diagrams, rather than on words or numbers.

Both features of figural response items have potential benefits for testing. For example, the student's constructed response might be a closer approximation of the desired skill than would be the selection of a multiple-choice option. Expression of understanding through figural representations might enable the measurement of knowledge that is difficult or impossible to express verbally or numerically.[1] In addition, people who tend to think pictorially might find expression of their knowledge facilitated.

Figural Response Projects

Paper-and-Pencil Delivery

At first, figural response items were printed on paper and answered with a pencil. Twenty-five such items were developed for the 1989-90 field test of the National Assessment of Educational Progress (NAEP). These science items were given to fourth, eighth, and twelfth grade students. To compare the properties of figural response items with multiple-choice items, each figural response question was matched with a multiple-choice counterpart. Items of the two formats were administered to equivalent groups and compared statistically.[2] Figural response items were found to be somewhat more difficult and were more predictive of overall ability than multiple-choice items, a pattern consistent with findings in other studies.[3]

In addition to comparing item format properties, this study pilot-tested a technology that involved high-resolution scanning of the students' pencil responses, and computer scoring of those responses.[4] While the NAEP items were intended to be scored by computer, they were first hand-scored by experts. Thus, figural response questions need not rely on automated scoring. They can be constructed and scored by local educators and thus supplement more traditional types of items, such as the multiple-choice format and essay writing.

Computer Delivery

Following the NAEP study, the project shifted to computer delivery of items, which enables an examinee to carry out operations not possible with paper and pencil. The project has two strands of development. In one strand, figural response items are being developed for architecture licensure testing (Figures 1 and 2). In the other strand, items are being created for assessment of proficiency in cell and molecular biology (Figures 3 and 4). Items in both architecture and biology use the same IBM-compatible computer interface for delivery (although the format is not limited to this platform). Test items are presented and answered on the computer using a mouse.

Computer-based responses to figural response items are made by using a set of tools activated by on-screen buttons. The tools are few in number, which helps keep the computer interface simple and accessible to users who lack extensive computer experience. Our "move object" tool enables an examinee to move an object on the screen, and the "rotate" tool allows an object to be rotated at 90-degree increments. Both move object and rotate are used in Figure 1 to position elements of the recreation center on the site. In Figures 3 and 4, the move object tool is used to move molecular groups and chromosomes, respectively.

Our "draw line" and "draw arrow" tools allow the user to create straight lines, with or without arrowheads. The draw line tool is used in Figure 2 to extend the survey of the property by drawing in two additional property lines. The draw tool can be used to create a free-form line, such as might be used to plot a function on a line graph. The "label" tool permits selection of a label from a list and placement of that label on a figure. Only the tools needed to answer a particular item are offered to the examinee. Again, all responses to the items are made by using a mouse.

Research

Our research program is guided by a central question: How does the figural

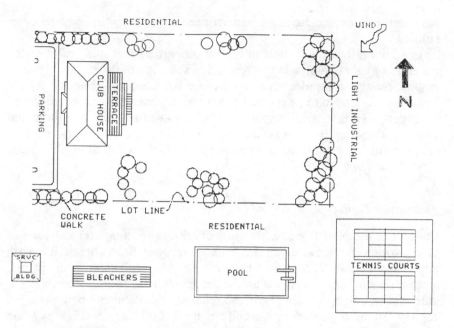

Figure 1. Sample figural response item on site planning. The item stem reads: "A recreational center site plan must accommodate a club house in its present position, as well as tennis courts, pool, bleachers, and a service building. Prepare the site plan according to the following objectives: (1) Preserve all trees. (2) Bleachers shall serve the tennis courts. (3) Pool shall be adjacent to the club house. (4) Service building shall relate to the club house and the parking lot." Tools needed: Move Object and Rotate.

response item format influence what a test actually measures? The research is intended to illuminate the format's construct validity — the meaning of what is measured. Some current or planned research directions are sketched below.

Psychometrics

If figural response items are to complement multiple-choice and other item types, their psychometric properties must be understood. These properties include difficulty and ability to predict overall proficiency of the examinee. As mentioned earlier, the NAEP figural response items were somewhat more difficult and predictive than their multiple-choice counterparts. Another study will determine whether the proficiencies measured by figural response and multiple-choice items are unidimensional; that is, whether they measure the same construct to a sufficient degree. If multiple-choice and figural response items are found to be unidimensional, they can exist

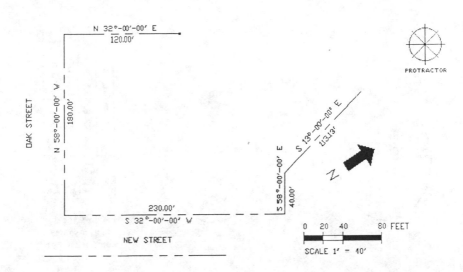

Figure 2. Sample figural response item on surveying. The item stem reads: "Given an excerpt of a Property Description and partial plotting of the described lot below, draw the boundary lines underlined in the partial description. . . . North 32 degrees -00'00" EAST 120.00' to a stake; thence parallel with said northeasterly side of Oak Street SOUTH 58 degrees -00'00" EAST 40.00' to a stake; thence at right angle to the northeasterly side of Oak Street, NORTH 32 degrees -00'-00" EAST 40.00' making an angle of . . ." Tools needed: Draw line, Erase, and Move Object.

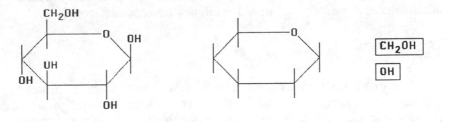

Figure 3. Sample figural response item on molecular configuration. The item stem reads: "Given the D-glucose below, construct its L-glucose stereoisomer using the template shown." Tools needed: Move object and Erase.

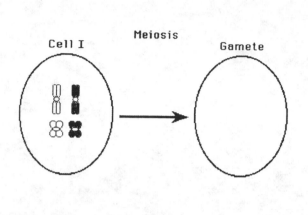
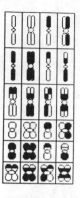

Figure 4. Sample figural response item on cell division. The item stem reads: "Cell 1 has the normal diploid chromosome complement of an organism. Using the chromosomes on the right, show the chromosome complement of one of the expected gametes from this organism. Assume there is no crossover involved. Tools needed: Move Object and Erase.

side-by-side in tests and count toward the same score; if not, they must be administered separately and would count toward separate scores.

Strategies

Our research must address the question of how item formats influence cognitive processing. Multiple-choice items can sometimes be answered by using specialized strategies.[5] For example, a response elimination strategy can eliminate implausible options. A test taker might conclude that options A and C are implausible; if two options (B and D) are left, there is a 50 percent chance of getting the item right by guessing alone. Another strategy involves "working backward" from the options — trying each with the question stem, and determining which one works. Test coaching programs emphasize competence in such strategies. Unfortunately, these strategies often have little to do with the ability that the test is intended to measure (for example, knowledge of biology). If scores can be elevated through the use of these strategies, "noise" is added to the measurements and the actual construct measured is to that extent impure. Advantages accrue to students who are "test wise." Worse yet, students might invest effort and/or money in learning test-taking strategies that are peculiar to the test item format. Constructed-response questions reduce or eliminate the utility of such strategies.

Diagnosis

Diagnosis is an often-stated and worthy goal of a new generation of tests. The addition of a diagnostic function can make testing more relevant to the processes of teaching and learning. Constructed-response items have been shown to be especially useful for diagnosis, since examinees are unconstrained and uncued in the expression of what they know.[6] "Unconstrained" means that responses are not limited to the options presented. The responses are also "uncued" in that options are not present to trigger retrieval or processing of information. Constructed responses therefore have greater fidelity to what is in the mind of the examinee.

Domain and Person Characteristics

As pointed out earlier, symbolic representations (such as figural, verbal, numeric) appear to interact with person characteristics. Some people prefer some representations more than others, or are better able to process them. Novel item formats, such as figural response, might facilitate the expression of knowledge and ability for some learners. Also, it appears that symbolic representations interact with subject-matter content. Some kinds of knowledge can be expressed easily in one format, but with great difficulty — or not at all — in another. Figural response items might be used to measure understandings that are untapped by most existing tests.

Learning

Basic characteristics of figural response tasks, their elicitation of constructed responses and use of figural material, might make them useful for promoting the learning of new material. Some of that value might come from the construction of the response. Information that is generated is more memorable than information that is merely visually inspected or copied.[7] In addition, visual information (or verbal information that can be translated into a visual representation) is generally more memorable than most verbal information.[8] Thus, the primary characteristics of figural response items might make them especially useful as tasks to faciliate learning.

Conclusion

Alternative forms of assessment are welcome and necessary for improving the relationship between assessment and education. Figural response items are one potentially important member of a new generation of testing methods. The visual nature of much of science and technology makes the use of figural response in these domains natural. As described in this paper, figural response items are being developed for use in large-scale assessment, but they can also be developed by teachers or school district staff for local testing. Figural response testing might, in combination with other new and old assessment methods, broaden the kinds of thinking called for by tests.

References

1. Larkin, J. H., and Simon, H. A., "Why a Diagram Is (Sometimes) Worth Ten Thousand Words," *Cognitive Science* **11** (1987), 65-99.
2. Martinez, M. E., "A Comparison of Multiple-choice and Constructed Figural Response Items," *Journal of Educational Measurement* (1991).
3. Traub, R. E., and MacRury, K., "Multiple-choice vs. Free-response in the Testing of Scholastic Achievement," in K. Ingenkamp and R. S. Jäger, eds., *Tests und Trends 8: Jarbuch der Pädagogischen Diagnostik.* (Weinheim und Basel, Germany: Beltz Verlag, 1990), 128-159.
4. Martinez, M. E., Ferris, J. J., Kraft, W., and Manning, W. H., "Automatic Scoring of Paper-and-Pencil Figural Responses," Research Report RR-90-23 (Princeton, NJ: Educational Testing Service, 1990).
5. Snow, R. E., "Aptitude Processes," in R. E. Snow, P.-A. Federico, and W. E. Montague, eds., *Aptitude, Learning, and Instruction, Volume 1: Cognitive Process Analyses of Aptitude* (Hillsdale, NJ: Lawrence Erlbaum Associates, 1980), 27-63.
6. Birenbaum, M., and Tatsuoka, K. K., "Open-ended versus Multiple-choice Response Formats—It Does Make a Difference for Diagnostic Purposes," *Applied Psychological Measurement* **11** (1987), 385-395.
7. Peynircioglu, Z. F., "The Generation Effect with Pictures and Nonsense Figures, *Psychologica* (1989), 153-160.
8. Paivio, A., *Imagery and Verbal Processes* (New York: Holt, 1971).

INDEX